Minerals, Metals and Sustainability

Minerals, Metals and Sustainability

Editor

Minh Chi Ngo

scitus
academics

Minerals, Metals and Sustainability
Edited by **Minh Chi Ngo**

ISBN: 978-1-68117-665-9
Library of Congress Control Number: 2016946532

616, Corporate Way, Suite 2, 4766,
Valley Cottage, NY 10989

www.scitusacademics.com

Notice

Printed in United States of America on Acid Free Paper

Preface

This book presents state of the art information on the sustainable use of finite mineral resources and also the waste allied with their manufacturing and utilization are subject to sustainability inspections. It offers a multidisciplinary approach that includes the physical and terrestrial sciences, social sciences, economics and environmental sciences. Today, the use of minerals is normal in daily products such as toothpaste, eye lenses and lifesaving equipment in hospitals. Additional areas for the extraction of minerals are identified and selected with the development of the mining industry and the development of advanced extraction techniques. Abandoned land mines, mining in open areas, drilling for all these activities also affects human health if unplanned and not environmentally friendly. This has given rise to concerns about the impact of mining on the ecological health of the people and the region. Different mining techniques have different effects on the planet. If we are to create a secure future for future generations, sustainable development will take the hour.

Minerals, Metals and Sustainability should be of great value to practical engineers, geologists and scientists; government service professionals in areas such as resources, environment and sustainability; and non-technical professionals working in the minerals industry or in the minerals sector.

Table of Contents

1

CHAPTER

Microbiological Advances in Biohydrometallurgy

Helen Watling

CSIRO Mineral Resources, Australian Minerals Research Centre, P.O. Box 7229, Karawara

ABSTRACT

The most exciting advances in biohydrometallurgy are occurring in the field of microbiology. The two main technologies employed in biohydrometallurgy, agitated tanks for the processing of refractory concentrates and heaps and dumps for the processing of low-grade ores, are technologically sound and widely practised at commercial scale, but their development began at a time when very little was known of the microorganisms that assisted metals extraction from sulfide ores. During and subsequent to those developments it has been shown that microbial communities in metals extraction are more diverse than originally thought, and extremely robust and adaptable to different and variable environments. Recent advances in genomics and proteomics, exploiting hugely increased computing power and speed, have made it possible to describe not only which microorganisms are present in bioleaching systems, but also what physiological functions are being exercised. The body of knowledge being acquired through the application of molecular biology methods will be used increasingly to monitor microbial behaviour, optimise conditions for more appropriate microbiological activity and/or infer the "microbiological health" of bioreactors (tanks and heaps).

Keywords: biomining; microbial communities; acidophiles; heap leaching; sulfide concentrates; sulfide ores; iron(II)-oxidation; sulfur-oxidation

1. INTRODUCTION

Biohydrometallurgy (biomining or mineral bioprocessing) is a sub-discipline of hydrometallurgy that exploits some attributes of micro-organisms (bio-catalysts) to facilitate and/or enhance the separation of elements from their ores or other materials. Traditional biohydrometallurgy involves the aqueous, inorganic chemistry of acidic sulfate solutions contacted with sulfide concentrates or ores containing valuable metals. As a method of extracting elements, biohydrometallurgy is an alternative to some intense chemical leaching methods [1,2,3,4]. However, its application to the processing of mineral concentrates compares unfavourably with pyrometallurgical processing because it is slower and because, usually, the infrastructure required for pyrometallurgical processing is already in place. According to Holmes and Debus [5], biological processing (of concentrates) would need to demonstrate a greater than 20% advantage over conventional pyrometallurgical processing to interest the mining industry. Similarly, Poulter*et al.* [6] remarked that industry reluctance to embrace the technology is partly due to the "inherent process and economic advantage of modern smelting technologies" and the "real and perceived risks associated with the introduction of a novel technology". On the other hand, biohydrometallurgy has achieved acceptance for the processing of low-grade, secondary copper sulfide ores on a very large scale, particularly in Chile, and accounts for approximately 20% of annual global copper production [7,8]. Likewise, pre-treatment of refractory gold concentrates in agitated tanks at commercial scale has a long and successful history from Fairview, South Africa [9] to Runro, the Philippines [10]. Looking to the future, this suite of well-developed and reliable biohydrometallurgical technologies can be applied, for example, to small deposits in remote locations perhaps with unfavourable terrain, to complex ores that are difficult to process, or to concentrates containing impurities that attract smelter penalties.

Man has benefited from the bio-assisted extraction of metals for much longer than the underlying microbiological foundation of the technology has been understood. Accounts from almost 2000 years ago describe the recovery of copper from mine water by cementation (Table 1) but the 'technology' could pre-date these accounts by many centuries, possibly since man started to mine ores and extract metals.

Table 1. Historical biohydrometallurgy: Approximately 2000 years of anecdotes and evidence up to the discovery of *Acidithiobacillus (At.) ferrooxidans* (collated from [11,12,13,14,15,16,17,18,19,20, 21,22,23,24,25,26,27,28]).

Chronology	Observation—Report
25–220 A.D.	The Roman Gaius Plinius Secundus (Pliny the Elder, 23–79 A.D.) wrote to the effect that chrysocolla is also artificially produced by slowly passing water through the mine during the winter until the month of June; subsequently the water is evaporated in June and July.
	Notes in old Chinese texts on leaching of rocks and the formation of gall ($CuSO_4$) springs; copper recovery was by cementation on iron.
	The scientist Galen described *in situ* leaching at a mine in Cyprus.
1086 A.D.	The Gall ($CuSO_4$)—copper process described: copper leaching and recovery using cementation on iron as a commercial process; the copper was used for coinage.
1500s	Copper leached from some ores (*in situ*) and mines (Spain, Hungary, Germany) and copper recovered by cementation on iron. A wood cut illustration in Agricola's *De re metallica* represents the manual collection of copper solutions in wooden 'vats' and transfer to evaporative ponds to concentrate the solutions.
1600s	The extraction of copper from mine water (Peru) described by Alvaro Alonso Barba de Garfias (priest and metallurgist); he was granted a patent to recover copper from mine water.
1800s	Large-scale copper heap leaching (Spain): Heaps of low-grade ore were left for 1–3 years to undergo *natural decomposition*.
1900–1920	Commercial copper production by leaching and copper cementation (Butte, MT, USA); in place (stope) leaching (Cananea, Sonora, Mexico); recovery of copper from underground mine water (Bisbee, AZ, USA).
1921–1940	The oxidation of sulfur by soil microorganisms described; *At. thiooxidans* (previously *Thiobacillus*) isolated and described.
	The concept of the 'biosphere' in which living organisms shaped the environment (biogeochemistry) expounded by Vladimir Vernadskii (1863–1945).
	Biological production of organic acids reported.
	Biooxidation of pyrite first reported and the suggested commercial application of biohydrometallurgy for zinc sulfide.
	Large scale dump leaching of open pit waste rock with recirculation of leach solutions (Bingham Canyon, UT, USA).
1941–1951	Roles of microorganisms in rock-weathering processes recognised.
	Acidithiobacillus ferrooxidans isolated from acid drainage of a coal mine and its role in oxidising Fe(II) and reduced inorganic sulfur compounds (RISC) in acid mine drainage (AMD) described.

Not surprisingly, metal extraction dominated commercial developments before the discovery that microorganisms were involved in the process. However, that discovery created a broad new research field within which the biogeochemical roles of acidophilic microorganisms were explored in relation to acidic drainage at mine sites and geothermally-heated sulfur-rich environments. Naturally, the greater understanding of microbial attributes was incorporated into established or new technologies. For the purposes of this targeted review of microbiological advances in biohydrometallurgy, it is convenient to discuss separately those advances relating to discoveries about the microorganisms, responses to processing environments, and diversity in managed heaps, columns and agitated tanks.

2. MICROBIOLOGICAL DISCOVERY

The discovery and description of *At. ferrooxidans* prompted a rapid increase in microbiological studies related to the oxidation of sulfide minerals. Arguably, *At. ferrooxidans* is the most studied acidophilic bacterium known, with more than 9000 papers in Web of Science, far more than any other acidophile. However, the perceived importance of *At. ferrooxidans* in bioleaching environments was largely a consequence of the relative ease of its enrichment and isolation, and its rapid growth in acidic iron(II) media, by which it out-competed other members of its microbial community. In the early years (post-1950), source materials of microorganisms tended to be spoils dumps at coal mines [29] or waste dumps at base metals mines [30]. Since then, many acidophiles have been isolated and described and many more are anticipated (Figure 1), constrained only by the number of research groups interested in acidophiles. Nevertheless, iron(II)- and sulfur-oxidising acidophiles are relatively few in number, especially those isolated from, or detected in, heap- or agitated-tank bioleaching reactors for the processing of sulfide minerals (refer to discussions and collations of microorganisms in [8,31,32,33,34,35].

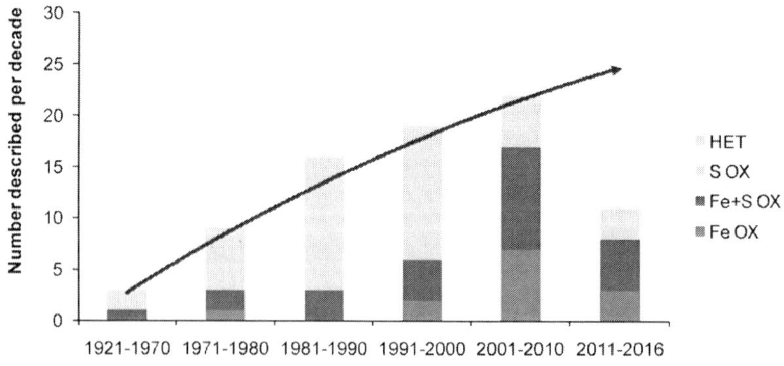

Figure 1. Trend with time of the numbers of acidophilic microorganisms (*Bacteria* and *Archaea*) isolated from environments relevant to biohydrometallurgy since the discovery of *At. thiooxidans*. They are grouped according to whether they are chemolithotrophs or mixotrophs that utilise iron(II) (Fe OX), both iron(II) and sulfur (Fe + S OX), sulfur but not iron(II) (S OX) or are heterotrophs (HET). Data to February 2016.

More general overviews have been focused on the broader range of microorganisms and microbial interventions associated with acid mine drainage (AMD) environments including collations of acidophilic prokaryotic micro-organisms on the basis of their physiological characteristics [36,37]. These microorganisms are of direct interest in the context of the microbial communities that can inhabit mineral sulfide heaps. Another group of acidophiles of strong interest are the hyperthermophilic archaea found in sulfur-rich geothermal regions (solfataras, volcanoes, hot springs or combusting coal dumps) [38,39,40,41] that might be used to enhance metal extraction from mineral sulfide concentrates in higher-temperature processes.

Before leaving microbiological discovery, it is worth recapping how biomining organisms assist metals extraction. The microbial capabilities of ferrous ion oxidation and RISC oxidation are the two key functions in the most commonly practised form of biohydrometallurgy, ferric sulfate—sulfuric acid leaching in agitated tanks and heaps/dumps. Microbial ferrous-ion oxidation regenerates the oxidant (ferric ion, Reaction (1)) required for sulfide mineral dissolution (e.g., covellite, Reaction (2)). At pH < 4, microbial ferrous ion oxidation is faster than chemical oxidation (e.g., Johnson [36] and references therein). RISC oxidation to sulfuric acid contributes to the acidic environment required for leaching (Reaction (3)).

$$4Fe^{2+} + 4H^{+} + O_2 + \text{microbial catalysts} \rightarrow 4Fe^{3+} + 2H_2O \quad (1)$$

$$CuS + 2Fe^{3+} \rightarrow Cu^{2+} + 2Fe^{2+} + S^{0} \quad (2)$$

$$2S^{0} + 3O_2 + 2H_2O + \text{microbial catalysts} \rightarrow 2H_2SO_4 \quad (3)$$

However, the geochemistry and biochemistry of sulfide mineral dissolution is more complex than is conveyed by Reactions (1) to (3). It encompasses the electrochemical reactions that take place at the mineral surface during sulfide dissolution as well as physiological processes that control the passage of compounds or ions across the microbial cell membrane and within the cell during growth. These multi-faceted topics are beyond the scope of the review but are discussed comprehensively [42] as well as in some focused reviews on individual species (e.g., *At. ferrooxidans* [43]), genera (e.g., *Sulfolobus* spp. [44]), or particular extreme conditions (e.g., microbial mechanisms of coping with high copper concentrations [34], living in acidic environments [39], or the adaptability required to thrive in biomining environments [45]).

3. MICROBIAL RESPONSES TO EXTREME AND VARIABLE HABITATS

The two main processes employed in bio-assisted extraction of metals, heap leaching of low-grade sulfide ores and agitated-tank leaching of sulfide concentrates, present very different habitats for the microbial communities that colonise them. Heaps of low-grade ore particles with a broad size distribution (up to 25 mm) are constructed on a large scale and are essentially dynamic, heterogeneous systems in terms of ore mineralogy and grade. Heat generation is a consequence of microbially-catalysed oxidation of sulfide minerals and heat management requires a suitable balance between heap aeration and irrigation [46]. In contrast, agitated tanks are closer to homogeneous in terms of concentrate grade, solution acidity, provision of O_2 and CO_2 and a selected and controllable temperature for the process. In tanks, there is an abundance of substrate (the sulfide minerals, ferrous ions and sulfur) that diminishes with passage through the train of tanks. Ideally, microorganisms are not exposed to severe fluctuations in conditions.

Much of our understanding of biomining microorganisms, their functions and capabilities, rests upon the collection of acidophiles from mine-impacted sites or geothermal regions that have been isolated and characterised. They form the foundation upon which subsequent studies of the microbial ecology of new sites can be described, monitored and expanded. In particular, the availability of "type strains" deposited in commercial culture collections has facilitated numerous research studies focused on microbial responses to different conditions relevant to bioleaching heap or agitated-tank processes. The four main parameters affecting microbial growth and activity are temperature, acidity, and cations and anions in process water, assuming a substrate is available. In the discussion that follows, unless otherwise noted, it can be assumed that studies are focused on the impact of a particular condition or contaminant on microbial ferrous-ion oxidation rates, this being the key reaction for successful sulfide dissolution and the concomitant extraction of the target metals.

3.1. Temperature

Primarily, microorganisms whether or not they are acidophiles tend to be "classified" according to the temperature range within which they grow, for example mesophiles with temperature optima of 20–40 °C; moderate thermophiles (40–60 °C) or extreme thermophiles (60–80 °C) [36]. Variable temperature is more likely to impact microorganisms in heaps than in agitated tanks. A number of temperature-related examples of advances in understanding taken from laboratory-, pilot- and commercial-scale studies of columns and/or heaps are reported in the literature. Rapid temperature increases and their impacts on microbial communities can be problematic in heap leaching, especially at start up when the sulfide content is at its greatest [47,48,49]. For example, the heat generated during the bioleaching of an ore in a dynamically-controlled, insulated column simulated heat generation in a 5000 tonne test heap of the same ore [47,50]. The catastrophic effect on the microbial population in that simulation as a consequence of the rapid heating to more than 70 °C is of particular interest (Figure 2). The difference in the rates of heat generation is attributed to the initial cell density in the column (5×10^9 cells per kg ore) being higher than that in the test heap (estimated to be 5×10^5 cells per kg ore).

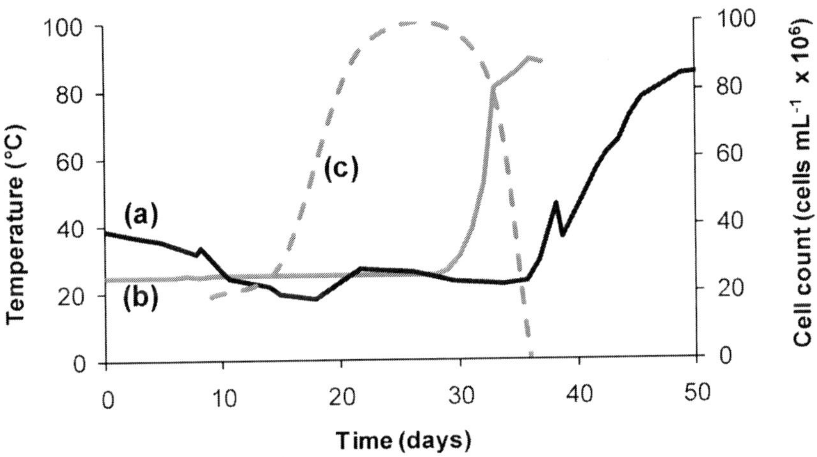

Figure 2. Heat generation profiles (solid lines) for (**a**) a 5000 tonne test heap of Ni-Cu-FeS ore and (**b**) the same ore, acid-agglomerated, inoculated and leached in an aerated, dynamically-controlled, insulated column; (**c**) estimates of cell numbers (broken line) in column effluent representing the microbial response to increased ore temperature.

The determination of preferred temperatures for growth (ranges and optima) of some single strains of acidophiles, based on their iron(II)- or RISC-oxidation rates, illustrates the rapid decline in the activities of microorganisms at temperatures greater than their respective optima [51,52]. It is generally held that hyperthermophiles, which grow well in sulfur-rich, persistently high-temperature environments such as solfataras and hot springs, would not naturally colonise heaped ores of low-sulfide content [35,51]. The results from a series of columns subjected to different temperature regimes and inoculated with mesophilic, moderately thermophilic and hyperthermophilic microorganisms as appropriate, showed a small benefit from inoculation with hyperthermophiles and led to the subsequent inoculation of a biooxidation heap for refractory gold ore with hyperthermophiles [53,54]. Recently "patches" of *Sulfolobus* spp. were found in the extremely large Escondida run-of-mine dump, indicative of the existence of persistent high-temperature regions in the dump [55], suggesting that thermophiles may colonise large sulfide dumps and heaps that generate and sustain high temperatures over a long period, removing the need for an inoculation strategy.

3.2. Acidity

In general, known acidophiles associated with AMD, solfataras and/or managed bioleaching reactors, while widely distributed across the pH range at temperatures favourable to their growth (Figure 3), tend to have a relatively narrow 'pH window' below which their activity (iron(II)- or RISC-oxidation) is greatly reduced [56]. However, differences in acid tolerance occur between species/strains. For example, in laboratory flask and reactor tests, a sudden drop in acidity from pH 1.5 to pH 1 caused the cessation of pyrite oxidation by *At. ferrooxidans* but did not affect the activity of a *Sulfobacillus*-like strain [57]. Very few acidophiles prefer or are active in habitats poised below pH 1. This was demonstrated in column tests where solution ORP fell, indicative of lower iron(II)-biooxidation activity, when the recycled, pH 1.7 solution, was acidified further to pH < 1 [58]. Similarly, the addition of acid to raffinate prior to recycle to a heap also caused decreased iron(II)- and RISC-oxidising activity in a heap microbial population [59]. Nevertheless, some acidophiles can adapt to very low-pH conditions; *Acidiplasma (Ap.) cupricumulans* (formerly*Ferroplasma cupricumulans*) grows in the range pH 0.4–1.8, *Ap. aeolicum* in the range pH 0–4 and*Picrophilus (P.) oshimae* and *P. torridus* can both grow at about pH 0 [60,61,62]. *Acidianus sulfidivorans* is another extremely acidophilic archaeon with preferred pH range 0.8–1.4 for growth [63].

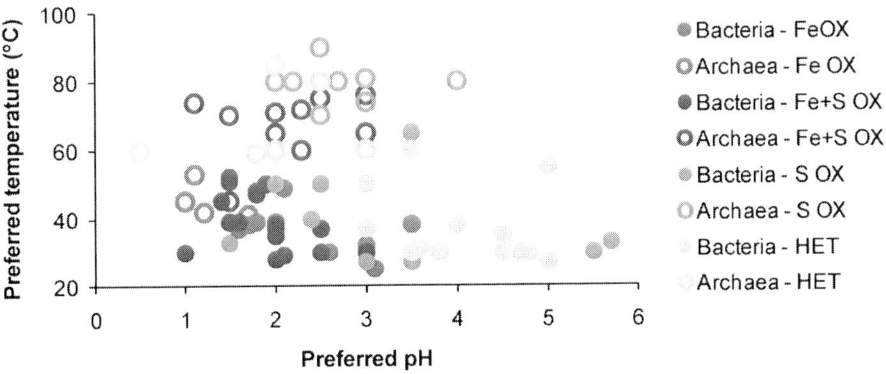

Figure 3. Preferred solution pH of microorganisms associated with AMD sites, geothermal regions and managed leaching heap or tank bioreactors, referenced against their preferred growth temperatures. Data obtained from their descriptions or from the recommended growth conditions obtained from commercial culture-collection database.

3.3. Cations

A series of laboratory studies have been conducted in which mainly mesophilic and moderately thermophilic microorganisms were adapted by successive culture to different metal cations, up to ($g \cdot L^{-1}$): Co, 15; Cu or Ni, 50; Zn, 98 but, overall, there are limited data on natural adaptation [64]. However, an important point to note is that, while tolerance to particular metals concentrations may increase the viability of a microbial population in a leaching operation, it does not necessarily result in increased metals extraction [65]. Indeed, slightly lower copper extractions were achieved in 40 days by the more Cu-tolerant strains of *S. thermosulfidooxidans* isolated from a commercial copper bioleaching heap (Figure 4).

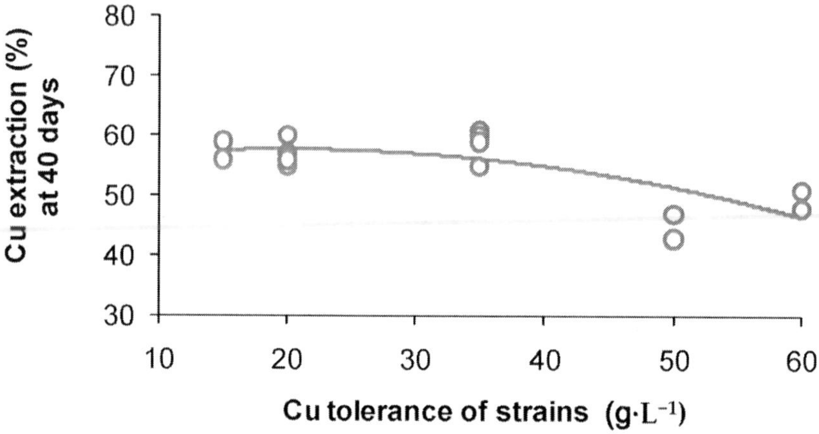

Figure 4. Bioleaching of chalcopyrite concentrate by isolates of *S. thermosulfidooxidans* exhibiting different copper tolerances. Tests were conducted with 3 wt % concentrate in basal salts medium (initial pH 1.8) in a shaking incubator (45 °C, 165 rpm). Initial cell density 5 × 10^6 cells·mL^{-1}.

While data from laboratory studies provide a helpful guide to microbial metals tolerances, more robust data can be obtained from long-term biological processing such as continuous agitated-tank or heap-leach processing. For example, two-years into the continuous pilot-scale operation of a biooxidation plant for the treatment of arsenopyrite gold concentrates, the retention time had reduced from 12 to 3.5 days, the mesophilic microbial population being active in solutions of 13 g·L^{-1} As [9]. Microbial adaptation to base metals in pilot- and demonstration agitated-tank plants is less well documented but microbial populations have been developed that tolerate >5 g·L^{-1} Co, 23 g·L^{-1} Ni with 38 g·L^{-1} Fe, or 36 g·L 1 Cu in cobalt, nickel and copper continuous plants, respectively [66,67,68,69]. From these data it can be inferred that adaptation has taken place, given that the concentrations are much higher than those encountered in most metal-rich acidic environments.

3.4. Anions

High concentrations of anions can be present in process waters associated with both heap and agitated-tank bioleaching, but in tank leaching sulfate will dominate. Ore heaps are more likely to contain mineral species enriched in chloride, nitrate or fluoride, which are known to inhibit microbial activity.

In bioleaching heaps, especially in Chile, the mineral atacamite ($Cu_2Cl(OH)_3$) presents a problem because of the ease with which it dissolves in acidic solutions and releases chloride ion to the circulating process water. In addition, Chile is an arid region with a scarcity of freshwater, leading to the use of seawater at some mines, a factor that has led to strong interest in the development of bacterial cultures active in seawater media. However, efforts to discover or develop acidophilic iron-oxidising strains or adapted cultures that tolerate chloride concentrations equivalent to seawater have had limited success. A survey of microorganisms from saline and acidic drains, lakes and sediments of Western Australia (52 samples) resulted in only five enrichment cultures that could tolerate between 26–41 $g \cdot L^{-1}$ Cl^- in growth media (pH 2) [70]. The authors commented that cultures that oxidised sulfur generally tolerated higher concentrations of salts than those that oxidised iron(II), consistent with previous data [71,72,73].

Overall, known biomining acidophiles do not tolerate chloride concentrations equivalent to seawater (approximately equivalent to 29 $g \cdot L^{-1}$ chloride salts) regardless of the conditions of growth. In several studies, mesophiles tolerated 2–20 $g \cdot L^{-1}$ NaCl, moderate thermophiles tolerated 10–30 $g \cdot L^{-1}$ NaCl and a mixed culture of thermophilic archaea tolerated up to 15 $g \cdot L^{-1}$ NaCl [70,71,72,74]. Data from descriptions of bacteria and archaea for salt tolerance referenced against their preferred pH for growth (Figure 5) indicated that acidophilic iron-oxidising strains with preferred pH 1–2 have limited tolerance to chloride. Strains with the dual capability of iron- and RISC-oxidation, some within slightly less acidic environments, exhibited greater chloride tolerance, particularly *Acidihalobacter prosperus* (formerly *Thiobacillus prosperus* [75]) and some*Alicyclobacillus*-like strains obtained from marine environments. In Figure 5, known biomining microorganisms mainly lie below the dotted line (equivalent chloride content to seawater). Halophilic bacteria, including the RISC-oxidising *Halothiobacillus* spp [76], and archaea that grow in environments at least 1.5 M NaCl, well above the chloride content of seawater, generally prefer to grow in environments with neutral pH or higher, above the extended acidophile or acid-tolerant range shown in Figure 5 and of lower acidity than would be useful in bio-assisted metals extraction. The discovery or development through adaptation of halo-tolerant acidophiles with iron(II)- and/or RISC-oxidising capability appears to be a major challenge that must be overcome if seawater is to become widely used in heap (bio)leaching.

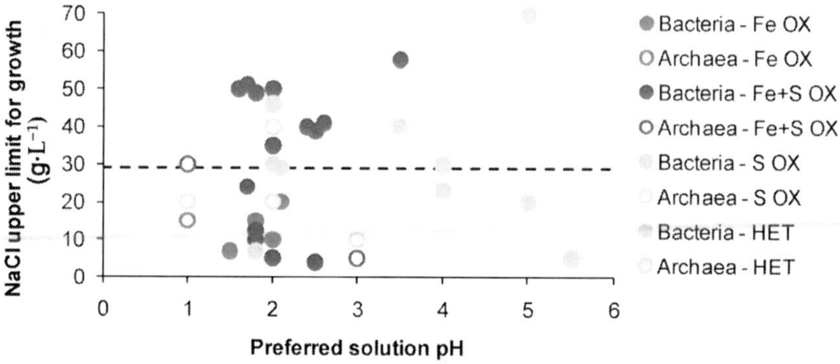

Figure 5. Maximum salt tolerance of acidophiles referenced against their preferred solution pH. The dotted line represents the concentration of chloride salts in seawater. (Data obtained from published descriptions of species or from the recommended growth conditions obtained from the DSMZ-Deutsche Sammlung von Mikroorganismen und Zellkulturen GmbH database).

Concentrations of fluoride higher than 0.5 $g{\cdot}L^{-1}$ are detrimental to microbial growth [69], potentially making fluoride in heap process water a greater challenge than chloride. A case in point was a chalcocite heap for which the failure to reach design recoveries was partly attributed to the failure to test the solubility of fluoride-containing minerals in the ore [77]. In that study, concentrations of iron-oxidizing microorganisms were lower than is typical for sulfide heap leaching, diminishing from the critical concentration of 10^5 cells·mL^{-1} [78] to 10^3 cells·mL^{-1} within a year. In an earlier but related study, finely ground particles of a variety of minerals inhibited iron(II) oxidation by *At. ferrooxidans* but fluorapatite ($Ca_5(PO_4)_3F$) particles (<74 μm) caused an anomalously severe response in suspensions < 2 wt % [79].

In a study of sulfur oxidation using *At. thiooxidans*, the toxicity of fluoride was shown to be pH dependent. It was strongly inhibitory at pH 2.3, where undissociated HF dominates the speciation, and is known to penetrate cell membranes [80,81]. Fluoride was less inhibitory at pH 4.5, where only 10% is present as undissociated HF, and not toxic at pH 7 where the undissociated MF molecule is dominant (M = Li, Na or K) [80]. Fluoride toxicity differs between bacterial species. In a recent comparison of iron(II)- and/or RISC-oxidation by five known bioleaching bacteria, the rank order of fluoride tolerance in media pH 1.5–2 was *At. ferrooxidans* > *At. thiooxidans* > *Leptospirillum ferrooxidans* > *At. caldus* > *Sulfobacillus (S.) thermosulfidooxidans* [82]. The fluoride tolerance of bacterial cultures can be increased substantially, in one case from 100 to 850 mg·L^{-1} through adaptive protocols [81,83].

The impact of the bioleach solution composition on the concentration of the HF species was investigated at 65 °C using *Sulfolobus metallicus* growing on pyrite (initial solution pH 1.5) and monitoring changes in ORP as a surrogate measure of microbial activity and growth [84]. The tests varied in the concentration of aluminium added to the growth medium, in the knowledge that Al(III) (and Fe(III)) form Al- or Fe-fluoride complexes [85,86,87]. A similar strategy of aluminium addition counteracted fluoride toxicity during the bioleaching of a low-grade, fluoride-containing ore using *S. thermosulfidooxidans* (50 °C, growth media in the range pH 1.4–2.1) [88], demonstrating that, if fluoride solubilisation is accompanied by strong aluminium or iron(III) solubilisation from gangue minerals, then the effects on the microorganisms may be substantially mitigated due to the formation of aluminium- or iron(III)-fluoro complexes.

As well as being a copper province, Chile is also a nitrate province. The very large Chilean deposits of salitre (saltpetre), mixtures of sodium and potassium nitrate, are a valuable export commodity. It is not surprising therefore that nitrate minerals are sometimes components of sulfide ores and that nitrate in process water may impact on microbial activity. Studies using *At. ferrooxidans* as the test species showed that nitrate was more inhibitory to iron(II) oxidation than to RISC oxidation [89,90]. However, in a recent comparison of the effects of nitrate ion on substrate utilisation by bacteria and archaea, it was found that the bacteria adapted to the presence of nitrate in growth media and resumed iron(II)-oxidation within a nine-week period in bioleaching tests [91]. Thus nitrate in ores may have only a transient impact on metals extraction in heaps. In contrast, in the same study, it was found that archaea did not adapt to the presence of nitrate. Based on this result, it was proposed that nitrate may be a useful means of controlling ORP during the higher-temperature oxidation of chalcopyrite in an agitated-tank process [91].

Surprisingly, the concentrations of sulfate ion, by far the predominant anion in most bioleaching process waters, are seldom reported and the effect of sulfate on microbial activity seems not to be of great concern. Sulfate concentrations can be more than 100 g·L^{-1} in agitated tanks [68,69] or up to 150 g·L^{-1} in the process water circulating through heaps [92,93], concentrations which are much higher than the 40 g·L^{-1} sulfate that caused an adapted mesophilic culture in iron(II) growth medium to halve its replication rate [94] or *At. caldus* in tetrathionate growth medium to increase the lag time two- to three-fold [74]. During a column bioleaching study (low-grade copper ore), planktonic cells in process water exiting a column declined rapidly by two orders of magnitude when concentrations of ferric and sulfate ions exceeded 30 and 75 $g \cdot L^{-1}$, respectively [95], and sulfur-oxidising bioactivity in the Escondida heap was reduced by 50% when sulfate concentrations in process water exceeded 120 $g \cdot L^{-1}$ [96].

4. BIODIVERSITY IN BIOLEACHING ENVIRONMENTS

The ease with which the microbial diversity of environmental samples can be examined has led to many recent descriptions of the orders, genera and sometimes species inhabiting bioleaching reactors (heaps, dumps, tanks) and/or AMD impacted environments. The collation of example data in Figure 6 illustrates the increased knowledge of bioreactor diversity as well as some effects of variable environments on that diversity, for example changes in community structure with time (H1, days 30 to 400; and T1, days 0 to 8), with depth in the heap (H2, depth 1 *cf.* 3 m) and with operating pH of the process water (H2 and H3, T2). In agitated tanks, the population of attached microorganisms differs from the planktonic community (T3, attached *vs.* planktonic) and the community structure differs with different solids loading (T3, 10% *cf.* 20%).

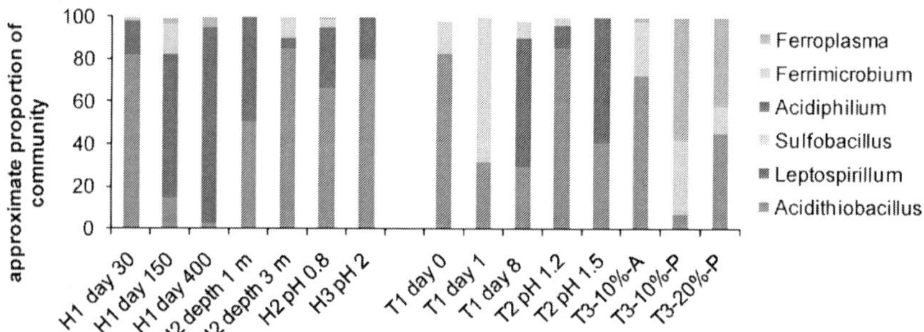

Figure 6. Microbiological diversity to genus level in heaps (H), and agitated tanks (T); A, attached; P, planktonic cells; % solids loading. Data collated from [97,98,99,100,101,102].

In general, abundances of *Acidithiobacillus* spp. in heaps including *At. ferrooxidans* and the sulfur-oxidising species *At. caldus*, *At. thiooxidans* and *At. albertensis* are greater early in leaching when there is a greater sulfur content [99,103]. *Leptospirillum* and *Ferroplasma* are more abundant later in the leach when the iron concentrations are higher and possibly inhibitory to *Acidithiobacillus* and other species [99,104,105]. Microbial communities change with time in both heaps and tanks, influenced by parameters such as substrate content and availability, solution composition and acidity, solids loadings in tanks, and temperature [92,102,103,104,106,107,108]. However, comparisons between studies tend to reveal a variety of effects as the communities respond to a combination of environmental parameters and, in many instances, to adapt to them with time [64].

It is in the nature of sulfide heaps that sulfur contents tend to be low and the heat generated within the bed during sulfide oxidation is transient [49], essentially because heap-management strategies are employed to maintain temperatures in the range suited to bacterial growth. In general, extreme thermoacidophiles with iron(II)- and sulfur-oxidising capabilities colonise habitats that are persistently hot and sulfurous [40,109,110]. As a consequence, it has been assumed that thermophiles such as *Acidianus*, *Metallosphaera* and *Sulfolobus* spp [111,112,113,114] would not naturally colonise sulfide heap operations and that thermophiles would need to be introduced to managed "hot heaps" to exploit increased metal extraction kinetics [49,54]. However, some recent studies on the biodiversity of sulfide heaps and mine-waste dumps included the putative identification of *Archaea* and some thermophiles have been detected, including *Acidianus*,*Acidiplasma*, *Ferroplasma*, *Metallosphaera*, *Sulfolobus* and *Thermoplasma* spp. [49,115,116,117,118,119].

The advances in both DNA sequencing and protein extraction, accompanied by huge advances in computing power and speed, launched the age of genomics and metagenomics, proteomics and meta-proteomics and metabolomics (Table 2).

Not surprisingly, the first genome sequence of a bioleaching microorganism was that of *At. ferrooxidans* [120]. Since then, the number of genomes of microorganisms of direct interest in bioleaching has increased significantly, largely due to a general interest in acidophiles and their activities in AMD environments. The strong overlap with bioleaching is understandable if bioleaching is viewed as an exemplar of extreme but controlled AMD. In a 2010 review [121], it was reported that there were 30 bacterial genomes and 26 archaeal genomes for extremely acidophilic microorganisms, eight metagenome projects of extremely acidic environments (four associated with the AMD at Iron Mountain, CA)

Table 2. Molecular biology terminology.

Term	Definition
Genomics	The branch of molecular biology concerned with the structure, function, evolution and mapping of genomes (the complete set of genes present in an organism).
Metagenomics	The study of the collected genetic material (genomes) from a mixed community of organisms.
Proteomics	The analysis of the structure, function and interactions of the proteins produced by the genes of a particular cell, tissue or organism; a proteome is the total complement of proteins present in a cell.
Metaproteomics	The study of all protein samples recovered directly from environmental sources; also called community proteomics or environmental proteomics.
Metabolomics	Systematic study of the unique chemical fingerprints that specific cellular processes leave behind; the study of their small-molecule metabolite profiles.

and complete sequences for 38 plasmids and 29 viruses from acidic environments. At the same time, it was concluded that these data were not sufficient "to provide a reasonably complete description of the genomic complexity and, by inference, of the full metabolic potential present in bioleaching operations" because (i) there are many gaps regarding the metabolic data; (ii) the microbial diversity of bioleaching and AMD environments is considerably greater than was previously appreciated; and (iii) often genomes were obtained using laboratory-maintained strains that may have accumulated genetic modifications.

Nevertheless, despite the body of research required to address the gaps and the greater diversity in natural microbial communities, the applications of the complementary techniques (Table 2) are already augmenting our understanding of the microbial processes involved in extracting metals from minerals. Numerous data on microbial functions and the conditions under which they are exercised best are being gathered. For example, their applications to AMD and bioleaching systems are resulting in better understanding of microbial physiology for individual organisms [40,122,123,124,125]. Bacterial, archaeal and eukaryotic diversity and their metabolic networks in AMD habitats have been described [126,127,128] and the ways in which metabolic activity levels and the partitioning of function are affected by community composition and the physico-chemical environment have been explored [129,130]. The extent to which each of the individual participants contributes to the process and how they evolve in time to keep the conglomerate healthy and therefore efficient during the entire process of bioleaching is being investigated [131]. Finally, the hypothesis that advanced knowledge of microbial processes can lead to better process design, control and optimization for metals extraction from minerals is supported by the interest shown by some mining companies in transcriptomic analysis of the genomic sequences of newly-isolated (commercially-sensitive) microorganisms with superior ability to extract copper from sulfides [132].

A valuable output from genomic studies is the generation of genomic sequences for microarray analysis [132]. The development of microarray technology over a number of years [133,134] has the potential to revolutionise the study of biodiversity in environmental samples. A review on the development of functional gene arrays was introduced as follows: "the recent advance of metagenomic technologies such as high throughput sequencing and functional gene arrays provides powerful high throughput tools for analysing microbial communities" [135]. A variety of microarrays have been constructed to contain probes for the genes involved in key microbial functions for specific applications [135]. For example, a microarray was developed based on most of the genes associated with acidophiles, including genes for carbon, nitrogen, sulfur and iron metabolism, DNA replication and repair, and metal resistance [136]. A suite of GeoChip microarrays, of which GeoChip 4 is the most comprehensive [137], have been used to characterize microbial communities in terms of functional diversity, composition, structure and metabolic activity/capability from a variety of habitats (e.g., soils, aquatic, contaminated sites, bioreactors).

Two modes of microarray application have been described. The use of microarrays based on the complete genomes of the members of a microbial community enables a nearly complete view of gene expression under different bioleaching conditions but it is time-consuming and expensive for the study of organisms isolated from the environment. Examples of specific studies based on complete genome sequences in combination with physiological and biochemical studies include for bacteria *Acidithiobacillus* spp [138,139,140] and for archaea, *Metallosphaera sedula* [141,142]. The shotgun DNA microarray, is a general approach by which gene expression can be studied in environmental isolates and can be applied to any organism regardless of how much of its genome is sequenced [132,143,144]. Applications for acidophiles include bacteria, *Leptospirillum* spp. [145,146] and *Acidithiobacillus* spp. [147] and archaea, *Acidianus* [148] and *Ferroplasma* [149]. Applications of microarray analysis of bioleaching and/or AMD environments include the study of bacterial population dynamics at a uranium-contaminated site [150], the monitoring of acidophile activity in AMD systems [151,152] and in the Escondida copper heap [99,153]. As no public domain research on heap or tank microbial populations funded by industry was found in this survey, it is assumed that such applications of advanced molecular-microbiological techniques for the discovery of superior bioleaching microorganisms are commercially sensitive.

5. SUMMARY

Until recently, the success of the two, widely-practised technologies employed in biohydrometallurgy, the pre-treatment of refractory gold concentrates in agitated tanks and the heap and dump leaching of low-grade, copper sulfide ores, was thought to depend on a few acidophilic microorganisms that oxidised iron(II) or sulfur compounds (chemolithotrophs). The technological concern was to optimise processing conditions to suit the microorganisms in the hope of faster and greater metals production.

In the late 20th century, the research emphasis was focused in two areas, discovering microorganisms with 'superior' metal-extraction capabilities and gaining a better understanding of how adaptable and resilient they were to extreme and variable mineral processing environments. The numbers of new species from natural and mine-impacted acidic environments increased steadily, with particular value being placed on extremely acidophilic, thermophilic and/or halophilic species with the requisite iron(II) and sulfur oxidising capabilities. Once characterised, their ready availability from commercial culture collections promoted an impressive number of studies on their responses to acid, metals and other soluble components, high ionic strength, high temperature and other processing parameters, depending on whether the research interest was rapid metals extraction from ores/concentrates or remediation of acid-drainage systems. The result is that what were once thought to be environments with low biodiversity have been shown to host diverse, robust, adaptable communities of acidophiles.

Now, in the 21st century, more than 50 genomes of acidophilic microorganisms have been published and analysed, leading to greatly increased knowledge of their physiological capabilities. Functional-gene microarrays are providing snapshots of microbial diversity in AMD environments and managed bioleaching reactors in terms of community structure and metabolic activity. The increasing capability of microarrays to describe not only what acidophiles are present but also what active contributions the acidophiles are making to metals extraction, is set to revolutionise both biodiversity studies of operating heaps and bioreactors as well as laboratory-based studies of the impacts of changing growth conditions on microbial activities. The body of knowledge being acquired through such studies can be used to predict microbial behaviour under varied conditions and from those data, to modify conditions for better metal extraction or for more appropriate microbial activity. Conversely, it should become possible to infer "microbial health" particularly in heaps based on measured physicochemical parameters and microbial diversity in process water, without having to disturb a production heap by sampling the solids.

ACKNOWLEDGMENTS

The Australian Government is thanked for funding through CSIRO Mineral Resources.

CONFLICTS OF INTEREST

The author declares no conflict of interests.

REFERENCES

1. McDonald, R.G.; Whittington, B.I. Atmospheric leaching of nickel laterites review. Part 1. Sulphuric acid technologies. Hydrometallurgy 2008, 91, 35–55.
2. McDonald, R.G.; Whittington, B.I. Atmospheric leaching of nickel laterites review. Part 2. Chloride and bio-technologies. Hydrometallurgy 2008, 91, 56–69.
3. McDonald, R.G.; Muir, D.M. Pressure oxidation leaching of chalcopyrite. Part I. Comparison of high and low temperature reaction kinetics and products. Hydrometallurgy 2007, 86, 191–205.
4. McDonald, R.G.; Muir, D.M. Pressure oxidation leaching of chalcopyrite. Part II. Comparison of medium temperature kinetics and products and effect of chloride ion.Hydrometallurgy 2007, 86, 206–220.
5. Holmes, D.S.; Debus, K.A. Biological opportunities for metal recovery. In Proceedings of the International Symposium on Biotechnology for Energy, Faisalabad, Pakistan, 16–21 December 1989; Malik, K.A., Naqvi, S.H.M., Aleem, N.I.H., Eds.; Nuclear Institute for Agriculture and Biology: Faisalabad, Pakistan, 1991; pp. 341–358.
6. Poulter, S.; Canterford, J.; Lunt, D. Biotechnology for sulphide minerals. A comparison with other leaching processes. In

Proceedings of the Biomine 99 and Waste Management in Metallurgical Operations Conference, Perth, Australia, 23–24 August 1999; Australian Mineral Foundation: Glenside, Australia, 1999; pp. 1–14.

7. Domic, E. A review of the development and current status of copper bioleaching operations in Chile: 25 years of successful commercial implementation. In Biomining; Rawlings, D.E., Johnson, D.B., Eds.; Springer: Berlin, Germany, 2007; pp. 81–95.

8. Watling, H.R. The bioleaching of sulphide minerals with emphasis on copper sulphides—A review. Hydrometallurgy 2006, 84, 81–108.

9. Van Aswegen, P.C.; Haines, A.K.; Marais, H.J. Design and operation of a commercial bacterial oxidation plant at Fairview. In Proceedings of the Randol Gold Conference, Perth, Australia, 14–17 March 1995; Randol International: Golden, CO, USA, 1988; pp. 144–147.

10. Van Niekerk, J. Factors affecting the selection of BIOX® as the preferred technology for the treatment of a refractory gold concentrate. Adv. Mater. Res. 2015, 1130, 191–196.

11. Hiskey, J.B. In situ leaching of copper: What's next? In Proceedings of the Hydrometallurgy '94, Cambridge, UK, 11–15 July 1994; Chapman & Hall: London, UK, 1994; pp. 43–67.

12. Pliny the Elder: the Natural History. Volume 37. Available online: http://penelope.uchicago.edu/Thayer/E/Roman/Texts/Pliny_the_Elder/home.html (accessed on 20 May 2016).

13. Walsh, J. Galen visits the Dead Sea and the copper mines of Cyprus. Bull. Geogr. Soc. Phila.1927, 25, 93–110.

14. Lung, T.N. The history of copper cementation on iron—World's first hydrometallurgical process from medieval China. Hydrometallurgy 1986, 17, 113–129.

15. Salkield, L.U. A Technical History of the Rio Tinto Mines: Some Notes on Exploitation from Pre-Phoenician Times to the 1950s; Institution of Mining and Metallurgy: London, UK, 1987.

16. Agricola, G. De re Metallica; Froben: Basel, Switzerland, 1556; Available online: www.gutenberg.org/ebooks/38015 (accessed on 20 May 2016).

17. Lipman, J.G.; Waksman, S.A.; Joffe, J.S. The oxidation of sulfur by soil microorganisms. Soil Sci. 1921, 12, 475–489.

18. Waksman, S.A.; Joffe, J.S. Microorganisms concerned in the oxidation of sulfur in the soil. II.Thiobacillus thiooxidans, a new sulfur-oxidizing organism isolated from the soil. J. Bacteriol.1922, 7, 239–256. [PubMed]

19. Rudolfs, W. Oxidation of iron pyrites by sulfur-oxidizing organisms and their use for making mineral phosphates available. Soil Sci. 1922, 14, 135–147.

20. Rudolfs, W.; Helbronner, A. Oxidation of zinc sulfide by microorganisms. Soil Sci. 1922, 14, 459–464.

21. Falck, R.; Kingma, V.T. The methodological and principles of production of organic acids by biological methods with help of aphanomyces. Ber. Deutsch. Chem. Ges. 1924, 57, 915–920.

22. Bernhauer, K. The problem of the formation of acids by Aspergillus niger. (Introductory announcement). Biochem. Z. 1924, 153, 517–521.

23. Colmer, A.R.; Hinkle, M.E. The role of microorganisms in acid mine drainage. Science 1947,106, 253–256.

24. Colmer, A.R.; Temple, K.L.; Hinkle, H.E. An iron-oxidizing bacterium from the acid mine drainage of some bituminous coal mines. J. Bacteriol. 1950, 59, 317–328. [PubMed]

25. Temple, K.L.; Colmer, A.R. The autotrophic oxidation of iron by a new bacterium:Thiobacillus ferrooxidans. J. Bacteriol. 1951, 62, 605–611. [PubMed]

26. Oldfield, J.D.; Shaw, D.J.B. VI Vernadskii and the development of biogeochemical understandings of the biosphere, c. 1880s–1968. Brit. J. Hist. Sci. 2013, 46, 287–310.

27. Krasilnikov, N.A. The role of micro-organisms in rock weathering. I. Microflora of rock stratum surfaces. Mikrobiologiya 1949, 18, 318–323.

28. Krasilnikov, N.A. The role of micro-organisms in rock weathering. II. The spread of micro-organisms from foci on rock surfaces. Mikrobiologiya 1949, 18, 492–497.

29. Leathen, W.W.; Braley, S.S., Sr.; McIntyre, L.D. The role of bacteria in the formation of acid from sulfuritic constituents associated with bituminous coal: II. Ferrous iron oxidizing bacteria. Appl. Microbiol. 1953, 1, 65–68. [PubMed]

30. Bryner, L.C.; Jameson, A.K. Microorganisms in leaching sulfide minerals. Appl. Microbiol.1958, 6, 281–287. [PubMed]

31. Rawlings, D.E. Characteristics and adaptability of iron- and sulfur-oxidizing microorganisms used for the recovery of metals from minerals and their concentrates.Microb. Cell Factories 2005, 4.

32. Norris, P.R. Acidophile diversity in mineral sulfide oxidation. In Biomining; Rawlings, D.E., Johnson, D.B., Eds.; Springer: Berlin, Germany, 2007; pp. 199–216.

33. Johnson, D.B. Biodiversity and interactions of acidophiles: Key to understanding and optimizing microbial processing of ores and concentrates. Trans. Nonferr. Met. Soc. China2008, 18, 1367–1373.

34. Orell, A.; Navarro, C.A.; Aranciba, R.; Mobarec, J.C.; Jerez, C.A. Life in blue: Copper resistance mechanisms of Bacteria and Archaea used in industrial biomining of minerals.Biotechnol. Adv. 2010, 28, 839–848.

35. Watling, H.R.; Watkin, E.L.J.; Ralph, D.E. The resilience and versatility of acidophiles that contribute to the bioassisted extraction of metals from mineral sulphides. Environ. Technol.2010, 31, 915–933.

36. Johnson, D.B. Extremophiles: Acidic Environments. In Encyclopedia of Microbiology; Schaechter, M., Ed.; Elsevier: Oxford, UK, 2009; pp. 107–126.

37. Johnson, D.B.; Hallberg, K.B. Carbon, iron and sulfur metabolism in acidophilic micro-organisms. Adv. Microb. Physiol. 2009, 54, 201–255.

38. Norris, P.R.; Burton, N.P.; Foulis, N.A.M. Acidophiles in bioreactor mineral processing.Extremophiles 2000, 4, 71–76.

39. Baker-Austin, C.; Dopson, M. Life in acid: pH homeostasis in acidophiles. Trends Microbiol.2007, 15, 165–171.

40. Auernik, K.S.; Cooper, C.R.; Kelly, R.M. Life in hot acid: Pathway analyses in extremely thermoacidophilic archaea. Curr. Opin. Biotechnol. 2008, 19, 445–453.

41. Kato, S.; Itoh, T.; Yamagishi, A. Archaeal diversity in a terrestrial acidic spring field revealed by PCR primer targeting archaeal 16S rRNA genes. FEMS Microbiol. Lett. 2011, 319, 34–43.

42. Ehrlich, H.L.; Newman, D.K.; Kappler, A. Geomicrobiology, 6th ed.; CRC Press: Boca Raton, FA, USA, 2015; p. 635, (and references therein).

43. Alexander, B.; Leach, S.; Ingledew, W.J. The relationship between chemiosmotic parameters and sensitivity to anions and organic

acids in the acidophile Thiobacillus ferrooxidans. J. Gen. Microbiol. 1987, 133, 1171–1179.

44. Wheaton, G.; Counts, J.; Mukherjee, A.; Kruh, J.; Kelly, R. The confluence of heavy metal biooxidation and heavy metal resistance: Implications for bioleaching by extreme thermoacidophiles. Minerals 2015, 5, 397–451.

45. Rawlings, D.E. Relevance of cell physiology and genetic adaptability of biomining microorganisms to industrial processes. In Biomining; Rawlings, D.E., Johnson, D.B., Eds.; Springer-Verlag: Heidelberg, Germany, 2007; pp. 177–198.

46. Dixon, D.G. Analysis of heat conservation during copper sulphide heap leaching.Hydrometallurgy 2000, 58, 27–41.

47. Hunter, C. BioHeap™ leaching of a primary nickel-copper sulphide ore. In Proceedings of the Nickel-Cobalt-8 Technical Proceedings, Perth, Australia, 20–22 May 2002; ALTA Metallurgical Services: Melbourne, Australia, 2002.

48. Readett, D.; Sylwestrzak, L.; Franzmann, P.D.; Plumb, J.J.; Robertson, W.R.; Gibson, J.A.E.; Watling, H. The life cycle of a chalcocite heap bioleach system. In Hydrometallurgy 2003; Young, C.A., Alfantazi, A.M., Anderson, C.G., Dreisinger, D.B., Harris, B., James, A., Eds.; TMS: Warrendale, PA, USA, 2003; pp. 365–374.

49. Plumb, J.J.; Hawkes, R.B.; Franzmann, P.D. The microbiology of moderately thermophilic and transiently thermophilic ore heaps. In Biomining; Rawlings, D.E., Johnson, D.B., Eds.; Springer-Verlag: Berlin, Germany, 2007; pp. 217–235.

50. Shiers, D.W.; Maree, M.D.; Collinson, D.M.; Watling, H.R.; Hosken, T.; Ingram, G.D. Use of a dynamically controlled column to assess the impact of temperature on copper extraction and microbial activity during copper sulfide bioleaching. In Proceeding of the Goldschmidt 2015 25th Anniversary, Praguz, Czech Republic, 16–21 August 2015.

51. Franzmann, P.D.; Haddad, C.M.; Hawkes, R.B.; Robertson, W.J.; Plumb, J.J. Effects of temperature on the rates of iron and sulfur oxidation by selected bioleaching Bacteria andArchaea: Application of the Ratkowsky equation. Min. Eng. 2005, 18, 1304–1314.

52. Watling, H.R.; Perrot, F.A.; Shiers, D.W. Comparison of selected characteristics ofSulfobacillus species and review of their occurrence in acidic and bioleaching environments.Hydrometallurgy 2008, 93, 57–65.

53. Brierley, J.A. Response of microbial systems to thermal stress in biooxidation-heap pretreatment of refractory gold ores. Hydrometallurgy 2003, 71, 13–19.

54. Logan, T.C.; Seal, T.; Brierley, J.A. Whole ore heap biooxidation of sulfidic gold-bearing ores. In Biomining; Rawlings, D.E., Johnson, D.B., Eds.; Springer: Berlin, Germany, 2007; pp. 113–138.

55. Soto, P.; Acosta, M.; Tapia, P.; Contador, Y.; Velásquez, A.; Espoz, C.; Pinilla, C.; Galleguillos, P.; Demergasso, C. From mesophilic to moderate thermophilic populations in an industrial heap bioleaching process. Adv. Mater. Res. 2013, 825, 376–379.

56. Plumb, J.J.; Muddle, R.; Franzmann, P.D. Effect of pH on rates of iron and sulfur oxidation by bioleaching microorganisms. Min. Eng. 2008, 21, 76–82.

57. Yahya, A.; Johnson, D.B. Bioleaching of pyrite at low pH and low redox potentials by novel mesophilic gram-positive bacteria. Hydrometallurgy 2002, 63, 181–188.

58. Tupikina, O.V.; Ngoma, I.E.; Minnaar, S.; Harrison, S.T.L. Some aspects of the effect of pH and acid stress in heap bioleaching. Min. Eng. 2011, 24, 1209–1214.

59. Zepeda, V.J.; Cautivo, D.; Galleguillos, P.A.; Salazar, C.N.; Velasquez, A.; Pinilla, C.; Demergasso, C.S. Effect of increased acid concentration on the microbial population inhabiting an industrial heap bioleaching plant. Adv. Mater. Res. 2013, 825, 348–351.

60. Hawkes, R.B.; Franzmann, P.D.; O'Hara, G.; Plumb, J.J. Ferroplasma cupricumulans sp. nov. a novel moderately thermophilic acidophilic archaeon isolated from an industrial-scale chalcocite bioleach heap. Extremophiles 2008, 10, 525–530.

61. Golyshina, O.V.; Yakimov, M.M.; Lunsdorf, H.; Ferrer, M.; Nimtz, M.; Timmis, K.N.; Wray, V.; Tindall, B.J.; Golyshin, P.N. Acidiplasma aeolicum gen. nov., sp. nov., a euryarchaeon of the family Ferroplasmaceae isolated from a hydrothermal pool and transfer of Ferroplasma cupricumulans to Acidiplasma cupricumulans comb. nov. Int. J. Syst. Evol. Microbiol. 2009, 59, 2815–2823.

62. Schleper, C.; Pühler, G.; Klenk, H.-P.; Zillig, W. Picrophilus oshimae and Picrophilus torridusfam. nov., gen. nov., sp. nov., two species of hyperacidophilic, thermophilic, heterotrophic, aerobic archaea. Int. J. Syst. Bacteriol. 1996, 46, 814–816.

63. Plumb, J.J.; Haddad, C.M.; Gibson, J.A.E.; Franzmann, P.D. Acidianus sulfidivorans sp. nov., an extremely acidophilic, thermophilic

archaeon isolated from a solfatara on Lihir Island, Papua New Guinea, and emendation of the genus description. Int. J. Syst. Evol. Microbiol.2007, 57, 1418–1423.

64. Watling, H.R. Adaptability of biomining organisms in hydrometallurgical processes. InBiohydrometallurgical Processes: A Practical Approach; Santos Sobral, L.G., Monteiro de Oliveira, D., Gomes de Souza, C.E., Eds.; CETEM/MCTI: Rio de Janeiro, Brazil, 2011; pp. 39–70.

65. Watkin, E.L.J.; Keeling, S.E.; Perrot, F.A.; Shiers, D.W.; Palmer, M.-L.; Watling, H.R. Metals tolerance in moderately thermophilic isolates from a spent copper sulfide heap, closely related to Acidithiobacillus caldus, Acidimicrobium ferrooxidans and Sulfobacillus thermosulfidooxidans. J. Ind. Microbiol. Biotechnol. 2009, 36, 461–465.

66. Battaglia, F.; Morin, D.; Ollivier, P. Dissolution of cobaltiferous pyrite by Thiobacillus ferrooxidans and Thiobacillus thiooxidans: Factors influencing bacterial leaching efficiency. J. Biotechnol. 1994, 32, 11–16.

67. D'Hughes, P.; Cezac, P.; Cabral, T.; Battaglia, F.; Truong-Meyer, X.M.; Morin, D. Bioleaching of a cobaltiferous pyrite: A continuous laboratory-scale study at high solids concentration.Min. Eng. 1997, 10, 507–527.

68. Heinzle, T.; Miller, D.; Nagel, V. Results of an integrated pilot plant operation using the BioNIC® process to produce nickel metal. In Proceedings of the Biomine 99 and Water Management in Metallurgical Operations 99, Perth, Australia, 24–25 August 1999; AusIMM: Melbourne, Australia, 1999; pp. 16–25.

69. Du Plessis, C.A.; Batty, J.D.; Dew, D.W. Commercial applications of thermophile bioleaching. In Biomining; Rawlings, D.E., Johnson, D.B., Eds.; Springer-Verlag: Berlin, Germany, 2007; pp. 57–80.

70. Rea, S.M.; McSweeney, N.J.; Degens, B.P.; Morris, C.; Siebert, H.M.; Kaksonen, A.H. Salt-tolerant microorganisms potentially useful for bioleaching operations where fresh water is scarce. Min. Eng. 2015, 75, 126–132.

71. Harahuc, L.; Lizama, H.M.; Suzuki, I. Selective inhibition of the oxidation of ferrous iron or sulfur in Thiobacillus ferrooxidans. Appl. Environ. Microbiol. 2000, 66, 1031–1037.

72. Zammit, C.M.; Mangold, S.; rao Jonna, V.; Mutch, L.A.; Watling, H.R.; Dopson, M.; Watkin, E.L.J. Bioleaching in brackish waters—Effect of

chloride ions on the acidophilic population and proteomes of model species. Appl. Microbiol. Biotechnol. 2012, 93, 319–329.

73. Johnson, D.B.; Grail, B.M.; Bonnefoy, V. New insights into salt-tolerance in acidophilic iron-oxidising bacteria. Adv. Mater. Res. 2015, 1130, 3–6.

74. Watling, H.R.; Shiers, D.W.; Zhang, G.J. Microbial behaviour under conditions relevant to heap leaching: Studies using the sulfur-oxidising, moderate thermophile Acidithiobacillus caldus. Hydrometallurgy 2012, 127–128, 104–111.

75. Cardenas, J.P.; Ortiz, R.; Norris, P.R.; Watkin, E.; Holmes, D.S. Reclassification of 'Thiobacillus prosperus' Huber and Stetter 1989 as Acidihalobacter prosperus gen. nov., sp. nov., a member of the family Ectothiorhodospiraceae. Int. J. Syst. Evol. Microbiol. 2015, 65, 3641–3644.

76. Kelly, D.P.; Wood, A.P. Reclassification of some species of Thiobacillus to the newly designated genera Acidithiobacillus gen. nov., Halothiobacillus gen. nov., andThermithiobacillus gen. nov. Int. J. Syst. Evol. Microbiol. 2000, 50, 511–516.

77. Brierley, J.A.; Kuhn, M.C. Fluoride toxicity in a chalcocite bioleach heap process.Hydrometallurgy 2010, 104, 410–413.

78. Murr, L.E.; Brierley, J.A. The use of large-scale test facilities in commercial leaching operations. In Metallurgical Applications of Bacterial Leaching and Related Microbiological Phenomena; Murr, L., Torma, A., Brierley, J., Eds.; Academic Press: New York, NY, USA, 1978; pp. 491–520.

79. Soljanto, P.; Rehtijarvi, P.; Tuovinen, O.H. Ferrous iron oxidation by Thiobacillus ferrooxidans: Inhibition by finely ground particles. Geomicrobiol. J. 1980, 2, 1–12.

80. Suzuki, I.; Lee, D.; McKay, B.; Harahuc, L.; Oh, J.K. Effect of various anions, pH, and osmotic pressure on oxidation of elemental sulfur by Thiobacillus thiooxidans. Appl. Environ. Microbiol.1999, 65, 5163–5168. [PubMed]

81. Peng, Z.; Yu, R.; Qiu, G.; Qin, W.; Gu, G.; Wang, Q.; Li, Q.; Liu, X. Really active form of fluorine toxicity affecting Acidithiobacillus ferrooxidans activity in bioleaching uranium. Trans. Nonferr. Met. Soc. China 2013, 23, 812–817.

82. Ma, L.; Li, Q.; Xiao, Y.; Wang, Q.; Yin, H.; Liang, Y.; Qiu, G.; Liu, X. Comparative study of fluoride tolerance of five typical bioleaching microorganisms. Adv. Mater. Res. 2013, 825, 214–218.

83. Wang, Q.; Liu, Y.; Hu, K.; Hu, E. Experimental study on acclimatization and culture of acid-tolerant and fluoride-tolerant bacteria. Min. Res. Dev. 2005, 25, 35–36.

84. Sundkvist, J.-E.; Sandström, Å.; Gunneriusson, L.; Lindstrom, E.B. Fluorine toxicity in bioleaching systems. In Proceedings of the 16th International Biohydrometallurgy Symposium, Cape Town, South Africa, 25–29 September 2005; Harrison, S.T.L., Rawlings, D.E., Petersen, J., Eds.; IBS: Cape Town, South Africa, 2005; pp. 19–28.

85. Corbillon, M.S.; Olazabal, M.A.; Madariaga, J.M. Potentiometric study of aluminium fluoride complexation equilibria and definition of the thermodynamic model. J. Solut. Chem. 2008, 37, 567–579.

86. Radic, N.; Bralic, M. Aluminium fluoride complexation and its ecological importance in the aquatic environment. Sci. Tot. Environ. 1995, 172, 237–243.

87. Soli, A.L.; Byrne, R.H. The hydrolysis and fluoride complexation behaviour of Fe(III) at 25 °C and 0.68 molal ionic strength. J. Solut. Chem. 1996, 25, 773–785.

88. Sicupira, L.; Veloso, T.; Reis, F.; Leao, V. Assessing metal recovery from low-grade copper ores containing fluoride. Hydrometallurgy 2011, 109, 202–210.

89. Razzell, W.E.; Trussell, P.C. Isolation and properties of an iron-oxidizing Thiobacillus. J. Bacteriol. 1963, 85, 595–603. [PubMed]

90. Harahuc, L.; Lizama, H.; Suzuki, I. Effect of anions on selective solubilisation of zinc and copper in bacterial leaching of sulfide ores. Biotechnol. Bioeng. 2000, 69, 196–203.

91. Shiers, D.W.; Ralph, D.E.; Watling, H.R. The effects of nitrate on substrate utilisation by some iron(II)- and sulfur-oxidising Bacteria and Archaea. Hydrometallurgy 2014, 150, 259–268.

92. Romero, J.; Yañez, C.; Vásquez, M.; Moore, E.R.B.; Espejo, R.T. Characterization and identification of an iron-oxidizing, Leptospirillum-like bacterium present in the high-sulfate leaching solution of a commercial bioleaching plant. Res. Microbiol. 2003, 154, 353–359.

93. Ruan, R.; Liu, X.; Zou, G.; Chen, J.; Wen, J.; Wang, D. Industrial practice of a distinct bioleaching system operated at low pH, high ferric concentration, elevated temperature and low redox potential for secondary copper sulfide. Hydrometallurgy 2011, 108, 130–135.

94. Shiers, D.W.; Blight, K.R.; Ralph, D.E. Sodium sulphate and sodium chloride effects on batch culture of iron oxidising bacteria. Hydrometallurgy 2005, 80, 75–82.

95. Watling, H.R.; Collinson, D.M.; Li, J.; Mutch, L.A.; Perrot, F.A.; Rea, S.M.; Reith, F.; Watkin, E.L.J. Bioleaching of a low-grade copper ore linking leach chemistry and microbiology. Min. Eng. 2014, 56, 35–44.

96. Cautivo, D.; Soto, P.; Zepeda, V.J.; Galleguillos, P.A.; Velasquez, A.; Pinilla, C.; Demergasso, C.S. Estimation of ionic load effect on the oxidizing activity of the microbial population in the heap bioleaching process at Escondida Mine. Adv. Mater. Res. 2013, 825, 219–222.

97. He, Z.; Xiao, S.; Xie, X.; Hu, Y. Microbial diversity in acid mineral bioleaching systems of Dongxiang copper mine and Yinshan lead-zinc mine. Extremophiles 2008, 12, 225–234.

98. Hao, C.; Wang, L.; Dong, H.; Zhang, H. Succession of acidophilic bacterial community during bio-oxidation of refractory gold-containing sulfides. Geomicrobiol. J. 2010, 27, 683–691.

99. Remonsellez, F.; Galleguillos, F.; Moreno-Paz, M.; Parro, V.; Acosta, M.; Demeragasso, C. Dynamic of active microorganisms inhabiting a bioleaching industrial heap of low-grade copper sulfide ore monitored by real-time PCR and olignucleotide prokaryotic acidophile microarray. Microbiol. Biotechnol. 2009, 2, 613–624.

100. Liu, X.; Chen, B.; Wen, J.; Ruan, R. Leptospirillum forms a minor portion of the population in Zijinshan commercial non-aeration copper bioleaching heap identified by 16S rRNA clone libraries and real-time PCR. Hydrometallurgy 2010, 104, 399–403.

101. Liu, X.; Wu, B.; Chen, B.; Wen, J.; Ruan, R.; Yao, G.; Wang, D. Bioleaching of chalcocite started at different pH: Response of the microbial community to environmental stress and leaching kinetics. Hydrometallurgy 2010, 103, 1–6.

102. Wang, Y.; Su, L.; Zeng, W.; Wan, L.; Chen, Z.; Zhang, L.; Qiu, G.; Chen, X.; Zhou, H. Effect of pulp density on planktonic and attached community dynamics during bioleaching of chalcopyrite by a moderately thermophilic microbial culture under uncontrolled conditions.Min. Eng. 2014, 61, 66–72.

103. Demergasso, C.; Galleguillos, F.; Soto, P.; Serón, M.; Iturriaga, V. Microbial succession during heap bioleaching cycle of low grade copper sulfides. Does this knowledge mean a real input for

industrial process design and control? Hydrometallurgy 2010, 104, 382–390.

104. Okibe, N.; Gericke, M.; Hallberg, K.B.; Johnson, D.B. Enumeration and characterization of acidophilic microorganisms isolated from a pilot plant stirred-tank bioleaching operation.Appl. Environ. Microbiol. 2003, 69, 1936–1943.

105. Kawabe, Y.; Inoue, C.; Suto, K.; Chida, T. Inhibitory effect of high concentrations of ferric ions on the activity of Acidithiobacillus ferrooxidans. J. Biosci. Bioeng. 2003, 96, 375–379.

106. Cameron, R.A.; Yeung, C.W.; Greer, C.W.; Gould, W.D.; Mortazavi, S.; Bedard, P.L.; Morin, L.; Lortie, L.; Dinardo, O.; Kennedy, K.J. The bacterial community structure during bioleaching of a low-grade nickel sulphide ore in stirred tank reactors at different combinations of temperature and pH. Hydrometallurgy 2010, 104, 207–215.

107. Mutch, L.A.; Watling, H.R.; Watkin, E.L.J. Microbial population dynamics of inoculated low-grade chalcopyrite bioleaching columns. Hydrometallurgy 2010, 104, 391–398.

108. Yu, R.; Shi, L.; Gu, G.; Zhou, D.; You, L.; Chen, M.; Qiu, G.; Zeng, W. The shift in microbial community under the adjustment of initial and processing pH during bioleaching of chalcopyrite concentrate by moderate thermophiles. Bioresour. Technol. 2014, 162, 300–307.

109. Segerer, A.H.; Burggraf, S.; Fiala, G.; Huber, G.; Huber, R.; Pley, U.; Stetter, K.O. Life in hot springs and hydrothermal vents. Orig. Life Evolut. Biosph. 1993, 23, 77–90.

110. Childs, A.M.; Mountain, B.W.; O'Toole, R.; Stott, M.B. Relating microbial community and physicochemical parameters of a hot spring: Champagne Pool, Wai-o-tapu, New Zealand.Geomicrobiol. J. 2008, 25, 441–453.

111. Stott, M.B.; Sutton, D.C.; Watling, H.R.; Franzmann, P.D. Comparative leaching of chalcopyrite by selected acidophilic Bacteria and Archaea. Geomicrobiol. J. 2003, 20, 215–230.

112. Zhang, L.; Peng, J.; Wei, M.; Ding, J.; Zhou, H. Bioleaching of chalcopyrite with Acidianus manzaensis YN25 under contact and non-contact conditions. Trans. Nonferr. Met. Soc. China2010, 20, 1981–1986.

113. Zhu, W.; Xia, J.; Yang, Y.; Nie, Z.; Peng, A.; Liu, H.; Qiu, G. Thermophilic archaeal community succession and function change

associated with the leaching rate in bioleaching of chalcopyrite. Bioresour. Technol. 2013, 133, 405–413.

114. Shiers, D.W.; Ralph, D.E.; Bryan, C.G.; Watling, H.R. Substrate utilisation by Acidianus brierleyi, Metallospaera hakonensis and Sulfolobus metallicus in mixed ferrous ion and tetrathionate growth media. Min. Eng. 2013, 48, 86–93.

115. Demergasso, C.; Galleguillos, P.; Escudero, L.; Zepeda, V.; Castillo, D.; Casamayor, E. Molecular characterization of microbial populations in a low grade copper ore bioleaching test heap. Hydrometallurgy 2005, 80, 241–253.

116. Fuchs, T.; Huber, H.; Teiner, K.; Burggraf, S.; Stetter, K.O. Metallosphaera prunae, sp. nov., a novel metal-mobilizing, thermoacidophilic Archaeum isolated from a uranium mine in Germany. Syst. Appl. Microbiol. 1995, 18, 560–566.

117. Xiao, S.; Xie, X.; Liu, J.; He, Z.; Hu, Y. Composition and structures of archaeal communities in acid mineral bioleaching system of Dongxiang copper mine and Yinshan lead-Zinc Mine, China. Curr. Microbiol. 2008, 57, 239–244.

118. Xie, X.; Xiao, S.; He, Z.; Liu, J.; Qiu, G. Microbial populations in acid mineral bioleaching systems of Tong Shankou copper mine, China. J. Appl. Microbiol. 2007, 103, 1227–1238.

119. Halinen, A.-K.; Beecroft, N.J.; Määttä, K.; Nurmi, P.; Laukkanen, K.; Kaksonen, A.H.; Riekkola-Vanhanen, M.; Puhakka, J.A. Microbial community dynamics during a demonstration-scale bioheap. Hydrometallurgy 2012, 125–126, 34–41.

120. Selkov, E.; Overbeek, R.; Kogan, Y.; Chu, L.; Vonstein, V.; Holmes, D.S.; Silver, S.; Haselkorn, R.; Fonstein, M. Functional analysis of gapped microbial genomes: Amino acid metabolism of Thiobacillus ferrooxidans. PNAS 2000, 97, 3509–3514.

121. Cárdenas, J.P.; Valdes, J.; Quatrini, R.; Duarte, F.; Holmes, D.S. Lessons from the genomes of extremely acidophilic bacteria and archaea with special emphasis on bioleaching microorganisms. Appl. Microbiol. Biotechnol. 2010, 88, 605–620.

122. Justice, N.B.; Norman, A.; Brown, C.T.; Singh, A.; Thomas, B.C.; Banfield, J.F. Comparison of environmental and isolate Sulfobacillus genomes reveals diverse carbon, sulfur, nitrogen, and hydrogen metabolisms. BMC Genom. 2014, 15.

123. Song, J.-L.; Jiang, C.-Y.; Liu, S.-J. Insight into the sulfur metabolism by thermoacidophilic archaeon Metallosphaera cuprina with

genomic, proteomic and biochemical tools. Adv. Mater. Res. 2015, 1130, 145–148.

124. Ullrich, S.R.; Poehlein, A.; Daniel, R.; Tischler, J.S.; Vogel, S.; Schlömann, M.; Mühling, M. Comparative genomics underlines the functional and taxonomic diversity of novel "Ferrovum" related iron oxidizing bacteria. Adv. Mater. Res. 2015, 1130, 15–18.

125. Goltsman, D.S.; Denef, V.J.; Singer, S.W.; VerBerkmoes, N.C.; Lefsrud, M.; Mueller, R.S.; Dick, G.J.; Sun, C.L.; Wheeler, K.E.; Zemla, A.; et al. Community genomic and proteomic analyses of chemoautotrophic iron-oxidizing "Leptospirillum rubarum" (Group II) and "Leptospirillum ferrodiazotrophum" (Group III) bacteria in acid mine drainage biofilms. Appl. Environ. Microbiol. 2009, 75, 4599–4615.

126. Hemme, C.L.; Deng, Y.; Gentry, T.J.; Fields, M.W.; Wu, L.; Barua, S.; Barry, K.; Tringe, S.G.; Watson, D.B.; He, Z.; et al. Metagenomic insights into evolution of a heavy metal-contaminated groundwater microbial community. ISME J. 2010, 4, 660–672.

127. Méndez-García, C.; Peláez, A.I.; Mesa, V.; Sánchez, J.; Golyshina, O.V.; Ferrer, M. Microbial diversity and metabolic networks in acid mine drainage habitats. Front. Microbiol. 2015, 6.

128. Chen, L.; Huang, L.; Méndez-Garcia, C.; Kuang, J.; Hua, Z.; Liu, J.; Shu, W. Microbial communities, processes and functions in acid mine drainage systems. Curr. Opin. Biotechnol.2016, 38, 150–158.

129. Banfield, J.F.; Verboekmoes, N.C.; Hettich, R.L.; Thelen, M.P. Proteogenomic approaches for the molecular characterization of natural microbial communities. OMICS 2005, 9, 301–333.

130. Zhang, X.; Niu, J.; Liang, Y.; Liu, X.; Yin, H. Metagenome-scale analysis yields insights into the structure and function of microbial communities in a copper bioleaching heap. BMC Genet. 2016, 17.

131. Valenzuela, L.; Chi, A.; Beard, S.; Orell, A.; Guiliani, N.; Shabanowitz, J.; Hunt, D.F.; Jerez, C.A. Genomics, metagenomics and proteomics in biomining microorganisms. Biotechnol. Adv. 2006, 24, 197–211.

132. Jerez, C.A. Metal extraction and biomining. In The Desk Encyclopedia of Microbiology; Schaechter, M., Ed.; Elsevier: Oxford, UK, 2009; pp. 762–775.

133. Stoughton, R.B. Applications of DNA microarrays in biology. Ann. Rev. Biochem. 2005, 74, 53–82.

134. Seidel, M.; Niessner, R. Automated analytical microarrays: A critical review. Anal. Bioanal. Chem. 2008, 391, 1521–1544.

135. He, Z.; Deng, Y.; Zhou, Y. Development of functional gene arrays for microbial community analysis. Curr. Opin. Biotechnol. 2012, 23, 49–55.

136. Yin, H.; Cao, L.; Qiu, G.; Wang, D.; Kellogg, L.; Zhou, J.; Dai, Z.; Liu, X. Development and evaluation of 50-mer oligonucleotide arrays for detecting microbial populations in acid mine drainages and bioleaching systems. J. Microbiol. Methods 2007, 70, 165–178.

137. Tu, Q.; Yu, H.; He, Z.; Deng, Y.; Wu, L.; Van Nostrand, J.D.; Zhou, A.; Voordeckers, J.; Lee, Y.-J.; Qin, Y.; et al. GeoChip 4: A functional gene-array-based high-throughput environmental technology for microbial community analysis. Mol. Ecol. Resour. 2014, 14, 914–928.

138. Quatrini, R.; Appia-Ayme, C.; Denis, Y.; Ratouchniak, J.; Veloso, F.; Valdes, J.; Lefimil, C.; Silver, S.; Roberto, F.; Orellana, O.; et al. Insights into the iron and sulfur energetic metabolism of Acidithiobacillus ferrooxidans by microarray transcriptome profiling.Hydrometallurgy 2006, 83, 263–272.

139. Quatrini, R.; Appia-Ayme, C.; Denis, Y.; Jedlicki, E.; Holmes, D.S.; Bonnefoy, V. Extending the models for iron and sulfur oxidation in the extreme acidophile Acidithiobacillus ferrooxidans. BMC Genom. 2009, 10, 394.

140. Valdés, J.; Pedroso, I.; Quatrini, R.; Holmes, D.S. Comparative genome analysis ofAcidithiobacillus ferrooxidans, A. thiooxidans and A. caldus: Insights into their metabolism and ecophysiology. Hydrometallurgy 2008, 94, 180–184.

141. Auernik, K.S.; Kelly, R.M. Physiological versatility of the extremely thermoacidophilic archaeon Metallosphaera sedula supported by transcriptomic analysis of heterotrophic, autotrophic and mixotrophic growth. Appl. Environ. Microbiol. 2010, 76, 931–935.

142. Auernik, K.S.; Maezato, Y.; Blum, P.H.; Kelly, R.M. The genome sequence of the metal-mobilizing, extremely thermoacidophilic archaeon Metallosphaera sedula provides insights into bioleaching-associated metabolism. Appl. Environ. Microbiol. 2008, 74, 682–692.

143. Haward, R.E.; Derisi, J.L.; Alfadhli, S.; Kaslow, D.C.; Brown, P.O.; Rathod, P.K. Shotgun DNA microarrays and stage-specific gene expression in Plasmodium falciparum malaria. Mol. Microbiol. 2000, 35, 6–14.

144. Parro, V.; Moreno-Paz, M. Gene function analysis in environmental isolates: The nif regulon of the strict iron oxidizing bacterium Leptospirillum ferrooxidans. PNAS 2003, 100, 7883–7888.

145. Tyson, G.W.; Lo, I.; Baker, B.J.; Allen, E.E.; Hugenholtz, P.; Banfield, J.F. Genome-directed isolation of the key nitrogen fixer Leptospirillum ferrodiazotrophum sp. nov. from an acidophilic microbial community. Appl. Environ. Microbiol. 2005, 71, 6319–6324.

146. Moreno-Paz, M.; Gómez, M.J.; Arcas, A.; Parro, V. Environmental transcriptome analysis reveals physiological differences between biofilm and planktonic modes of life of the iron oxidizing bacteria Leptospirillum spp. in their natural microbial community. BMC Genomics2010, 11.

147. Vera, M.; Krok, B.; Bellenberg, S.; Sand, W.; Poetsch, A. Shotgun proteomics study of early biofilm formation process of Acidithiobacillus ferrooxidans ATCC 23270 on pyrite. Proteomics2013, 13, 1133–1144.

148. Laska, S.; Lottspeich, F.; Kletzin, A. Membrane-bound hydrogenase and sulfur reductase of the hyperthermophilic and acidophilic archaeon Acidianus ambivalens. Microbiology 2003, 149, 2357–2371.

149. Golyshina, O.V. Environmental, biogeographic, and biochemical patterns of archaea of the family Ferroplasmaceae. Appl. Environ. Microbiol. 2011, 77, 5071–5078.

150. Brodie, E.L.; DeSantis, T.Z.; Joyner, D.C.; Baek, S.M.; Larsen, J.T.; Andersen, G.L.; Hazen, T.C.; Richardson, P.M.; Herman, D.J.; Tokunaga, T.K.; et al. Application of a high-density oligonucleotide microarray approach to study bacterial population dynamics during uranium reduction and reoxidation. Appl. Environ. Microbiol. 2006, 72, 6288–6298.

151. Garrido, P.; Gonzalez-Toril, E.; Garcia-Moyano, A.; Moreno-Paz, M.; Amils, R.; Parro, V. An oligonucleotide prokaryotic acidophile microarray: Its validation and its use to monitor seasonal variations in extreme acidic environments with total environmental RNA. Environ. Microbiol. 2008, 10, 836–850.

152. Shen, L.; Yin, H.; Hu, Q.; Liu, X.; Qiu, G. Microarray-based characterization of microbial community functional structure and heterogeneity associated with acid mine drainages.Adv. Mater. Res. 2015, 1130, 40–44.

153. Acosta, M.; Galleguillos, P.; Ghorbani, Y.; Tapia, P.; Contador, Y.; Velásquez, A.; Espoz, C.; Pinilla, C.; Demergasso, C. Variation in microbial community from predominantly mesophilic to thermotolerant and moderately thermophilic species in an industrial copper heap bioleaching operation. Hydrometallurgy 2014, 150, 281–289.

2

CHAPTER

Heterogeneous Nb-Based Nuclei for the Grain Refinement of Al-Si Alloys

L. Bolzoni[1,2] and N. Hari Babu[2]

[1]*Waikato Centre for Advanced Materials (WaiCAM), The University of Waikato, Private Bag 3105, Hamilton 3240, New Zealand.*

[2]*Institute of Materials and Manufacturing, Brunel University London, Uxbridge, Middlesex UB8 3PH, UK.*

ABSTRACT

Nb-based intermetallics are, generally, low-density high-temperature materials used for structural applications or cryogenic superconductors. In this work, we report the development of an Al(96)-Nb(2)-B(2) master alloy where in situ-formed micrometric Nb-based intermetallics (i.e. NbB_2 and Al_3Nb) are used for a completely different purpose: to promote the refinement of Al-Si alloys by taking advantage of enhanced heterogeneous nucleation. Nb-based intermetallics have the right characteristics, like low density, stability at high temperature and good lattice match, to be used as heterogeneous nucleation substrates. It was found that the addition of these Nb-based intermetallics permits the significant refinement of the microstructural features of the Al-Si alloy studied. The enhanced heterogeneous nucleation makes the grain size of the material far less dependent on the cooling rate, which is one of the critical parameters influencing the variation of the properties of the alloy.

INTRODUCTION

Niobium is a transition metal characterized by some specific properties such as superconductivity at the cryogenic temperature of 9.2 K. Due to its outermost electronic configuration, niobium is able to form different compounds. The two most common Nb-based intermetallics are probably, niobium borides and niobium aluminides. Specifically, NbB_2 has been extensively studied and it has been reported that it exhibits superconductive transition behaviour depending on the stoichiometry of the compound.1–3 Niobium aluminides, specifically Al_3Nb, were developed due to promising attributes for high-temperature applications such as high melting point and low density,4,5 high strength over a wide range of temperatures and good oxidation resistance. These properties made them strong candidates for advanced aeropropulsion systems and for other high-temperature components, either as a coating or as structural materials. Two of the previously mentioned aspects (i.e. high-temperature stability and low density) are fundamental prerequisites for the development of potential heterogeneous nucleation substrates for aluminium alloys, characteristics which should be coupled with good lattice match between the structure of the substrate and the nucleating phase. The most common example of heterogeneous nucleation substrates added to a molten metal to promote the heterogeneous nucleation and, thus, the refinement of its microstructure is the addition of Al-Ti-B master alloys to Al and wrought Al alloys.6–9These master alloys are composed of Ti-based compounds (i.e. TiB_2 and Al_3Ti), and it is believed that nucleation is promoted because TiB_2 acts as the actual heterogeneous nucleation site, on top of which a thin atomic-scale layer of Al_3Ti is formed to reduce and accommodate the lattice strain.10,11 This is because the lattice mismatch between Al and Al_3Ti is lower than that between Al and TiB_2.12 Nevertheless, Al-Ti-B master alloys are not extensively used in the case of cast Al-Si alloys because titanium and silicon react between themselves hindering the efficient refinement of the alloy (this occurrence is known as Si poisoning).13,14 Based on the promising characteristics of the Nb-based intermetallics (i.e. NbB_2 and Al_3Nb), we analysed crystallographic databases and found that Nb-based intermetallics, which have an isomorphous structure and similar lattice parameters to the respective Ti-based compounds, would be successful candidates as heterogeneous nucleation substrates for Al-Si alloys.15 In our first works,16,17 in which the Nb-based intermetallics were formed by adding Nb powder and KBF_4 flux to Al-Si alloys, we demonstrated the grain refining potency of Nb-based intermetallics on Al-Si alloys. The grain refinement is not hindered even though niobium can form silicides, as these intermetallic phases are stable at much higher temperatures with respect to their Ti-based counterparts. This means that their kinetics are much slower, in particular at processing temperatures typically used in Al foundries, and, thus, no poisoning was detected.16,17 In this work, we focused on the development of an Al(96)-Nb(2)-B(2) master alloy, which has been found to be easier to add to the melt, starting from a different source of raw materials [i.e. Nb powder and Al(95)-B(5) master alloy]. The aim of this work is to report and discuss the employment of the (in situ-formed micrometrics) Nb-based intermetallics for an alternative application rather than as a structural material. Specifically, the grain refining effectiveness of the Nb-based intermetallics is assessed on an Al-Si casting alloy solidified using different cooling rates.

MATERIALS AND METHODS

The raw materials employed in this study are (values given refer to the weight of the elements except where otherwise stated): pure aluminium (purity >99.5), niobium powder (purity >99.8), Al(95)-B(5) master alloy and a hypoeutectic Al-Si alloy (Al-8.4Si-0.3Mg-0.13Fe). The required amounts of aluminium, niobium and Al(95)-B(5) were prepared in order to fabricate an Al(96)-Nb(2)-B(2) master alloy. Aluminium was melted at 1063 K (790°C) for 3.6 ks (i.e. 1 h) and, subsequently, the Al(95)-B(5) master alloy was added to the melt and the temperature was increased to 1123 K (850°C). After 3.6 ks (1 h) at 1123 K (850°C), niobium powder was added and left to dissolve for 7.2 ks (2 h) with intermediate stirring. The microstructure of the Al(96)-Nb(2)-B(2) master alloy was analysed using optical and scanning electron microscopy (SEM). Understanding the fundamental effect of the addition of Nb-based intermetallics on the solidification behaviour was assessed by means of thermal analysis (i.e. recording the cooling curves using K-type thermocouples). For consistency, the same setup was used for all the undercooling experiments in which the thermocouple was placed in the middle of the 10-cm-diameter crucible at a height of 5 cm from the bottom of the crucible. For the grain refining experiments, 5% of Al(96)-Nb(2)-B(2) master alloy (equivalent to 0.1% of Nb) was added to the Al-8.4Si-0.3Mg-0.13Fe alloy melted at 1063 K (790°C) and left in contact with the melt for 1.8 ks (i.e. 30 min). Afterwards, the alloy was left to cool and poured from a temperature of 1013 ± 3 K (740 ± 3°C). Reference samples (without the addition of any master alloy) were also cast under the same processing conditions for the sake of comparison. In order to simulate the cooling conditions of many industrial processes and, thus, to solidify the alloy under a varying range of cooling rates, a wedge-shaped (copper) mould and a cylindrical (steel) mould were used. Characterisation was performed at both macro- and micro-level. Macro-visual analysis gave a first insight of the effect of the addition of the grain refinement whereas the microstructural analysis permitted the quantification of the grain size reduction as well as the effect of the addition of Nb-based intermetallics on the other microstructural features. Macro- and micro-etching (i.e. anodising for grain size measurements) were performed, respectively, by means of Tucker's and Barker's solutions. The analysis of the microstructures was conducted using a Carl Zeiss Axioskop 2 MAT optical microscope and the linear intercept method was used to determine grain sizes.

RESULTS AND DISCUSSION

Characterisation of the Al(96)-Nb(2)-B(2) Master Alloy

Once produced, the Al(96)-Nb(2)-B(2) master alloy was sectioned and analysed in order to confirm the formation and presence of, as well as to characterise, the (Nb-based) intermetallic compounds. Optical microscopy revealed that the Al(96)-Nb(2)-B(2) master alloy has a rather uniform distribution of intermetallic phases dispersed and embedded in the aluminium matrix as can be seen in Fig. 1a. Moreover, it indicated that there were at least two different types of intermetallics which differ in both colour and size, the black ones being larger than the dark grey ones.

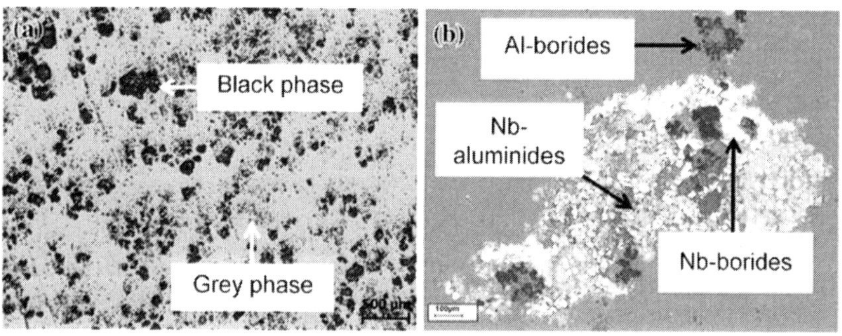

Figure 1. Details of the nature and distribution of the intermetallic particles present in the Al(96)-Nb(2)-B(2) master alloy: (a) bright field optical micrograph and (b) SEM picture

Examining the master alloy with the SEM permitted the confirmation of the formation and nature of the Nb-based intermetallics (i.e. NbB_2 and Al_3Nb) due to the reaction of niobium with boron or aluminium, respectively (see Fig. 1b). Furthermore, it was found that aluminium borides (with non-stoichiometric composition AlB_x, where $x = 2–12$ on the basis of the semi-quantitative EDS chemical analysis) were also present. They are thought to derive from the employment of the Al(95)-B(5) master alloy used as a source of boron because of their less favourable enthalpy of formation in comparison to niobium borides.18 Nonetheless, it is also possible that some of them may have formed during the fabrication of the Al(96)-Nb(2)-B(2) master alloy. Although aluminium borides (AlB_x) have some refining effect on Al-Si alloys, their performance is limited if titanium is present as an impurity in the alloy.19 This is because aluminium borides tend to transform to titanium borides due to the greater stability of the latter. The Al-Si alloy considered in this study has a relatively high titanium impurity level (~0.1) and, therefore, the great majority of AlB_x will transform to Ti-borides, which cause Si poisoning. Consequently, the refining effect of aluminium borides (AlB_x) and Ti-borides on the performances of the Al(96)-Nb(2)-B(2) master alloy can be ignored.

Fundamental Effect of Nb-Based Intermetallics on Nucleation

Thermal analysis, i.e. the measurement of the cooling curve during the solidification of a metal, has been proposed and used as an estimation of the effect of the addition of heterogeneous nucleation substrates and, thus, the effectiveness of a grain refiner.20,21The solidification curve of the Al-Si alloy considered in this study (reference) is the classical solidification curve with a valley, whose minimum and recalescence temperatures are 887.0 K (614.0°C) and 888.4 K (615.4°C), as can be seen in Fig. 2. The undercooling (ΔT) needed for aluminium atoms to group into clusters and form stable nuclei which subsequently grow into grains is the difference between the two previously mentioned temperatures which, therefore, equals 1.4 K (see Table I).

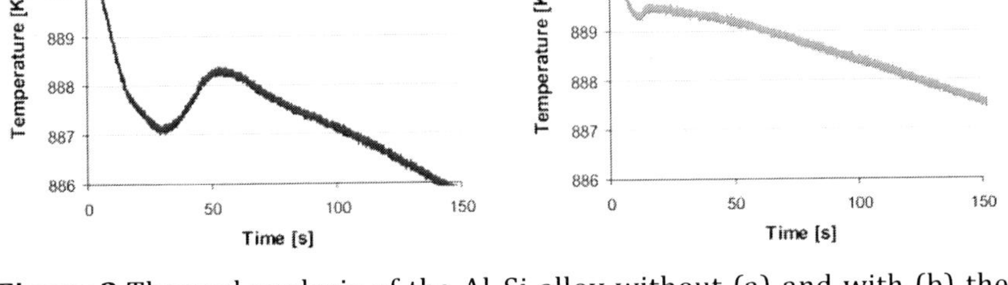

Figure 2. Thermal analysis of the Al-Si alloy without (a) and with (b) the addition of the Al(96)-Nb(2)-B(2) master alloy

Table I. Results of the analysis of the cooling curves shown in Fig. 2

Material	Valley minimum temperature [K (°C)]	Valley maximum temperature [K (°C)]	Δ *T* (K)
Reference	887.0 (614.0)	888.4 (615.4)	1.4
Al(96)-Nb(2)-B(2) master alloy addition	889.3 (616.3)	889.6 (616.6)	0.3

As can be seen in Fig. 2b, the introduction of the Nb-based intermetallic particles displaces the cooling curve to a higher temperature, which can be expected when grain refiners are added.20,22,23 From the analysis of the cooling curves (data reported in Table I), it can be seen that the addition of the Al(96)-Nb(2)-B(2) master alloy induces the reduction of the undercooling from 1.4 K to 0.3 K. The reduction of the energy barrier needed for the nucleation of the α-Al grains confirms the hypothesis that the Nb-based intermetallics are good potent heterogeneous nucleation sites for the nucleation and growth of aluminium grains. Thermal analysis was carried out with the aim of understanding the effect of Nb-based intermetallic particles and a reduction was expected on the basis of the interrelationship which exists between the lattice mismatch parameter (f) and the undercooling ($\Delta T = \alpha \cdot f$, where α is a variable that depends on physical properties such as heat of fusion, molar volume of crystals, etc.).24 It is worth recalling that the fact that Nb-based intermetallics have good lattice match with the Al nucleating phase was one of the reasons why the Al(96)-Nb(2)-B(2) master alloy was developed.

Assessment of the Effect of Nb-Based Intermetallics on Microstructural Features

As just explained, the reduction of the undercooling confirmed that enhanced heterogeneous nucleation is the main solidification mechanism and, consequently, because of the greater number of nucleation events, a reduction in the final grain size is expected. The comparison of the results of the microstructural analysis of the Al-Si alloy without (Fig. 3) and with the addition of the Al(96)-Nb(2)-B(2) master alloy (Fig. 4) is self-explanatory. Nevertheless, there are some points that should be stressed.

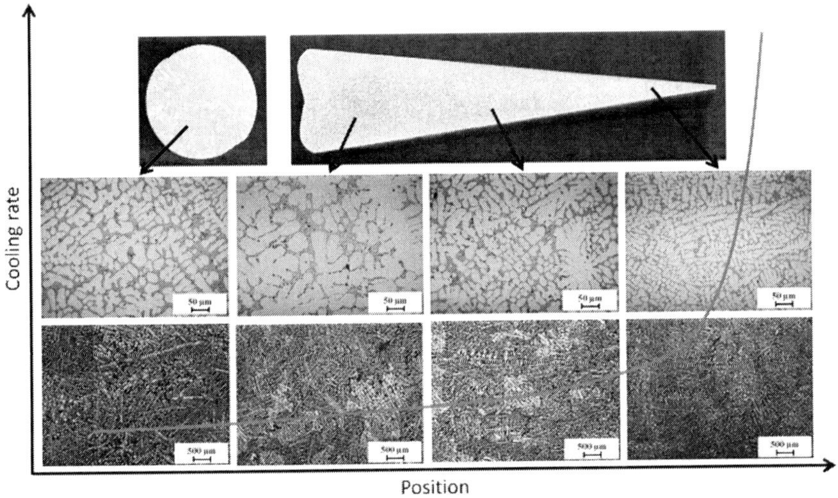

Figure 3. Microstructural analysis results for the Al-Si alloy without the addition of the Al(96)-Nb(2)-B(2) master alloy

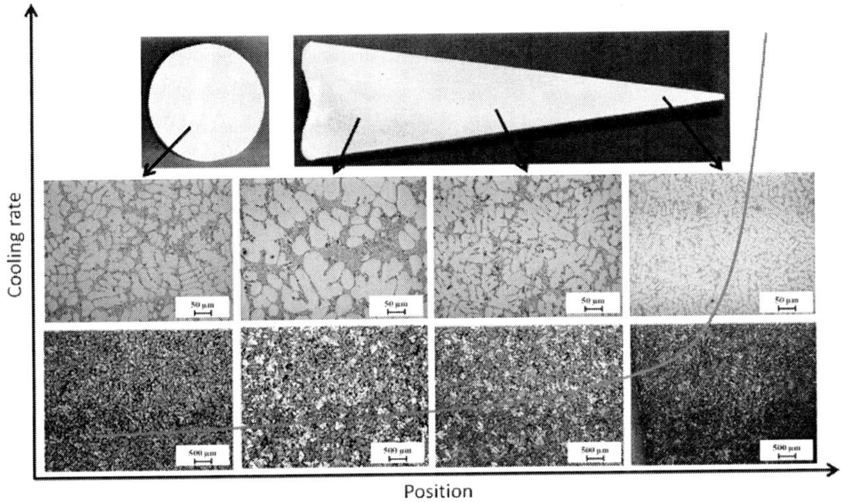

Figure 4. Microstructural analysis results for the Al-Si alloy with the addition of the Al(96)-Nb(2)-B(2) master alloy

In the case of the reference (Fig. 3), as the cooling rate decreases the size of the aluminium grains (polarised light-coloured micrographs) increases significantly passing from hundreds of microns to millimetres. The fine grain microstructure at the tip of the wedge-shaped sample is due to the very fast heat extraction rate (~100 K/s at the point where the micrograph was taken) and short solidification time of the thin section. The distribution of the second phase (silicon intermetallic) moves from rather homogeneously distributed amongst the fine dendritic arms (fast cooling) to rather inhomogeneous entrapment between the secondary arms of the dendrites. No comment can be made about the morphology of the secondary Si eutectic phase as the Al-Si alloy comes already modified (i.e. with the addition of strontium to change the morphology from plate-like to fibrous).25 These variations indicate that the nucleation process mainly depends on the way the heat is extracted from the solidification front and solute rejection. It is worth mentioning that columnar grains and the columnar/equiaxed transition zones are present in the reference material, especially at relatively slow cooling rates (i.e. cylindrical sample). After the addition of the Al(96)-Nb(2)-B(2) master alloy, it is clear that the grain structure is finer in each section (i.e. regardless of the cooling rate) and that the increment in the grain size with the decrement of the heat extraction rate is far less pronounced (Fig. 4) with respect to the reference material (Fig. 3). That is, the introduced Nb-based intermetallics actually act has heterogeneous nucleation sites for the α-Al grains, enhancing their formation (grain refinement), making the solidification process less dependent on the heat extraction from the solidification front and solute rejection. A direct consequence of the finer structure of the aluminium grains is a more uniform distribution of the secondary phase due to the more even scattering of silicon when rejected from the solidification front or in the solute-rich pools remaining in interdendritic regions. Trace elements introduced into the melt via the use of the Al(96)-Nb(2)-B(2) master alloy also contribute to growth restriction during solidification. Nevertheless, their effect is not as critical, as in the case of refinement of pure aluminium, due to their relatively low contribution in comparison to the effect of amount alloying elements of the Al-8.4Si-0.3Mg-0.13Fe alloy. It is worth mentioning that the addition of the Al(96)-Nb(2)-B(2) master alloy did not affect the Si modification because the eutectic phase is still characterised by a fibrous morphology after grain refinement (Fig. 4). Finally, the enhanced heterogeneous nucleation promoted by the Nb-based intermetallics suppresses the columnar/equiaxed transition zone in addition to the formation of columnar grains, especially at low cooling rates. The grain refining effectiveness of the Nb-based intermetallics as a function of the cooling rate is shown in Fig. 5.

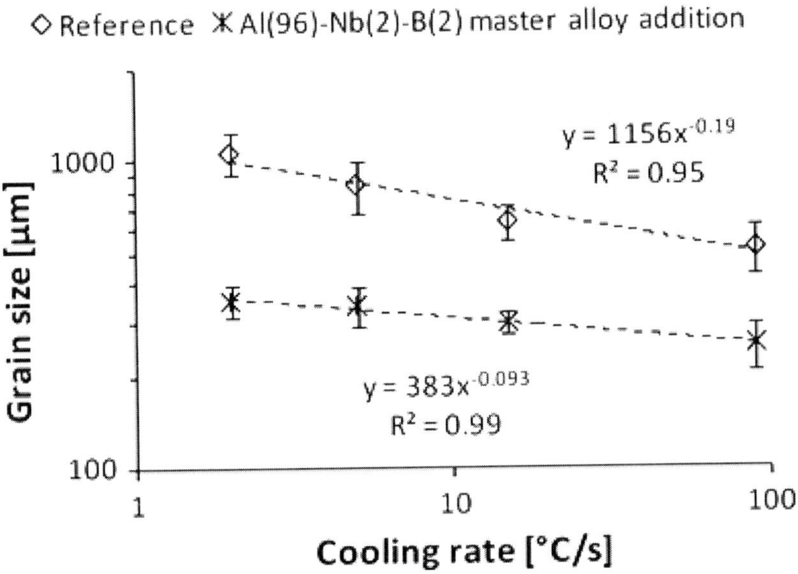

Figure 5. Variation of the grain size as a function of the cooling rate

The trends of the grain size with respect to the cooling rate, shown in Fig. 5, are as expected from the microstructural analysis, and it was quantitatively proven that the difference between the Al-Si alloy grain size without and with the addition of the Al(96)-Nb(2)-B(2) master alloy decreases with increasing the cooling rate. This does not mean that the effectiveness of the Nb-based intermetallics in promoting heterogeneous nucleation decreases, rather that the heat extraction mechanism becomes more important for higher cooling rates. The plotting of the data in logarithmic scales (Fig. 5) gives as a result two linear straight lines and the experimental data from the microstructural analysis can reasonably be represented by power-law equations ($R^2 \geq 0.95$). On the basis of this equation, the grain size is proportional to the cooling rate although there are two experimental parameters that must be determined (i.e. grain size = $\beta \cdot$(cooling rate^n) where β is the pre-exponential constant). It is worth mentioning that both β and n are composition-dependent. In the literature, it can be found that the distance between the arms of the primary dendrite depends on, apart from other physical parameters, the growth velocity and the temperature gradient.26 Assuming that there are no significant variations in the physical parameters (such as the Gibbs–Thomson coefficient) due to the addition of the Al(96)-Nb(2)-B(2) master alloy, the variation of the grain size can actually be estimated as proposed, with power-law equations.27 The lower β and n parameters of the alloy refined by means of the addition of the Al(96)-Nb(2)-B(2) master alloy is a further indication that the grain size and, thus, the nucleation/growth depends less on the thermal processing parameters and in particular the cooling rate.

CONCLUSION

In this study, micrometric Nb-based intermetallics were formed in situ in an Al(96)-Nb(2)-B(2) master alloy in order to take advantage of their intrinsic characteristics such as low density and high-temperature stability. This work demonstrates that Nb-based intermetallics can successfully be used to refine the microstructural features of Al-Si alloys because they act as heterogeneous nucleation substrates and promote the formation and growth of α-Al grains. This is possible because Nb-based intermetallics present low lattice mismatch (30.6 and 4.2 for NbB_2 and Al_3Nb, respectively) with the crystal structure of the aluminium grains. From the experimental results, it is concluded that the addition of Nb-based intermetallics makes the solidification process and the corresponding microstructural features less sensitive to the cooling conditions as well as being much finer and more uniformly distributed.

ACKNOWLEDGEMENT

The authors are thankful for the financial support from the Technology Strategy Board (TSB) through the TSB/101177 Project and to the Engineering and Physical Sciences Research Council (EPSRC) through the EP/J013749/1 and EP/K031422/1 Projects.

REFERENCES

1. A.S. Cooper, E. Corenzwit, L.D. Inginotti, B.T. Matthias, and W.H. Zachariasen, *Proc. Natl. Acad. Sci. USA* 67, 313 (1970).

2. Yamamoto, C. Takao, T. Masui, M. Izumi, and S. Tajima, *Phys. C* 383, 197 (2002).

3. H. Takagiwa, E. Nishibori, N. Okada, M. Takata, M. Sakata, and J. Akimitsu, *Sci. Technol. Adv. Mater.* 7, 22 (2006).

4. C.-P. Reipt and G. Sauthotit, *Intermetallics* 3, 159 (1993).
5. C.-P. Reipt and G. Sauthotit, *Intermetallics* 4, 377 (1996).
6. G.K. Sigworth, *Metall. Trans. A* 15A, 277 (1984).
7. M.M. Guzowski, G.K. Sigworth, and D.A. Sentner, *Metall. Trans. A* 18A, 603 (1987).
8. T. Sritharan and H. Li, *J. Mater. Process. Technol.* 63, 585 (1997).
9. A.L. Greer, P.S. Cooper, M.W. Meredith, W. Schneider, P. Schumacher, J.A. Spittle, and A. Tronche, *Adv. Eng. Mater.* 5, 81 (2003).
10. B.J. Mc Kay and P. Schumacher, *TMS 2005*, 155–164 (2005).
11. P. Schumacher and B.J. Mc Kay, *J. Non-Cryst. Solids* 317, 123 (2003).
12. G.E. Totten and D.S. Mac Kenzie, *Handbook of Aluminium vol. 2: Alloy Production and Materials Manufacturing* (New York: Marcel Dekker Inc., 2003), p. 81.
13. S.A. Kori, V. Auradi, B.S. Murty, and M. Chakraborty, *Mater. Forum* 29, 387 (2005)
14. T.E. Quested, A.T. Dinsdale, and A.L. Greer, *Mater. Sci. Technol.* 22, 1126 (2006).
15. L. Bolzoni, M. Nowak, and N. Hari Babu, *J. Alloys Compd.* 623, 79 (2015).
16. M. Nowak, L. Bolzoni, and N. Hari Babu, *Mater. Des.* 66, 366 (2015).
17. L. Bolzoni, M. Nowak, and N. Hari Babu, *Mater. Des.* 66, 376 (2015).
18. Ihsan Barin, *Thermochemical Data of Pure Substances* (Weinheim: VCH Verlagsgesellschaft mbH, 1993), p. 1080
19. Y. Birol, *J. Alloys Compd.* 513, 150 (2012).
20. B.S. Murty, S.A. Kori, and M. Chakraborty, *Int. Mater. Rev.* 47, 3 (2002).
21. D. Apelian, G.K. Sigworth, and K.R. Whaler, *AFS Trans.* 92, 297 (1984).
22. M. Easton and D.H.S. John, *Metall. Mater. Trans. A* 30A, 1625 (1999).
23. J.A. Spittle, *Foundry Trade J.* 181, 308 (2008).
24. D. Turnbull and B. Vonnegut, *Ind. Eng. Chem.* 44, 1292 (1952).
25. S.-Z. Lu and A. Hellawell, *JOM* 47, 38 (1995).

26. J.A. Dantzig and M. Rappaz, *Solidification* (Lausanne: EPFL Press, 2009), p. 324

27. C. Merton, *Flemings, Solidification Processing* (New York: McGraw-Hill, 1974).

3

CHAPTER

Additive Manufacturing of High-Entropy Alloys by Laser Processing

V. Oceli´K,[1,2] N. Janssen,[1] S.N. Smith,[1] and J.Th.M. De Hosson[1]

[1]Department of Applied Physics, Zernike Institute for Advanced Materials, University of Groningen, 9474 AG Groningen, The Netherlands

ABSTRACT

This contribution concentrates on the possibilities of additive manufacturing of high-entropy clad layers by laser processing. In particular, the effects of the laser surface processing parameters on the microstructure and hardness of high-entropy alloys (HEAs) were examined. AlCoCrFeNi alloys with different amounts of aluminum prepared by arc melting were investigated and compared with the laser beam remelted HEAs with the same composition. Attempts to form HEAs coatings with a direct laser deposition from the mixture of elemental powders were made for AlCoCrFeNi and AlCrFeNiTa composition. A strong influence of solidification rate on the amounts of face-centered cubic and body-centered cubic phase, their chemical composition, and spatial distribution was detected for two-phase AlCoCrFeNi HEAs. It is concluded that a high-power laser is a versatile tool to synthesize interesting HEAs with additive manufacturing processing. Critical issues are related to the rate of (re)solidification, the dilution with the substrate, powder efficiency during cladding, and differences in melting points of clad powders making additive manufacturing processing from a simple mixture of elemental powders a challenging approach.

INTRODUCTION

Alloying is accepted as a classic method to improve mechanical properties and performance, as for instance has been well executed in steel, aluminum, or magnesium alloys. In the field of "materials design," Yeh et al.[1,2] made a breakthrough by putting forward the concept of "high-entropy alloys" (HEAs). These alloys are claimed to belong to a new class of materials that present novel phase structures and properties. In the literature, HEAs are defined to have at least five metal constituents in equimolar or near-equimolar ratios.[2] Although HEAs with a hexagonal crystal structure have been reported,[3] previous studies have shown that most HEAs consist of simple face-centered cubic (fcc) or body-centered cubic (bcc) solid solutions instead of intermetallic or other complex phases due to the high-entropy effect.[1] The underlying mechanism behind the HEAs is a minimization of the Gibbs free energy through a balance between entropy and enthalpy.[4] The basic idea in developing HEAs is that a negative Gibbs free energy of mixing can be achieved, even when the enthalpy of mixing is positive or close to zero, through an increasing mixing entropy contribution with increasing temperature. Ideally this is so in the case of random solid solution (the so-called regular solution in the quasi-chemical approximation). This trick is often spoiled and hampered by the appearance of intermetallic compounds, and the basic idea should be considered as an interesting materials design guideline, not as a "necessary" and neither as a "sufficient" condition in materials design.

Nonetheless, a lot of research work has shown that HEAs possess many promising properties, such as high hardness, strength and ductility,[5] superior thermal stability, excellent high-temperature oxidation resistance, high wear[6–9] and corrosion resistance,[10] and interesting magnetic properties.[11] Apart from bulk material consolidation methods such as casting and sintering, HEAs can also be deposited as a surface coating. Various techniques of welding,[9, 12] thermal spraying,[13] selective laser melting,[14] direct laser deposition[8, 15–18] and additive manufacturing[19–22] are already reported in the literature as experimental methods to produce HEA coatings.

Laser processing is particularly advantageous in synthesizing HEA coatings due to its ability to reach a high instantaneous temperature that leads to stabilizing the HEA phase due to high quench rates associated with the technique.[16] In particular, rapid quenching becomes important as HEA may undergo various phase transformations while slowly cooling due to reduction in the thermodynamic driving force at lower temperatures. Furthermore, rapid quenching restricts diffusion of the elements and restricts nucleation and growth of the brittle intermetallic compounds, thereby retaining the HEA phase.[16] Another advantage of direct laser deposition is that coatings could be deposited in a few steps, which eliminates the influence of the substrate and allows gradual chemical compositional changes through the coating thickness.[23] Special additive manufacturing techniques like selective laser melting[14] allow also for building three-dimensional (3D) dimensional objects directly from FeCoCrNi HEA pre-alloyed powder, exhibiting high strength and ductility that are comparable with engineering materials like stainless steels. Direct deposition techniques like laser or TIG cladding are suitable to deposit HEA from a mixture of elemental powders. Such a mixture has been used to clad TiVCrAlSi HEA on Ti-6Al-4V substrate.[15] We have noticed also disadvantages of these direct deposition methods of HEAs, namely, the evaporation of certain elements, different rates of intake of the powders, as well as dilution of the coating materials by the substrate during laser cladding. To avoid complications with powder feeding, a mixture of commercial pure elemental powders could be preplaced on the surface and then remelted by the laser beam.[16,17,24]

Typically, a slow laser beam scanning speed (<10 mm/s) is used to allow a proper mixture of elemental phases inside the melt pool formed under the laser beam.[17,25–27] However, these scanning speeds still result in much higher solidification rates than during classic arc melting and casting. In the case when pre-alloyed HEA powders are used,[22] a wider range of scanning speeds could be used to study the influence of solidification rates on the microstructure and properties of laser-deposited HEAs. It is expected that a decrease of grain size of AlCoCrFeNi deposits with increasing scanning speed will increase the microhardness.

The microstructure, phases, and properties of the HEA system Al_xCoCrFeNi (x = 0–1.8) were recently studied by Wang et al.[28] at room and elevated temperatures. A rather strong influence of Al on the phases at room temperature as well as at elevated temperatures was investigated. At a low Al content (x = 0–0.3), an fcc solid solution phase is formed, followed by a mixture of fcc and bcc phases for higher amounts of Al and a pure bcc for Al content higher than x = 1.2. The microstructure is also influenced by homogenization at elevated temperatures, which usually follows the casting.

Comparative study of arc melted and laser beam fabricated Al_xCoCrFeNi (x = 0.3, 0.6, and 0.85) HEAs[19] shows similar microstructures for low and high Al content alloys with a strong ⟨001⟩ fiber texture along the direction of solidification. A very first review of HEA coatings fabricated mainly by laser cladding and their application prospects was published by Zhang et al.[29]

The goal of our experimental work is a comparative study of the microstructure and hardness among HEAs prepared by conventional arc melting with a laser-beam-assisted surface coating technique. Laser beam remelting of Al_xCoCrFeNi HEAs should demonstrate the strong influence of the (re)solidification rate on the microstructure and properties of HEAs, i.e., different for conventional casting and laser surface treatments. Direct laser deposition of HEA coatings from elemental powders mixture should prove an ability of this method for fast and versatile HEA surface coatings fabrication.

EXPERIMENTAL PROCEDURE

The HEAs studied in this work were prepared by vacuum arc melting and by laser processing. Ingots with nominal compositions of Al_xCoCrFeNi (x values in molar ratios,x = 0.3, 0.5, and 0.7) alloys were prepared by arc-melting, homogenized at 1250°C, and forged at the same temperature with reduction of 50%. Energy dispersive x-ray spectroscopy (EDS) analysis performed after sample preparation reveals different compositions in comparison with the nominal compositions. In the following, the specimens will be marked according their real chemical compositions: 0.4, 0.7, and 1.0, respectively.

The second set of specimens was prepared by surface laser remelting of the samples prepared as mentioned earlier. The surface was sandblasted to increase absorption of the laser beam and then remolten using the 1.07-μm wavelength radiation generated by 3-kW IPG Fiber laser. A laser beam spot with a diameter of 1.5 mm (Gaussian distribution) was used to scan the surface with speeds of 5 and 10 mm/s and laser power of 300 and 450 W, respectively.

A third group of samples was prepared by direct laser beam deposition on the SS 305 steel bar substrate using the laser beam scanning speed of 5 mm/s and laser power between 600 and 650 W. A premixture of pure elemental powders (particles size between 5 μm and 120 μm) with equimolar nominal compositions of AlCoCrFeNi and AlCrFeNiTa was fed by a Sulzer Metco Twin10C powder feeder using a side cladding nozzle and forming a single layer coating with a thickness of about 500–800 μm. To avoid modification of the chemical composition due to dilution from the molten steel, substrate multiple layer coatings were created so that the effects of dilutions at the outer layer are minimized.

Samples for microstructural investigations and hardness measurements were cut and prepared by standard mechanical grinding and lapping with suspensions containing 9-μm, 3-μm, and 1-μm diamond particles and polishing with oxide polishing suspension (OPS) colloidal silica. A Philips XL30 Field Emission Gun scanning electron microscope (SEM) and Tescan Lyra Dual beam system were used for microstructural observations in secondary electron (SE) and backscattered electron (BSE) imaging mode. The constituent phases were determined using a combination of EDS and electron backscatter diffraction (EBSD) using an EDAX Octane Super EDS detector and TSL EBSD system equipped with a Hikari Super CCD camera. Team 4.2 data collection and analysis software was used; EBSD data were further analyzed by TSL OIM Analysis 7.3 software.

The XRD analyses were done using an x-ray diffractometer (XRD, Bruker D8 Advance) with Cu Kα radiation. The samples were scanned over a 2θ range from 20° to 120° with a step size of 0.02° and a step time of 10 s. The XRD spectra were analyzed using Match software.

RESULTS AND DISCUSSION

Arc-Melted Al_xCoCrFeNi HEAs

The peaks in the XRD spectra were assigned to the crystallographic planes and phases in the samples prepared by arc melting. The assignment agreed well with literature.28 Closer inspection of the position of (111) fcc phase peak revealed a peak shift with increasing aluminum concentration. This convolutes to a shift in the fcc lattice parameter from $a_{Al\_0.4}$ = 3.560 Å to $a_{Al\_0.7}$ = 3.592 Å and to $a_{Al\_1.0}$ = 3.603 Å. This phenomenon is expected since aluminum atoms are much larger than the other constituent atoms. A similar feature is observed for the bcc lattice parameter as Table I clearly demonstrates. The lattice parameters of both phases were calculated by averaging the lattice parameters derived from the various peaks in the XRD spectra.

Table I. Lattice parameters (Å) for the fcc and bcc phase in arc-melted Al_xCoCrFeNi samples calculated from corresponding peak positions in XRD spectra with their standard deviations

	$a_{Al\_0.4}$	std	$a_{Al\_0.7}$	std	$a_{Al\_1.0}$	std
fcc	3.581	0.030	3.596	0.003	3.605	0.003
bcc	2.880	0.004	2.899	0.010	2.890	0.011

std standard deviation.

The reason why the peak shift is significantly larger from Al_0.4 to Al_0.7 than for Al_0.7 to Al_1.0 was explained by Wang et al.28 to be an upper limit of aluminum concentration in the fcc and bcc phases. At that limit, the chemical compositions of the two phases does not change anymore; i.e., only the relative amount of fcc with respect to bcc changes.

As Fig. 1 clearly demonstrates, there are different amounts of fcc and bcc phase present in arc-melted samples. Al_0.4 contains 99.9%, Al_0.7 contains 97.0%, and Al_1.0 contains 65.3% of fcc phase, respectively. Figure 2 summarizes the microstructural study of the Al_1.0 sample by a combination of EDS and EBSD mapping technique from a larger area ($250 \times 550\ \mu m^2$ with step of $0.4\ \mu m$). The [001] inverse pole figure map (Fig. 2a) shows no preferential grains orientation for neither the bcc nor the fcc phase. However, different deformation mechanisms could be clearly detected in the fcc and bcc phases. The fcc phase is full of Σ3 twins (60°@ ⟨111⟩), shown as yellow lines in all maps depicted in Fig. 2. Figure 2c shows the ratio between parent and daughter fraction of fcc grains. A total of 88.5% of fcc grains are twinned with 35% of twinned volume. On the other side, the bcc phase is characterized by large, highly deformed grains. Figure 2d shows the map of local average misorientation, indicating a high local deformation through dislocation glide and climb mechanisms. High plastic deformation found in all three arc-melted samples is a consequence of the 50% forging applied in the last step of their preparation.

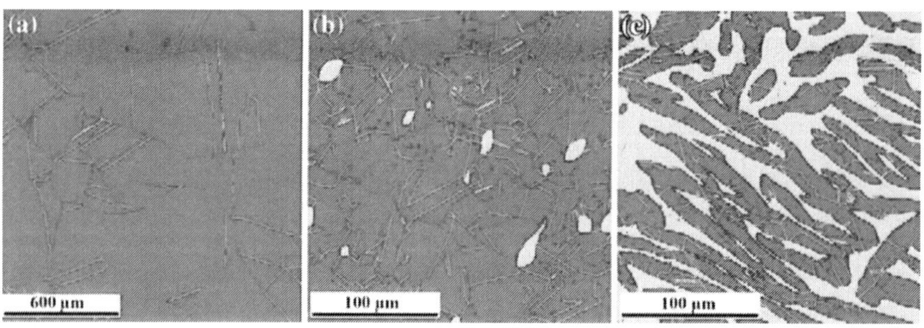

Figure 1.EBSD phase map of (a) Al_0.4, (b) Al_0.7, and (c) Al_1.0 arc-melted samples showing fcc (red) and bcc (green) phases. Grain boundaries (>5°, black lines) and Σ3 twin boundaries (yellow lines) in fcc phase are also shown

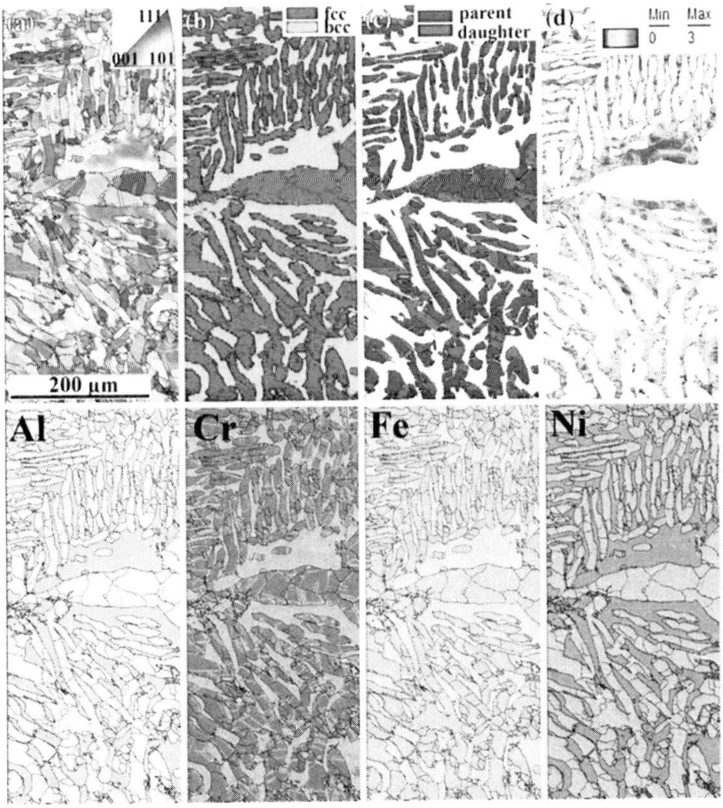

Figure 2 Microstructural characteristics of Al_1.0 alloy studied by combination of EBSD and EDS mapping of about 1 M points with step of 0.4 μm. (a) [001] Inverse pole figure with 5° grain boundaries (black lines) and Σ3 twin boundaries (yellow lines) in fcc phase; (b) EBSD phase map; (c) EBSD map of parent and daughter Σ3 twins in fcc phase; (d) local average misorientation map in bcc phase (blue 0°, red 3°). Bottom: Al, Cr, Fe, and Ni concentration EDS maps showing Al, Ni-rich bcc, and Cr-, Fe-rich fcc phases. Grain boundaries (>5° in black) and Σ3 twins in fcc phase (yellow) are also shown

The EDS maps in Fig. 2 show clearly elemental segregation. The bcc phase is rich of aluminum and nickel, whereas the fcc phase is rich of chromium and iron. Table IIsummarizes elemental compositions of both phases in all three samples measured by EDS. The relative amounts of Co, Cr, Fe, and Ni are almost the same for all three alloys. The composition of the fcc phase in Al_0.7 and Al_1.0 is also very close to each other, but the composition of the bcc phase in these two is different. The bcc phase in Al_1.0 is less rich of Al and Ni, and more rich of Co, Cr, and Fe.

Table II verall chemical composition (at.%) of the arc-melted samples and compositions of the phases therein determined by EDS analysis (note that the Al_0.4 consists of the fcc phase only)

	Al_0.4	Al_0.7			Al_1.0		
	fcc	Overall	fcc	bcc	Overall	fcc	bcc
Al	9.3	14.6	13.7	33.0	19.8	13.3	27.2
Co	22.9	21.6	21.9	16.5	20.5	22.2	18.5
Cr	22.7	21.5	21.8	11.2	20.0	23.2	16.1
Fe	22.7	21.4	21.9	12.7	20.0	23.1	16.2
Ni	22.4	20.9	20.7	26.6	19.7	18.2	22.0

The grain size distributions of arc-melted samples are plotted in Fig. 3. The grain sizes of the different samples vary greatly. In $Al_{0.4}CoCrFeNi$, the grain size is larger than 500 μm in diameter. In $Al_{0.7}CoCrFeNi$, small equiaxed grains of bcc appear surrounded by much bigger equiaxed fcc grains, although in this particular alloy, the fcc grains are much smaller than in Al_0.4. With AlCoCrFeNi, the fraction of bcc is even higher. The bcc grains are bigger and the fcc grains smaller than in Al_0.7. The bcc grains form large connected grains in between the fcc phase.

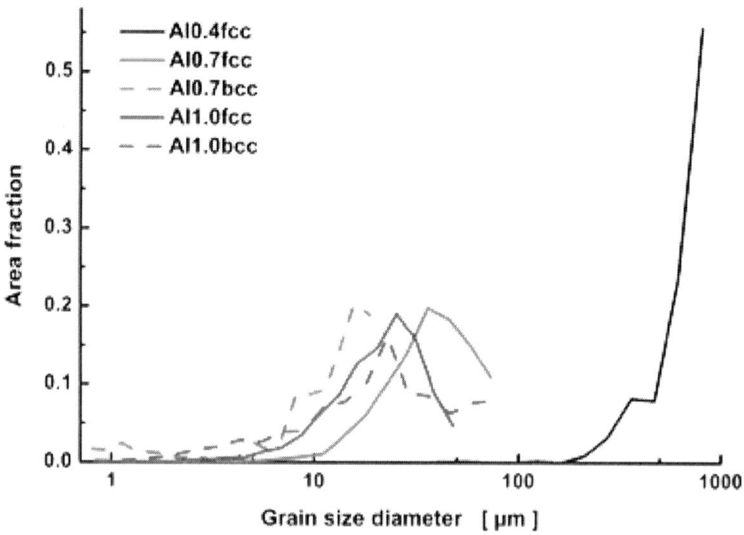

Figure 3. Grain size distributions of the fcc and bcc phases of the arc-melted samples; twin boundaries and grains at the edge of the scans are excluded

Vickers hardness tests showed differences in hardness between the samples, The hardness is almost doubled going from Al_0.4 (148 HV) to Al_1.0 (278 HV). This increase in hardness is attributed to the increase of the fraction of bcc phase in the alloy. The Al_0.4 has almost no bcc phase. The phase fraction of bcc in Al_0.7 is only 3%, and a small increase of 40 HV is measured with respect to Al_0.4. However, the hardness increases of 90 HV going from Al_0.7 to Al_1.0, where the phase fraction of bcc is approximately 35%. The increase of the interface area between fcc and bcc phases and the reduction of grain size probably also contributes by reducing the mean free path length of moving dislocations.

Laser Beam Remelted Al_xCoCrFeNi HEAs

To investigate the microstructure and properties of HEAs with the same compositions, but faster solidification, remelted tracks were made on the original arc-melted samples using a moving laser beam. Two tracks were remelted in each of the three samples. One track had a laser power of 300 W with a beam speed of 5 mm/s and one track had a laser power of 450 W at a beam speed of 10 mm/s, hereinafter referred to as Track1 and Track2, respectively. Both tracks were approximately 1.2 mm wide.

The remelted tracks in $Al_{0.4}$CoCrFeNi do not show any change with respect to the nonremelted alloy. The phase and orientations of the grains inside the remelted area are not modified. Since the grain size was bigger than the depth of the remelted track, this suggests that during the resolidification inside the molten zone, the grains keep the orientation of the grains that were there originally. This was the case for both remelting speeds. The overall compositions of the remelted zones in the rest of the samples shown in Table III resemble the compositions of the nonremelted samples quite well. This can be seen by comparing Table II with Table III.

Table III. Overall composition (at.%) of the remelted tracks and compositions of their phases determined by EDS analysis

	Al_0.7						Al_1.0	
	Track1			Track2			Track1	Track2
	Overall	fcc	bcc	Overall	fcc	bcc	bcc	bcc
Al	14.9	13.4	19.7	15.3	14.2	19.6	19.9	20.1
Co	21.5	22.3	19.1	21.4	22.1	19.2	20.4	20.5
Cr	21.3	21.4	21.2	21.2	21.1	21.2	20.0	19.9
Fe	21.4	22.5	18.4	21.3	22.1	18.7	20.1	20.0
Ni	20.9	20.4	21.6	20.8	20.5	21.3	19.6	19.5

The phases observed in the remelted samples are also the same as the phases seen in the original arc-melted samples and so are the phase fractions outside the remelted zones. However, the phase fractions and distribution of fcc and bcc phases inside the remelted zones differ substantially, as well as the elemental compositions of fcc and bcc phases. In the microstructure of a remelted Al_0.7 sample, there are some remarkable microstructural changes, as depicted in Fig. 4. The grain size of the fcc phase in the remelted zone is reduced to about 30 μm in diameter, and only very small islands of the bcc phase could be detected on an overall scan shown in Fig. 4a. The detailed scan inside the remelted zone (Fig. 4b) reveals a dendritic structure of fcc grains and the presence of bcc phase only inside interdendritic areas with the size of individual islands of about 2 μm in diameter. However, crystallographic orientation of these islands is the same inside one fcc dentritic grain, which means that the bcc phase solidified at the end of solidification as an "eutectic phase." The amount of the bcc phase in the remelted zone has increased to 7.5% in comparison with only 3% in the nonremelted Al_0.7 sample. EDS maps for this detailed scan are shown at the bottom of Fig. 4. They clearly discover elemental partitioning for Al and Fe, while the other three elements were partitioned much less, as Table III also demonstrates.

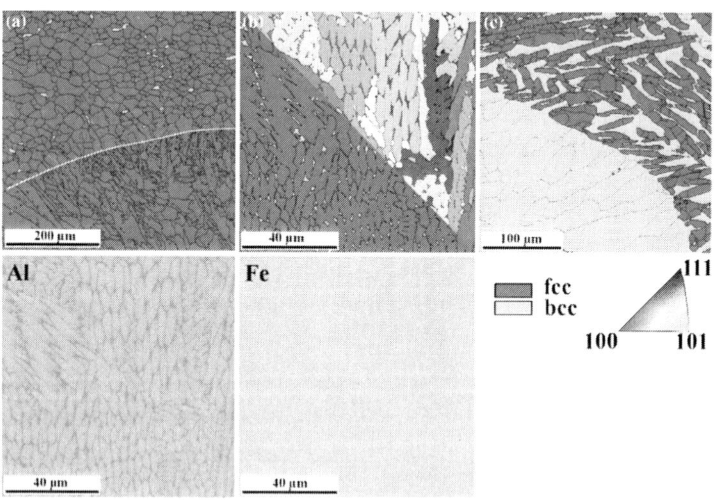

Figure 4. Microstructural characteristics of Al_0.7 and Al_1.0 alloy after laser beam remelting studied by a combination of EBSD and EDS mapping. (a) EBSD phase map of the interface between substrate and remelted zone of Al_0.7 sample. Interface is marked by white line, and laser remelted zone at the bottom; (b) combination of [001] inverse pole figure map (top-right) and phase map (bottom-left) inside remelted zone of Al_0.7 sample; (c) EBSD phase map of the interface between substrate and remelted zone of Al_1.0 sample, and laser remelted zone at the bottom; bottom: Al and Fe concentration EDS maps inside of remelted zone shown in (b)

The mechanism that favors the formation of the bcc phase in rapidly solidified volumes becomes apparent in remelting of the Al_1.0 sample (see Fig. 4c). In this alloy the entire remelted zone has turned into big columnar bcc phase grains (up to 100 μm long and tens of microns wide), elongated in the solidification direction. The phase fraction of bcc is 99.9%. This difference with the phases present in the original vacuum arc-melted sample (bcc phase fraction of 35%) finds its origin in the high solidification rate after laser remelting. We estimated cooling rates in our remelting experiments using a simple laser melting model,30 with 3.9×10^3 K/s and 3.3×10^4 K/s, for a laser beam velocity of 5 mm/s and 10 mm/s, respectively. It is clear that these cooling rates do not allow the full partitioning required for the formation of two phases in this HEA. When the overall composition becomes close to composition of the bcc phase, the fast solidification allows only the formation of this bcc phase.

In Table III, the compositions of the phases inside the remelted tracks are also shown. In both Al_0.7 and Al_1.0, no apparent difference was observed between the tracks with different speeds, neither in composition nor microstructure. This means that these combinations of laser power and scanning speed yield a similar solidification behavior. The fact that we do not observe any microstructural changes in the so-called "heat-affected zone" (area where the temperature during remelting was just below T_m) confirms the high thermal stability of both Al_0.7 and Al_1.0 alloys prepared by the arc-melting process.

Compared with arc-melted samples, the hardness of Al_0.7 has increased by almost 20%. This is due to the increased surface area between grains and phases because of the finer dendritic microstructure and a small increase in the amount of the bcc phase. In the case of Al_1.0, the laser remelted track is twice as hard (270 HV) as the non-remelted sample. This is due to the fact that the phase of the remelted zone is now solely bcc, which is the harder phase, compared with fcc. From here on, we may conclude that the synthesis procedure itself, affecting the fraction and spatial distribution of phases, is a very important parameter determining the hardness of AlCoCrFeNi HEAs.

Laser-Deposited AlCoCrFeNi and AlCrFeNiTa HEAs

A couple of attempts were made to laser clad equimolar AlCoCrFeNi and AlCrFeNiTa compositions using a standard side powder feeding setup.31 The clads were made with a scanning speed of 5 mm/s on a stainless steel bar substrate with a 50 mm diameter. In all cases, a 50% laser track overlap was used to create continuous coatings, and up to three layers of the coatings were built to eliminate compositional change by dilution32,33 from the substrate. Equimolar mixtures of elemental powders with purity higher than 99.9% were fed using argon as a carrier gas into the melt pool created by a defocused laser beam (diameter of Gaussian beam of about 2.3 mm). The typical laser power to clad one layer of the coating with a thickness of about 0.7–0.8 mm was ~600–650 W. We experienced that getting to an equimolar composition inside the HEAs coatings was a challenge. Starting with an equiatomic powder mixture does not mean that one ends up with the required composition. There are a few factors affecting the powder feeding efficiency for different elemental powders. Due to the different shapes, sizes, and densities of the powder particles, the probability for these particles to join the melt pool is also different. Obviously, also the various melting temperatures of the elements employed may influence the final solidification composition.

Therefore, a few iterations of elemental powders mixture are necessary to achieve the final clad layer composition being close to the required target. The composition profile by EDS of one of our best AlCoCrFeNi laser clad coatings is shown in Fig. 5. Here the coating profile is seen in a longitudinal cross-section parallel to the laser beam propagation direction. The whole coating contains three clad layers and is about 2.3 mm thick. A clear dilution effect could be observed in the first clad layer, where Fe content substantially increased in a vicinity of the substrate. Quite a homogeneous elemental distribution is present in the second and third clad layer, although the amount of nickel is far from the expected 20 at.%. Figure 5 also shows the longitudinal IPF map of the same layers. The map clearly demonstrates the columnar grain growth of bcc phase in the 〈001〉 direction perpendicularly to the solidification front.34 The second and third layer each contains 99.9% of the bcc phase, and the average hardness of this HEA coating was 492 ± 12 HV.

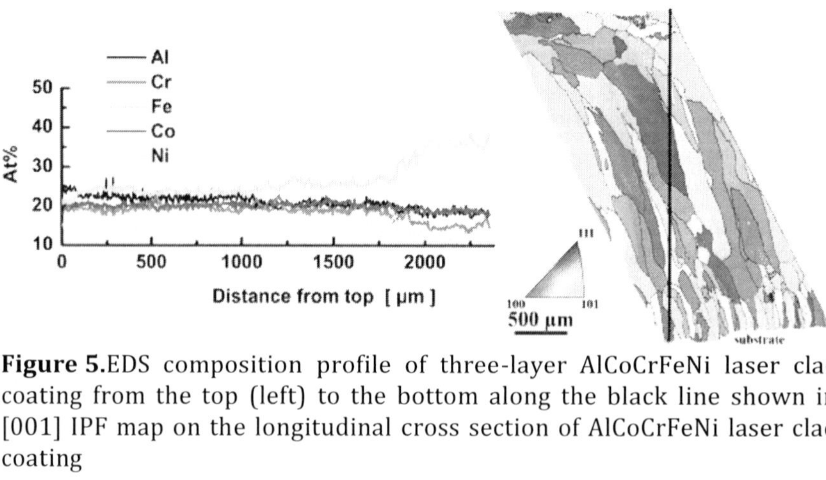

Figure 5.EDS composition profile of three-layer AlCoCrFeNi laser clad coating from the top (left) to the bottom along the black line shown in [001] IPF map on the longitudinal cross section of AlCoCrFeNi laser clad coating

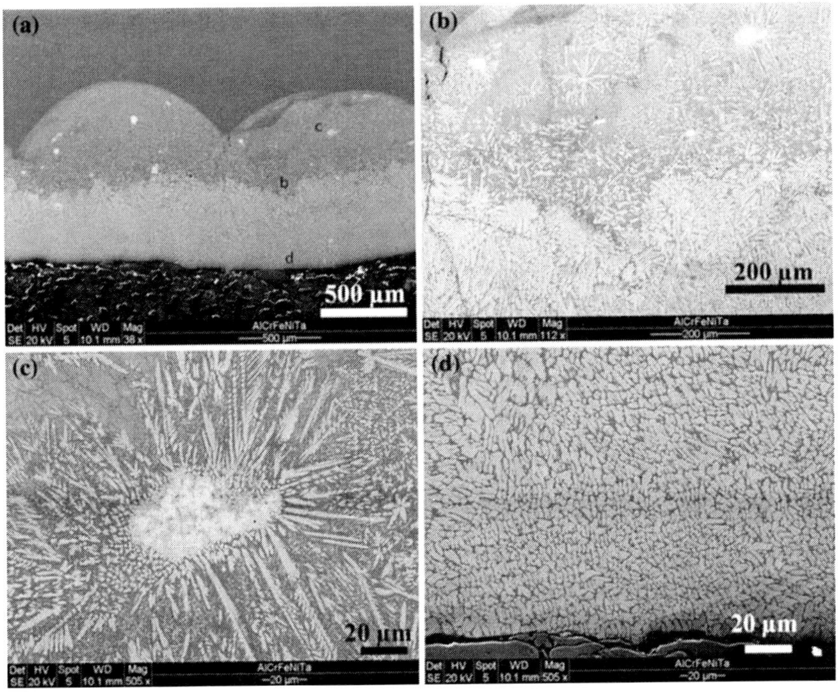

Figure 6. Secondary electron SEM images from AlCrFeNiTa laser clad coating. (a) An overview of the double-layered coating on transversal cross section. Note that the steel substrate is at the top. (b) Microstructure on the interface between layers. (c) Ta-rich area in the microstructure of the first layer. (d) Microstructure on the top of the second layer

Another attempt was made to clad a modified HEA composition. In this composition, Co was substituted by tantalum. Ta is a refractory element with an extremely high melting (3290 K) point. This melting point is even higher than the boiling point of aluminum (2792 K). The addition of tantalum was argued so as to give the alloy enhanced strength and structural stability at elevated temperatures due to an intensified sluggish diffusion. The composition was cladded with a scanning speed of 5 mm/s. Using a laser power of 600 W, a single and double layer coating were made, both with 50% overlap between adjacent tracks. The single layer clad showed about 70% dilution.

As a result, the single layer coating and the first layer of the double layer coating contain a very high amount of iron (up to 54 at.%) originating from the substrate. Even the second layer has still a high concentration of iron (~34 at.%). Another observation that stands out is the incredibly low

Al concentration in the top layer (7 at.%). This is probably due to the evaporation from the very hot melt. Based on the SEM images displayed in Fig. 6 in combination with EDS analysis, we conclude that the AlCrFeNiTa laser clad coating consists of multiple phases. Throughout the coating, sporadic bright white spots appear. These spots are almost pure (about 94 at.%) tantalum particles supposedly not molten in the laser cladding (Fig. 6a and c). On the top side of the clad (Fig. 6d), an equiaxed dendritic structure appears, clearly showing a two phase microstructure. The Vickers hardness values for this laser clad AlCrFeNiTa in the top layer are three times higher (860 HV) than in the AlCoCrFeNi arc-melted sample.

CONCLUSION

It is concluded that a high-power laser is a versatile tool to synthesize interesting high entropy alloys with additive manufacturing processing. From the laser remelting experiment, we see that the solidification rate is a very important factor for determination of the presence and distribution of phases in AlCoCrFeNi HEAs. Fast solidification substantially increases the preference for the bcc phase with a corresponding substantial increase in hardness. The high thermal stability of AlCoCrFeNi base HEA has been confirmed by the absence of a heat-affected zone in the vicinity of the remelting track. It goes without saying that the success of direct laser deposition of HEAs coatings using a feeding of mixture of elemental powders is affected by several materials and processing parameters, e.g., in-take of particular powders, melting temperatures, laser beam velocity, and power density.

ACKNOWLEDGEMENTS

Professor Peter K. Liaw and Dr. Z. Tang from the Department of Materials Science and Engineering, The University of Tennessee, are gratefully acknowledged for the preparation of arc-melted AlCoCrFeNi samples.

REFERENCES

1. J.W. Yeh, S.K. Chen, S.J. Lin, J.Y. Gan, T.S. Chin, T.T. Shun, C.H. Tsau, and S.Y. Chang, *Adv. Eng. Mater.* 6, 299 (2004).

2. B.S. Murty, J.-W. Yeh, and S. Ranganathan, *High Entropy Alloys* (Burlington, MA: Butterworth-Heinemann, 2014).
3. M. Feuerbacher, M. Heidelmann, and C. Thomas, *Mater. Res. Lett.* 3, 1 (2015).
4. F. Otto, Y. Yang, H. Bei, and E.P. George, *Acta Mater.* 61, 2628 (2013).
5. K.B. Zhang, Z.Y. Fu, J.Y. Zhang, J. Shi, W.M. Wang, H. Wang, Y.C. Wang, and Q.J. Zhang, *J. Alloy. Compd.* 502, 295 (2010).
6. C.-J. Tong, M.-R. Chen, S.-K. Chen, J.-W. Yeh, T.-T. Shun, S.-J. Lin, and S.-Y. Chang,*Metall. Mater. Trans. Phys. Metall. Mater. Sci.* 36, 1263 (2005).
7. S. Varalakshmi, R. Appa, M. Kamaraj, and B.S. Murty, *J. Mater. Sci.* 45, 5158 (2010).
8. X. Ji, H. Duan, H. Zhang, and J. Ma, *Tribol. Trans.* 58, 1119 (2015).
9. Y.C. Lin and Y.H. Cho, *Surf. Coat. Technol.* 203, 1694 (2009).
10. Y.Y. Chen, T. Duval, U.D. Hung, J.W. Yeh, and H.C. Shih, *Corros. Sci.* 47, 2257 (2005).
11. Y. Zhang, T. Zuo, Y. Cheng, and P.K. Liaw, *Sci. Rep.* 3, 1455 (2013).
12. W. Huo, H. Shi, X. Ren, and J. Zhang, *Adv. Mater. Sci. Eng.* 2015, 1 (2015).
13. A.S.M. Ang, C.C. Berndt, M.L. Sesso, A. Anupam, P.S.R.S. Kottada, and B.S. Murty,*Metall. Mater. Trans. A* 46, 791 (2015)
14. Y. Brif, M. Thomas, and I. Todd, *Scr. Mater.* 99, 93 (2015).
15. C. Huang, Y. Zhang, J. Shen, and R. Vilar, *Surf. Coat. Technol.* 206, 1389 (2011).
16. S. Katakam, S.S. Joshi, S. Mridha, S. Mukherjee, and N.B. Dahotre, *J. Appl. Phys.* 116, 104906 (2014).
17. X.-W. Qiu and C.-G. Liu, *J. Alloy. Compd.* 553, 216 (2013).
18. X.-W. Qiu, Y.-P. Zhang, L. He, and C. Liu, *J. Alloy. Compd.* 549, 195 (2013).
19. J. Joseph, T. Jarvis, X. Wu, N. Stanford, P. Hodgson, and D.M. Fabijanic, *Mater. Sci. Eng. A* 633, 184 (2015).
20. Kunce, M. Polanski, and J. Bystrzycki, *Int. J. Hydrogen Energy* 38, 12180 (2013).

21. Kunce, M. Polanski, and J. Bystrzycki, *Int. J. Hydrogen Energy* 39, 9904 (2014).

22. Kunce, M. Polanski, K. Karczewski, T. Plocinski, and K.J. Kurzydlowski, *J. Alloy. Compd.* 648, 751 (2015).

23. B.A. Welk, R.E.A. Williams, G.B. Viswanathan, M.A. Gibson, P.K. Liaw, and H.L. Fraser, *Ultramicroscopy* 134, 193 (2013).

24. H. Zhang, Y. Pan, and Y.-Z. He, *Mater. Des.* 32, 1910 (2011).

25. H. Zhang, Y. Pan, Y. He, and H. Jiao, *Appl. Surf. Sci.* 257, 2259 (2011).

26. W. Wu, L. Jiang, H. Jiang, X. Pan, Z. Cao, D. Deng, T. Wang, and T. Li, *J. Therm. Spray Technol.* 24, 1333 (2015).

27. T.M. Yue, H. Xie, X. Lin, H.O. Yang, and G.H. Meng, *J. Alloy. Compd.* 587, 588 (2014).

28. W.-R. Wang, W.-L. Wang, and J.-W. Yeh, *J. Alloy. Compd.* 589, 143 (2014).

29. H. Zhang, Y. Pan, Y.-Z. He, J.-L. Wu, T.M. Yue, and S. Guo, *JOM* 66, 2057 (2014).

30. N.A. Berjeza, S.P. Velikevitch, V.I. Mazhukin, I. Smurov, and G. Flamant, *Appl. Surf. Sci.* 86, 303 (1995).

31. V. Ocelík and J.T.M. De Hosson, *Advances in Laser Materials Processing* (Boca Raton, FL: Woodhead Publishing Limited, 2010), pp. 426–458.

32. Y. Shon, S.S. Joshi, S. Katakam, R. Shanker Rajamure, and N.B. Dahotre, *Mater. Lett.*142, 122 (2015).

33. Hemmati, V. Ocelik, and J.T.M. De Hosson, *Mater. Lett.* 84, 69 (2012).

34. V. Ocelík, I. Furár, and J.T.M. De Hosson, *Acta Mater.* 58, 6763 (2010).

4

CHAPTER

Precipitation Reactions in Age-Hardenable Alloys During Laser Additive Manufacturing

Eric A. Ja¨ Gle ,[1,4] Zhendong Sheng,[1,2] Liang Wu,[1] Lin Lu,[1] Jeroen Risse,[3] Andreas Weisheit,[3] and Dierk Raabe[1]

[1]*Max-Planck-Institut fu¨r Eisenforschung GmbH, Max-Planck-Strasse 1, 40237 Du¨sseldorf, Germany.*

[2]*Institut fu¨r Eisenhu¨ttenkunde, Rheinisch-Westfa¨lische Technische Hochschule Aachen, Intzestrasse 1, 52072 Aachen, Germany.*

[3]*Fraunhofer Institut fu¨r Lasertechnik, Steinbachstrasse 15, 52074 Aachen, Germany*

ABSTRACT

We describe and study the thermal profiles experienced by various age-hardenable alloys during laser additive manufacturing (LAM), employing two different manufacturing techniques: selective laser melting and laser metal deposition. Using scanning electron microscopy and atom probe tomography, we reveal at which stages during the manufacturing process desired and undesired precipitation reactions can occur in age-hardenable alloys. Using examples from a maraging steel, a nickel-base superalloy and a scandium-containing aluminium alloy, we demonstrate that precipitation can already occur during the production of the powders used as starting material, during the deposition of material (i.e. during solidification and subsequent cooling), during the intrinsic heat treatment effected by LAM (i.e. in the heat affected zones) and, naturally, during an ageing post-heat treatment. These examples demonstrate the importance of understanding and controlling the thermal profile during the entire additive manufacturing cycle of age-hardenable materials including powder synthesis.

INTRODUCTION

Many classes of alloys owe their high strength to the presence of finely dispersed second phase particles (i.e. phases different from the matrix phase). Since they form by precipitation phase transformations, they are called precipitates, and the materials featuring them, precipitation-strengthened alloys,[1] Examples are most Al alloys,[2–4]many Ni-based alloys[5–7] and some steels.[8–11] During conventional processing, these materials undergo two subsequent heat treatments. First, in a homogenization treatment in the single-phase region of the phase diagram, i.e. at relatively high temperature, all elements are brought into solid solution. This is followed by a rapid quenching, designed to limit or fully suppress any precipitation during cooling. Precipitates occurring after quenching from the solutionized state are called primary precipitates. Subsequently, the material is annealed at a lower temperature, where the remaining solutes that are now in a supersaturated state are allowed to precipitate. In this age-hardening step, the desired fine dispersion of particles forms. These particles are called secondary precipitates and are typically only up to a few nanometres in size.

During laser additive manufacturing (LAM), the same alloy experiences a very different, unique thermal profile as it is subjected to the subsequent processing steps (see Fig. 1).[12]First, the alloy needs to be rendered into powder form in an atomization process. For this, the material is melted and the liquid is atomized via gas or water into small droplets that cool down quickly and thus assume a solid powder particle state.[13] This cooling from the melt is very quick and precipitation is normally suppressed. Therefore, an additional homogenization heat treatment step is usually deemed unnecessary. During LAM, the powder is re-heated to temperatures above the melting temperature. Heating to and cooling from this temperature are also very quick, because of the small size of the heat source (the laser beam), which generates a melt pool that is small in size relative to the underlying substrate or the underlying layers, respectively ("self-quenching effect"). Again, precipitation is typically suppressed during this step. Additionally to the melting and deposition temperature "spike" or "peak", the already-deposited material experiences additional peaks as more material is deposited (in an adjacent track, or an overlying layer, i.e. the adjacent material is in the heat-affected zone (HAZ) of the melt pool). This series of subsequently imposed temperature peaks can be referred to as "intrinsic heat treatment". The maximum temperatures during the peaks are at first quite close to the melting temperature (and indeed some portions of every layer are re-melted during the deposition of the ensuing layer), but the peak temperature drops quickly from layer to layer. Due to insufficient heat conduction to the substrate and/or an installed heating system, the base temperature between the peaks may also be elevated. Finally, an ageing heat treatment analogous to conventional processing may be performed.

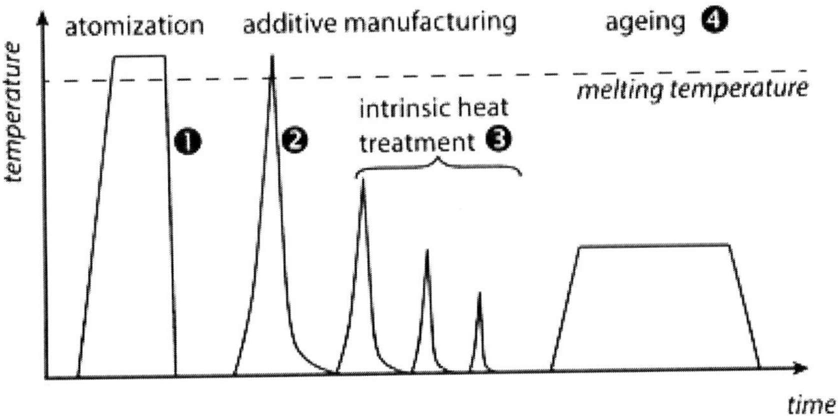

Figure 1. Schematic image of the complex temperature–time profile experienced by an alloy in the course of the production of an additively manufactured part. The numbers designate specifc processing steps when precipitation, desired and undesired, may occur (see text for details)

During these processing steps, either desired or undesired precipitation may occur during (at least) four of the different processing steps (cf. the numbers in Fig. 1):

(1)During atomization, either if the quenching rate is not high enough to suppress precipitation, or when the time spent in the liquid state is not sufficient to completely dissolve pre-existing, coarse precipitates

(2)During the actual LAM processing, i.e. during cooling from the liquid state after deposition. Again, this might occur if the cooling rate is not fast enough.

(3)During the intrinsic heat treatment, i.e. during temperature peaks.

(4)During the regular ageing heat treatment applied to the final part.

Of the above list, in traditional processing only possibility number 4 would be considered as desired option for precipitation strengthening, as the particles that result from too slow cooling (in steps 1 and 2) are either too coarse, or not dispersed homogeneously throughout the microstructure (e.g., concentrated at grain boundaries), or both. Secondary precipitation during the intrinsic heat treatment is an interesting possibility for desired precipitation that would allow shortening or even completely avoiding subsequent ageing treatment. Primary precipitates forming during solidification may be beneficial for grain refinement.

The control of precipitation during additive manufacturing is important, because solution heat treatment after LAM, which might be used to re-dissolve unwanted precipitates is not desirable or in some cases even impossible. Solutionizing increases the cost and complexity of the entire manufacturing process and subsequent quenching might induce warpage of the treated part, spoiling the (near-) net-shape nature of the LAM process. In some alloys such as the supersaturated Al-Sc-alloy discussed below, solutionizing cannot be used at all to re-dissolve solutes.

In this paper, examples of all four types of precipitation occurring in LAM-produced alloys are given, after a brief description of the common experimental techniques. The alloys under investigation are a maraging steel, an Al-Sc alloy and a nickel-based superalloy.

EXPERIMENTAL

Laser Additive Manufacturing

Two different LAM processes were employed in this work, namely selective laser melting (SLM) and laser metal deposition (LMD).12 In SLM, a thin powder layer is spread on a substrate plate using a powder distribution system and selectively melted using a scanning laser beam with a diameter in the range of 100 μm to 1000 μm. After each layer, the build platform is lowered and new powder is applied. Typical layer heights DLDL are 20 μm to 150 μm, laser powers PLPL are in the range of 50 W to 1000 W and scanning speeds, vSvS, may exceed 2000 mm/s. In LMD, on the other hand, powder is fed into the melt pool, generated by a laser beam, by a carrier gas being conducted through a co-axial or off-axial nozzle. The entire deposition head consisting of laser optics and powder-feeding nozzle is then moved relative to the fixed substrate plate creating, track by track and layer by layer, a bulk volume. Typical layer heights are 200 μm to 1000 μm, i.e. much larger than in SLM, but travel speeds are two orders of magnitude slower than in SLM (tens of mm/s). The beam diameter can vary in a wide range (100 μm to several 1000 μm). Depending on the beam size, the laser power employed in LMD ranges from a few 100 W to several kW. These different process parameters obviously strongly influence the time–temperature profile experienced by the material. Solidification and cooling rates are generally lower in LMD as compared to SLM (10^3–10^5 versus 10^4–10^6 K/s). In Table I, the main process parameters used to produce the materials analyzed in this work are compiled. The volume energy density, EV, introduced by the laser beam is defined as:

$$E_V = \frac{P_L}{y_H v_S D_L} \quad (1)$$

Here, the hatch distance, ΔyHΔyH, defines the distance between adjacent scan vectors (tracks).

Table I. LAM processes and corresponding main process parameters used for making the materials in this study

Material	Scalmalloy®	Maraging steel		Inconel® 738LC
Process	SLM	SLM	LMD	SLM
Volume energy density, E_V (J mm^{-3})	222.2	198.4	211.6	90.0
Scan speed, v_S, (mm s^{-1})	300	150	10	1000
Layer height, D_L, (μm)	30	30	420	20

Analysis Techniques

Specimens were deposited using SLM and LMD in the shape of cubes with side lengths of 10–20 mm. They were cut along the build direction (cross-section view) and metallographically prepared. Most of the analyses presented in this work were obtained by using SEMs (scanning electron microscopes; JEOL 6500F and Zeiss Merlin) equipped with EDS (energy dispersive spectroscopy) detectors. For the analysis of fine precipitates, atom-probe tomography (APT) was employed. For this method, samples are lifted from polished cross-sections by a FIB (focused ion beam)-liftout technique (using an FEI Helios NanoLab 600i dual-beam microscope) and also sharpened by FIB milling. The samples are measured in a Cameca LEAP 3000 HR X local-electrode atom probe, using laser pulsing at a repetition rate of 250 kHz, a pulse energy of 0.4 nJ and a base temperature of 60 K (with the exception of the Al-Sc-alloy, for which voltage pulsing at a pulse fraction of 15% is applied at a base temperature of 40 K).

EXAMPLES

Precipitates in the Raw Powder: A Supersaturated Al-Sc-Alloy

Al–Sc alloys have been shown to possess an outstanding combination of strength, ductility and corrosion resistance.14 A new class of Al-Sc-(Mg-Zr)-alloys called Scalmalloy® has been developed for use in additive manufacturing.3 In this alloy, the Sc content is well above the maximum equilibrium solubility of ~0.3wt.%. Therefore, to obtain a homogeneous supersaturation of Sc in the Al matrix, rapid solidification (and further, rapid cooling) needs to be employed. It has been shown that this way, the entire Sc content can be used in the strengthening precipitation reaction yielding $Al_3(Sc,Zr)$ particles.3

We investigated specimens produced by SLM from two different batches of Scalmalloy®powder. The first batch had been atomized by a standard gas atomization technique (ECKA Granules, Fürth, Germany) while the second batch was manufactured using the electrode induction-melting gas atomization (EIGA) technique (TLS Technik, Bitterfeld, Germany). In this crucible-free technique, the bottom part of a rotating bar is melted by induction heating and the resulting flow of liquid metal becomes atomized in a nozzle system using inert gas.

In otherwise identically SLM-produced specimens, the material made with EIGA-atomized powder showed particles of 10–50 μm diameter (see Fig. 2b for an example) distributed homogeneously. Analysis by SEM–EDS shows that the precipitates contain approximately 19 at.% Sc, 7 at.% Zr and 74 at.% Al (with traces of Si and Mn), compatible with the phase $Al_3(Sc,Zr)$. These particles are already visible in the powder material before processing by LAM (see Fig. 2a). The size of the particles as well as the absence of the particles in the material produced by standard gas atomization suggest that the particles precipitated during the EIGA process, or, more likely, were already present in the material prior to atomization and were not (fully) dissolved during the comparatively short time in the liquid state during this process.

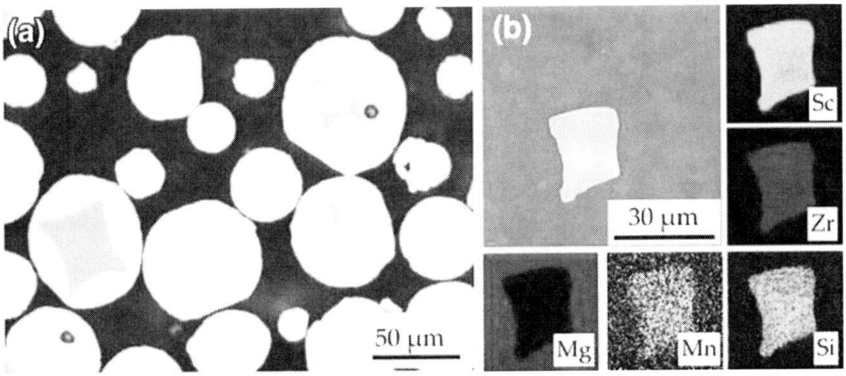

Figure 2. (a) Light optical micrograph of polished Al-Sc-alloy powder produced by electrode induction-melting gas atomization and used as starting material in LAM. Large precipitates in some of the powder particles are visible. (b) SEM image of an Al–Sc-alloy produced by SLM. A coarse $Al_3(Sc,Zr)$-precipitate is appearing bright in the backscattered electron detector signal. EDS mappings of various elements are arranged around the SEM image

Using the measured volume fraction of particles (~0.3%) and their measured composition, the amount of Sc removed from the matrix by being bound inside these large particles can be estimated: 0.09 wt.%. This small amount of scandium accounts for only a moderate drop in the strength of the material. Fractographic analysis of tensile test specimens made from this material reveal that the large precipitates also do not act as crack initiation sites and hence have only a small deteriorating impact on mechanical strength and ductility.

Precipitation during material deposition: Al-Sc alloy and Ni-base superalloy

Microsegregation occurs during solidification of most alloys (containing elements with partitioning coefficient not equal to one), provided the cooling rates are not high enough to completely trap solute into the growing crystal.15 We observed microsegregation both in a maraging steel (18Ni-300, see next example) and in a Ni-base superalloy (Inconel 738LC®). Evidently, even the relatively fast cooling rates prevailing in the SLM process (estimated by simulations to be 10^4–10^6 Ks^{-1}, depending on process parameters)16 are not sufficient for (complete) solute trapping. As the remaining liquid phase becomes strongly enriched in solutes during solidification, the formation of intermetallic phases may become energetically favorable.

In samples of Inconel 738LC® produced by SLM, APT measurements revealed the presence of carbides (TiC), borides (Cr- and Ni-rich) and an intermetallic phase (Zr-, Si-, Ta-, and Mo-rich), mostly in the intercellular and interdendritic regions where the final liquid film solidified (see Fig. 3). Interestingly, the boride phase and the intermetallic phase are always co-located, with the boride often forming an incomplete shell around the intermetallic phase. Note that Scheil-type thermodynamic simulations (using the software ThermoCalc in conjunction with the database TTNI8) do indeed predict both the formation of carbide (TiC) and boride (TiB_2), but also predict the formation of the coherent γ'-phase (which is the phase that emerges during age-hardening, but is not found in our experiments) and other intermetallic phases whose composition does not match the one observed in our experiments.

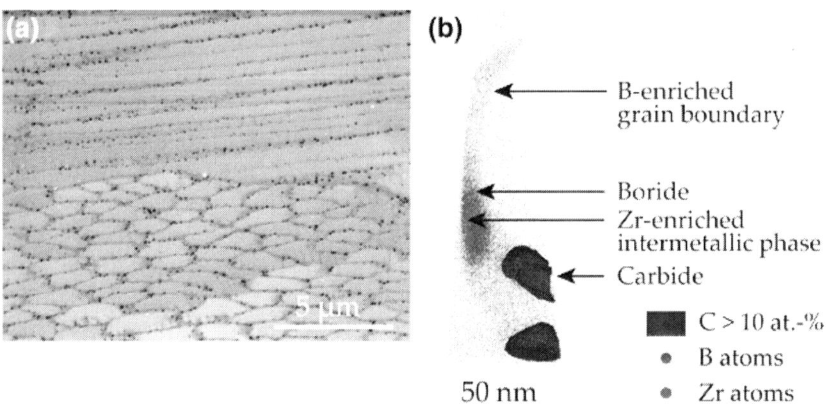

Figure 3.(a) An image recorded using a focused ion beam showing precipitates (dark) along the intercellular/interdendritic boundaries in the as-SLM-produced nickel-base superalloy Inconel 738LC®. They form due to microsegregation during the solidification of this alloy. (b) An APT dataset of such a boundary showing boron enrichment on the boundary as well as carbide, boride and intermetallic phase precipitation

In samples of Scalmalloy® produced by SLM, various precipitate phases are visible both inside small grains and along their grain boundaries (see Fig. 4). APT measurements allow the determination of the chemical composition of the precipitates, which are compatible with Al_3(Sc,Zr)-, Al_6Mn-, and Mg-rich precipitates. Al_6Mn precipitation is known to occur during the ageing of 5083 alloy, which has a similar composition as Scalmalloy® except for the addition of Sc and Zr; however, the concentration observed in the center of the Mg-rich precipitate is significantly higher (>50 at.%) than would be expected for the commonly observed Al_3Mg_2-phase (β-phase).17,18 Instead, the composition is compatible with the $Mg_{17}Al_{12}$-phase.

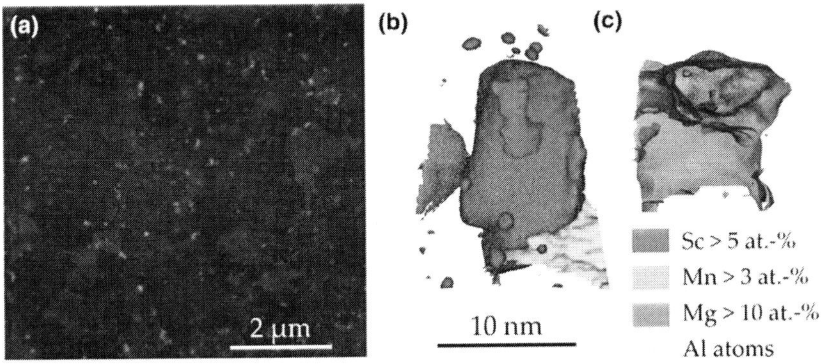

Figure 4 (a) SEM image showing fine precipitates in a highly supersaturated Al–Sc-alloy that formed most probably during cooling after solidification (SLM). (b) and (c) Two APT datasets showing that in fact three different types of precipitate are formed: $Al_3(Sc,Zr)$, Al_6Mn and $Mg_{17}Al_{12}$

Due to the presence of the precipitates both inside the grains and on the grain boundaries, the exact mechanism of precipitation is not clear. They could emerge in the solid state during cooling after solidification or they could be the result of re-heating during the deposition of an adjacent track/layer (see next example) and afterwards stimulate nucleation during partial re-melting of the layer as the next one is deposited above.

Precipitation During Intrinsic Heat Treatment: Maraging Steel

A characteristic feature of LAM processes, in particular the SLM process, are the high cooling rates experienced by the processed material owing to the small size of the melt pool and effective heat conduction through the substrate and already deposited material. It can therefore be expected that no precipitation in the bulk material during cooling after solidification takes place. This can indeed be shown by analyzing the material using statistical analysis of APT datasets.

We investigated a maraging steel (18Ni-300/1.2709/Böhler V720®). Maraging steels obtain their high strength and toughness by a microstructure consisting of martensite and finely dispersed intermetallic phases (here: Ni_3Ti, Ni_3Mo and Fe_7Mo_6 8) that form during a precipitation hardening heat treatment.9,10,19 In Fig. 5a, the radial distribution function (RDF) of pairs of titanium atoms is shown. The curves are obtained from APT datasets measured using conventionally produced material, material produced by SLM8,20 and produced by LMD. In the conventionally synthesized material in the as-produced state, i.e. before precipitation hardening, the RDF assumes a value close to one for all Ti–Ti pair distances. This means that it is equally likely to find a Ti atom in the close vicinity of another Ti atom as it is finding it further away. Also in as-produced material made by SLM, there is practically no deviation from unity visible. In LMD-produced material (taken from the middle of the sample), however, there is a clear deviation to values larger than one at small Ti–Ti distances. Apparently, in this material, the Ti atoms are likely to already be found in close vicinity to each other without precipitation heat treatment. This is an indication of the onset of precipitation, because the Ti atoms exhibit high mobility (compared to Mo atoms) and therefore form Ni_3Ti precipitates first.

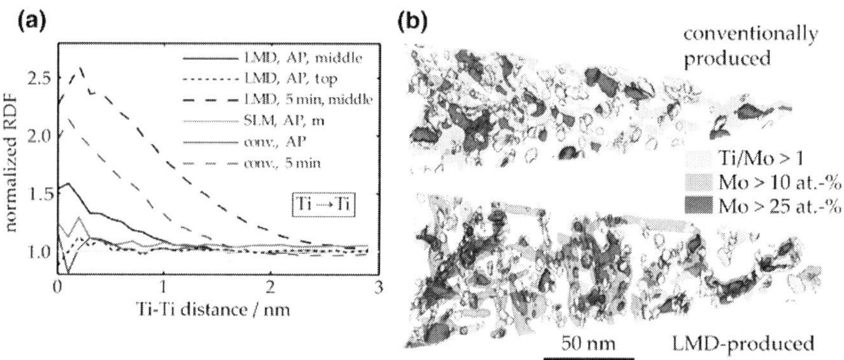

Figure 5.(a) The radial distribution function of Ti atoms in a maraging steel (18Ni-300) as determined by APT for different processing conditions (conventionally produced, SLM- and LMD-produced) and heat treatments (AP: as-produced, 5 min: ageing at 480°C, middle and top refer to the respective positions in build direction from where the APT specimens were taken. While the middle layers are exposed to additional heating cycles when depositing further layers, the top layer did not experience such an additional heat exposure). (b) APT datasets comparing the shape, size and number of precipitates present in conventionally produced and LMD-produced maraging steel after post-manufacturing ageing heat treatment (8 h at 480°C, i.e. until peak hardness). No significant differences are visible. The three iso-concentration surfaces delineate Ni_3Ti-precipitates (cyan), Ni_3Mo-precipitates (light red) and Fe_7Mo_6-precipitates (dark red), respectively

To prove that this clustering (early stage of precipitation) does not occur during cooling after solidification but rather during the intrinsic heat treatment, i.e. re-heating of the material during the deposition of additional layers, we also investigated samples taken from the very top of the specimen. In the uppermost layer, and in particular in the last track of this layer, no intrinsic heat treatment takes place. Indeed, the RDF corresponding to this situation is again nearly equal to one for all pair distances (cf. the black, dotted line in Fig. 5a).

Precipitation During Ageing Heat Treatment: Maraging Steel

Finally, further precipitation reactions occur during post-manufacturing heat treatments. Even though some of the initial solute content may not be available for precipitation any more, as explained in the previous examples, the remainder of solute atoms reacts in the same way as in conventionally produced materials. A potential difference exists in the different defect density of LAM- and conventionally produced material. Conventionally produced material is solution heat treated before precipitation and hence in recrystallized state, which is relatively poor in defects (e.g., dislocations, dislocation cells, and low-angle grain boundaries). Residual stresses generated during the LAM process can be released by plastic deformation, generating a higher density of dislocations. This presence of dislocations could influence the nucleation rate and the spatial distribution of nucleation sites. We observed precipitation in a LMD-produced and post-heat treated (8 h at 480°C, i.e. until peak hardness is reached) maraging steel and compared the morphology, number density and chemical composition of the emerging particles with the ones observed in a conventionally produced and identically heat-treated specimen (see Fig. 5b). Despite the different processing routes, no significant differences between the samples could be observed (note that in other regards, e.g., in the presence of retained and reversed austenite, grain size and morphology, and texture, significant differences between the samples do exist).

CONCLUSION

We studied desired and undesired precipitation reactions before, during and after laser additive manufacturing in a variety of alloys that gain their strength by precipitation hardening. In particular, precipitates emerging during the following steps were described:

(1)Precipitation of $Al_3(Sc,Zr)$ during the production of starting material powder in a supersaturated Al-Sc alloy.

(2)Precipitation of carbide, boride and a Zr-rich intermetallic phase during the LAM process, i.e. during solidification in a Ni-based superalloy. Similarly, precipitation of $Al_3(Sc,Zr)$, Al_6Mn and $Mg_{17}Al_{12}$ from a supersaturated Al-Sc alloy during cooling after deposition was observed.

(3)Early stages of Ni_3Ti precipitation (Ti–Ti clustering) during the intrinsic heat treatment occurring in LMD of a maraging steel.

(4)Precipitation of Ni_3Ti, Ni_3Mo and Fe_7Mo_6 during the ageing heat treatment after LAM in a maraging steel.

These findings reveal that LAM can be used to produce supersaturated alloys that show precipitation during age hardening annealing, as intended. In addition, LAM even provides the potential to dispense with the need of a post-heat treatment by exploiting its intrinsic heat treatment. However, care has to be taken in the production of the powder and in the choice of process parameters to make sure that no undesired precipitation reactions occur in the process chain. Hence, a detailed knowledge of the thermal profile experienced by the material before and during LAM, as obtained, e.g., by process simulations, is necessary.

ACKNOWLEDGEMENTS

Open access funding provided by Max Planck Society (Max Planck Institute for Iron Research). The authors would like to thank F. Palm, Airbus Group Innovations, for providing the SLM-produced Scalmalloy® specimens, J. van Humbeeck, KU Leuven, for providing the SLM-produced maraging steel specimens and S. Kleber, Böhler Uddeholm GmbH for providing the conventionally produced maraging steel (Böhler V720). The support by M. Kuzmina, P. Kürnsteiner and S. Ocylok with experiments is gratefully acknowledged.

REFERENCES

1. E.J. Mittemeijer, *Fundamentals of Materials Science* (Heidelberg: Springer, 2011).

2. P.A. Rometsch, H. Zhong, K.M. Nairn, T. Jarvis, and X. Wu, *Scr. Mater.* 87, 13 (2014).

3. K. Schmidtke, F. Palm, A. Hawkins, and C. Emmelmann, *Phys. Procedia* 12, 369 (2011).
4. E.O. Olakanmi, R.F. Cochrane, and K.W. Dalgarno, *Prog. Mater. Sci.* 74, 401 (2015).
5. K. Kunze, T. Etter, J. Grässlin, and V. Shklover, *Mater. Sci. Eng. A* 620, 213 (2015).
6. L. Rickenbacher, T. Etter, S. Hövel, and K. Wegener, *Rapid Prototyp. J.* 19, 282 (2013).
7. D. Tytko, P.P. Choi, J. Klöwer, A. Kostka, G. Inden, and D. Raabe, *Acta Mater.* 60, 1731 (2012).
8. E.A. Jägle, P. Choi, J. Van Humbeeck, and D. Raabe, *J. Mater. Res.* 29, 2072 (2014).
9. W. Sha, A. Cerezo, and G. Smith, *Metall. Trans. A* 24, 1221 (1993).
10. R.F. Decker, J.T. Eash, and A.J. Goldman, *Trans. ASM* 55, 58 (1962).
11. D. Raabe, D. Ponge, O. Dmitrieva, and B. Sander, *Scr. Mater.* 60, 1141 (2009).
12. D. Gu, W. Meiners, K. Wissenbach, and R. Poprawe, *Int. Mater. Rev.* 57, 133 (2012).
13. Lawley, *JOM* 33, 13 (1981).
14. J. Røyset and N. Ryum, *Int. Mater. Rev.* 50, 19 (2005).
15. D.A. Porter and K.E. Easterling, *Phase Transformations in Metals and Alloys*, 2nd ed. (Berlin: Springer-Science+Business Media, 1992).
16. D. Buchbinder, *Selective Laser Melting von Aluminiumgusslegierungen* (Aachen: Shaker Verlag, 2013).
17. Y. Zhu, D.A. Cullen, S. Kar, M.L. Free, and L.F. Allard, *Metall. Mater. Trans. A Phys. Metall. Mater. Sci.* 43, 4933 (2012).
18. R. Goswami, G. Spanos, P.S. Pao, and R.L. Holtz, *Mater. Sci. Eng. A* 527, 1089 (2010).
19. M.N. Rao, *Int. J. Mater. Res.* 97, 1594 (2006).
20. K. Kempen, E. Yasa, L. Thijs, J.-P. Kruth, and J. Van Humbeeck, *Phys. Procedia* 12, 255 (2011).

5

CHAPTER

A Study of Digging Productivity of an Electric Rope Shovel for Different Operators

Mohammad Babaei Khorzoughi [1,*] and Robert Hall [2]

[1]*Norman B. Keevil Institute of Mining Engineering, University of British Columbia, Vancouver, BC V6T 1Z4, Canada*

[2]*School of Mining and Petroleum Engineering, University of Alberta, Edmonton, AB T6G 2W2, Canada*

ABSTRACT

A performance monitoring study of an electric rope shovel operating in an open pit coal mine was conducted. As the mining industry moves toward higher productivity, profitability and predictability, the need for more reliable, productive and efficient mining shovels increases. Consequently, it is critical to study the productivity of these machines and to understand the effect of different operational parameters on that. In this paper a clustering analysis is performed to classify shovel digging effort and behaviour based on digging energy, dig time and payload per pass. Then the influence of the operator on the digging efficiency and productivity of the machine is analyzed with a focus on operator technique during digging. A statistical analysis is conducted on different cycle time components (dig time, swing time, return time) for different operators. In addition to time components, swing and return angles as well as loading rate and mucking rate are observed and analyzed. The results of this study help to understand the effect of different operators on the digging productivity of the shovel and then to set the best operator practice.

Keywords: electric rope shovels; productivity; digging energy; operator; digging time

1. INTRODUCTION

Loading efficiency has a critical role in the success of the mining process as the loading equipment is the source of ore supply [1] or waste removal. An efficient loading practice can help to increase production and reduce cost. For instance, according to Scott and Mackee [2], one million dollars can be saved in a surface coal mine with a 1% improvement in loading efficiency. Therefore, it is essential to study and monitor the performance of loading equipment.

The electric rope shovel is one type of mining equipment primarily used in most large, high volume operations as a loading unit. The performance of electrical rope shovels may vary with the muck-pile characteristics, operator practice and skills, and machine type and conditions. Previous research attempts show that shovel performance is directly influenced by muck-pile characteristics [3,4]. However, in addition to muck-pile characteristics, operator proficiency and skill play a significant role in the productivity of loading equipment [5,6,7,8,9,10,11,12].

This paper investigates the digging efficiency and productivity of a shovel and the effect of different operators on its performance. The shovel is currently operating in an open pit coal mine in Canada and the data has been collected during a week of field trials in the summer of 2015. The machine used in the present study is a P&H4100XPB shovel with a nominal dipper capacity of approximately 90 metric tonnes or approximately 48.4 cubic meters of material volume. The shovel's main functions during loading haul trucks include digging, swinging and dumping. The digging component is a combination of hoist and crowd actions. These functions are mainly accomplished by two hoist, one crowd and two swing motors on-board the shovel. Additionally, for an electric rope shovel a normal and productive cycle includes: digging, swinging, waiting (usually for the first pass), dumping and returning back to the face.

It should be noted that the onboard monitoring system on the shovel provides real-time feedback to the operator. The system was implemented to improve operator productivity and ensure consistent shovel performance. With this system, the operators adjust their digging tactics to current digging condition to achieve the highest productivity.

2. BACKGROUND

Shovel productivity and performance are strongly influenced by operator proficiency. A well-trained operator is essential to achieve maximum productivity [10]. Especially, with the current cost pressure on the mining industry, it is important to have high productivity. The influence of operator practice and skill should be a significant factor in any productivity assessment. Some research studies have reported the effect of operators on shovel performance.

Hendricks [5] monitored the performance of four electric rope shovel operators. He concluded that operators adjust their digging tactic to compensate for variations in muck-pile digging conditions; however, each operator operates within a particular range of dipper trajectories.

Jesset [6], as part of his research, established a framework that might help to set the best operator practice to improve shovel productivity and reduce loading duty. By statistically comparing measured data for different operators, Jessett [6] concluded that operator's style affects shovel productivity and duty loading.

In contrast to blasted muckpile digging, Patnayak *et al.* [8] reported the influence of operating practice on the shovel performance in oil sand digging. For the purpose of this study, performance parameters, recorded from the shovel, were compared for four teams of operators. They believed that "the operating characteristics of each team of operators will overshadow the influence of material diggability at a given shovel location" [8] (p. 133). Comparing average shift hoist and crowd motor power during different shifts and for different teams, Patnayak *et al.* [8] concluded that the consumed hoist power depends on the way that a team operates the shovel while the crowd power is independent of an operators' team digging tactic. They also showed that the hoist energy per unit volume of payload can be a measure of the operators' team performance.

As reported by Hendricks [5] and Patnayak *et al.* [8], energy consumption and digging effort of shovels are significantly influenced by operating characteristics. Similarly, other attempts have been made in the past to address the effect of operator practice and skill on equipment energy consumption and its performance. Widzyk-Capehart and Lever [13] similar to Jessett [6] stated that operator style has a significant effect on shovel productivity. Similarly, Onederra *et al.* [7] , based on the result of their case study, showed that operator proficiency is critical in shovel performance which was indicated through production rate variability. Vukotic [10] established a methodology to evaluate rope shovel operators and then to minimize energy consumption and maximize production rate. He developed a model to analyze operator's performance in different parts of the shovel loading cycle based on the energy consumption and production rate. Bernold [14] compared operator's digging performance by analyzing digging forces through a backhoe simulator. He estimated operator's performance on the basis of total energy per digging cycle, total path distance per digging cycle and bucket average velocity. Komljenovic *et al.* [15] developed a performance indicator for dragline operators. This indicator was defined as the ratio of dragline hourly production rate and hourly energy consumption. Awuah-Offei and Frimpong [16] introduced hoist rope and crowd arm speeds as critical parameters in evaluating operator's performance. In this study, a simulation of a rope shovel was conducted.

3. APPROACH

3.1. Field Studies

In this study data was collected during a one-week period using a commercially available fleet management and health monitoring system on-board the shovel. This monitoring system records data from sensors mounted on the shovel and processes them to estimate a comprehensive set of key shovel performance indicators (KPIs) per cycle. These KPIs include dig time, swing time, return time, swing angle, return angle, payload, and equivalent digging energy. This data is stored in a MySQL database and can be queried. In addition to the mentioned system, an embedded computer system, Octagon (Octagon Systems, Westminster, CO, USA) (Figure 1), has been installed in the shovel house to capture on-board shovel signals such as electrical motor responses and joystick reference signals. The Octagon computer records the data using an OPC (Open Platform Communications) interface. There is OPC bridge software installed on the two computers on the shovel. The software makes a connection to these bridges to collect the signals from digital side of the programmable logic controller (PLC). The signals pertaining to this paper include the hoist and crowd joysticks references as well as the hoist rope retraction and crowd arm extension. The sampling rate for these signals has been set at 20 Hz. Finally, in parallel to all of these data sets, the entire operation shift has been recorded using a USB camera mounted in the operator cab. These digital video records help to interpret collected data.

Figure 1. Octagon computer installed in the shovel house.

During the monitoring trial, four different operators worked the shovel; throughout the paper they are identified as "Operator A", "Operator B", "Operator C" and "Operator D". The shovel was loading 315 tons trucks (930 E) from both sides (left and right) in 3 to 5 passes. Digging conditions were also assessed based on the operator comments, the blast engineer comments and the authors' field observations. Figure 2 shows the digging sequence of the shovel with different colours representing different dates during the field trial. As this figure displays, the shovel was mainly digging two blast patterns (1905-04 and 1905-06) and the edges of another one (1905-05) (Default blast design parameters for mentioned patterns include: Burden: 9.5 m; Spacing: 11 m; Bit Size: 13 inch; Bench Height: 15 m; Sub-drilling: 2 m). Two zones have also been identified in Figure 1 as easier (coarse but loose material) and harder digging (hard toe and coarse and tight material) conditions. Operator A was working on the easier condition and Operator B was working on the harder one. It should be noted that Operators C and D were working during the night shifts and the authors couldn't make a conclusive statement about digging conditions while these operators were working.

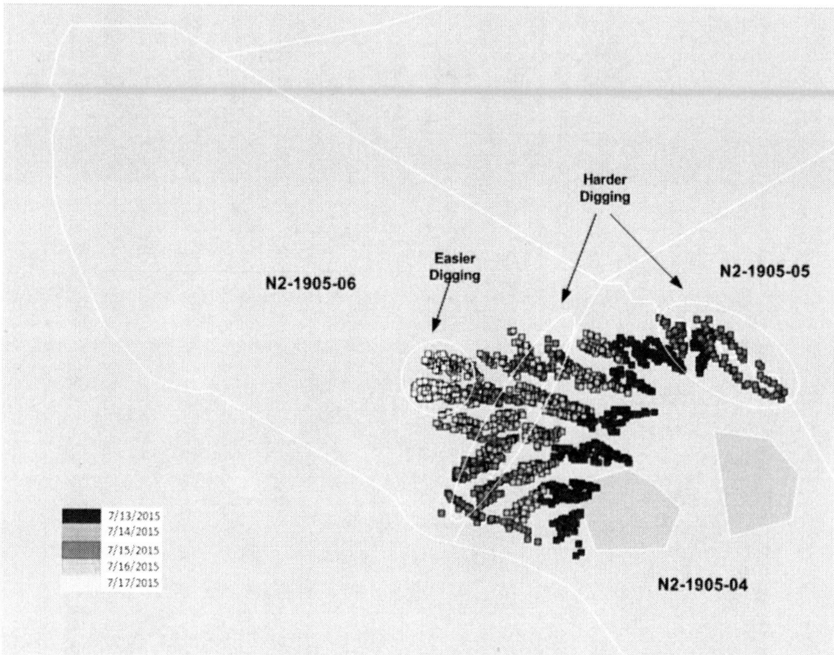

Figure 2. Shovel dig playback during the field trial (provided by mine).

3.2. Statistical and Clustering Analysis

As the first step in the analysis of the data collected during field studies, different shovel key performance indicators recorded by the commercial monitoring system on-board the shovel are statistically compared. Especially, their coefficient of variation (COV) are compared to assess the variability in productivity of the machine from cycle to cycle and for different operators. This is done to understand the effect of the operator on shovel performance. Additionally, the analysis of COVs in conjunction with other data analysis can show if the new monitoring technologies are successful in increasing production and reducing variability of machine productivity across different operators.

As the next step, to further investigate the effect of different operational parameters on the shovel performance, a clustering analysis is performed based on digging energy, dig time and payload per pass. The goal of this clustering is to classify shovel digging behavior and effort which depend on operator practice and skills, digging condition and machine type and conditions. In this paper, to solve the clustering problem, K-means clustering technique is used because of its simplicity and speed in dealing with large datasets.

K-means [17] is an unsupervised learning algorithm that partitions a set of *n* data points in Rd (Rd is the data space of d dimensions) into *k* clusters. Given an integer *k* and a set of *n* data points X⊂Rd, K-means algorithm aims at minimizing an objective function (*J*), in this case sum of the square of the distance from data points to the clusters centers (centroid), so that *k* cluster centers $C = [c_1, c_2, \ldots, c_k] \in Rd$ are defined.

$$J = \sum_{x \in X}^{n} d(x,c)^2 \qquad (1)$$

where c∈C and d(x,c) denotes the Euclidean distance between data points and each center. The details of K-means algorithm have been presented in [17].

In this paper, the number of clusters has been set as k=4 to allow for the results from this work to be integrated into a diggability algorithm that is a subject of ongoing research. Additionally, Euclidian distance, the most common metric used for clustering, has been chosen to assign data points to each cluster.

Finally, K-means++ algorithm [18] has been used to initialize cluster centers. According to Arthur and Vassilvitskii [18], this algorithm improves the speed and quality of clustering of K-means.

In this paper to ascertain that the K-means algorithm will result in a solution that is a global minimum, 10 replicates with different starting points (according to K-means++ algorithm) have been used. The results of productivity analysis, as well as clustering analysis are presented next.

4. ANALYSIS OF RESULTS

4.1. Productivity

A total of 4791 shovel cycles were monitored during the course of the field trial. As discussed inSection 3.1 a commercial monitoring system was used to record different shovel activities such as digging, swinging and returning. Additionally, the monitoring system detects and records activities that are not associated with loading the truck such as cleaning up the face. In this paper, such activities are not included in the analysis. As the current study mainly focuses on the digging part of each loading cycle, Figure 3 illustrates a histogram and a box plot of dig time values. The histogram indicates that dig times are positively skewed and they range from 2 to 48 s. The box plot also describes the spread of data and highlights outliers. In this paper an outlier is defined as a value that is more than 1.5 times the interquartile range away from the top or bottom of the box in the boxplot.

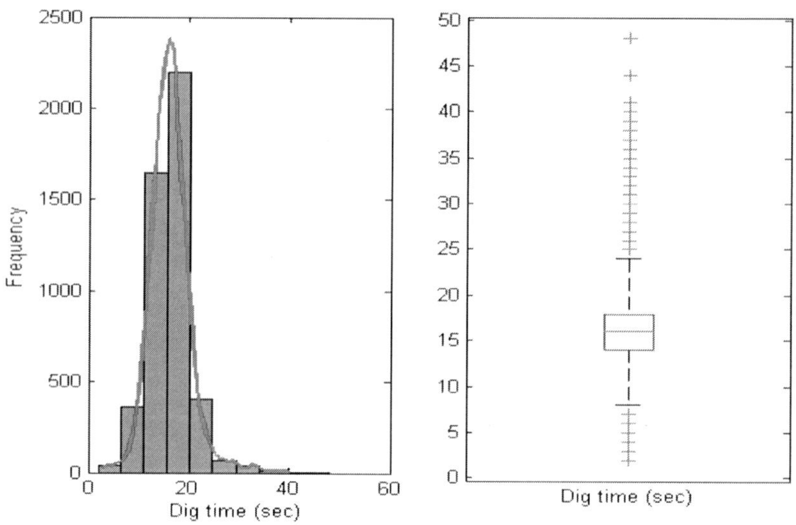

Figure 3. Dig time distribution.

Outliers were removed from the data and statistics for different KPIs were calculated. The KPIs include payload, cycle time components, swing and return angles, equivalent digging energy, loading rate and mucking rate. In this paper, productive cycle time is defined as the interval between two consecutive dumps excluding waiting times. Dump time and the positioning of the bucket before digging have been included in return time (The shovel monitoring system has been coded in this way). Equivalent digging energy, loading rate and mucking rate are also given by Equation (2) to Equation (4), respectively:

$$\text{Equivalent Digging Energy} = \int_{t_0}^{t_1} F_b \times \text{Crowd Rate}\, dt \quad (2)$$

$$\text{Loading Rate} = \frac{\text{Payload}}{\text{Dig time}} \tag{3}$$

$$\text{Mucking Rate} = \frac{\text{Payload}}{\text{Productive Cycle Time}} \tag{4}$$

where t_0 and t_1 are the start and the end of digging respectively, F_b is the bail force (bail pull) and crowd rate is the rate of change of crowd arm angle with respect to horizon. It should be noted that all these KPIs are being estimated by the commercial monitoring system on-board the shovel using a suit of sensors such as load cells and gyro sensors. As Equation (2) shows the digging energy estimated by the monitoring system on-board the shovel is an equivalent mechanical energy (tons×rad) during the digging part of the cycle. The equivalent digging energy is measured at the bail and is a measure of mechanical energy transferred to the bucket teeth. This energy is provided by electrical DC motors on-board the shovel. A summary of the aforementioned KPIs for all of the operators is presented in Table 1.

Table 1 shows that the largest component of the cycle is the dig phase which on average accounts for about 50% of the productive cycle time. Swing time and return time each accounts for 25% of the productive cycle time which is lower than dig time portion. This could be due to low swing and return angles (<90°).

Among the parameters presented in Table 1, waiting time has the highest variation (highest COV) which is independent of machine performance. Furthermore, the largest variations within each cycle occur in the swing and return phases which are mainly controlled by the operator. Payload also has the lowest variation which demonstrates that operators try to adapt their digging technique to current digging conditions to achieve the desired payload which is the highest target load per pass without exceeding the truck capacity; therefore, it is mainly consistent from cycle to cycle. It should be noted that current loading practice is not an automated process and machine monitoring systems only help the operators to more consistently fill the bucket. The variations in payload can also be a representation of variations in productivity of the machine. Generally, operations are interested in lower variations in their shovel productivity which should be reflected in a narrow distribution of truck loads.

Table 1. Key shovel performance indicators.

Operator	Statistics	Payload (tons)	Dig Time (s)	Swing Time (s)	Swing Angle (°)	Return Time (s)	Return Angle (°)	Productive Cycle Time (s)	Waiting Time (s)	Equivalent Digging Energy (tons × rad)	Loading Rate (tons/s)	Mucking Rate (tons/s)
Operator A 589 Cycles	Mean	98.3	16.4	8.2	68.8	8.6	62.3	33.3	18.9	266,265.5	6.2	3.0
	COV	0.17	0.20	0.31	0.36	0.24	0.36	0.15	1.94	0.36	0.26	0.19
	Min	13	8	2	9	1	0	21	0	36,461	1	0
	Max	133	24	18	180	27	173	53	373	588,114	15	6
Operator B 1629 Cycles	Mean	104.1	16.5	8.5	66.8	8.1	61.8	33.1	18.9	231,875.7	6.5	3.2
	COV	0.14	0.18	0.33	0.32	0.33	0.43	0.16	2.18	0.32	0.20	0.16
	Min	11	8	2	1	0	0	18	0	6717	1	0
	Max	139	24	30	166	36	175	67	411	617,241	15	6
Operator C 1633 Cycles	Mean	98.6	15.7	8.5	70.8	8.5	64.6	32.7	17.0	204,946.2	6.5	3.1
	COV	0.16	0.20	0.32	0.32	0.30	0.38	0.15	2.03	0.34	0.22	0.19
	Min	17	8	1	2	1	0	16	0	19,629	1	0
	Max	137	24	23	172	46	161	70	540	491,753	15	6
Operator D 671 Cycles	Mean	97.0	15.3	7.6	66.8	8.5	58.8	31.4	12.8	253,197.2	6.4	3.1
	COV	0.21	0.21	0.31	0.37	0.31	0.37	0.16	2.46	0.37	0.22	0.19
	Min	11	8	2	4	1	0	19	0	22,938	1	1
	Max	139	24	23	165	32	131	59	476	557,642	17	5
All Data 4522 Cycles	Mean	100.3	16.0	8.3	68.5	8.4	62.5	32.7	17.4	230,100.6	6.4	3.1
	COV	0.16	0.20	0.32	0.34	0.31	0.40	0.16	2.14	0.36	0.22	0.18
	Min	11	8	1	1	0	0	16	0	6717	1	0
	Max	139	24	30	180	46	175	70	540	617,241	17	6

Table 1 also shows that although different operators have different average equivalent digging energy per cycle, they have similar average loading rates as well as mucking rates. Among the four operators, Operator A has the lowest digging rate and mucking rate while he has the highest average equivalent digging energy per cycle. It should be noted that Operator A encountered one of the easier digging conditions during the field trial. Operator B has the highest loading and mucking rates while he was digging one of the harder digging conditions during the field trial. The digging practice of these two operators will be compared in Section 4.3 using recorded signals from PLC to better understand the effect of different operational parameters on the shovel performance. Operators C and D have not been chosen for comparison due to technical issues that caused the PLC signals not to be recorded while they operated the shovel.

In order to further investigate the effect of operator on the machine productivity, especially digging component, a clustering analysis is performed and operators' techniques during digging are compared. The results are presented next.

4.2. Clustering of Shovel Cycles

To classify shovel digging behavior and effort, first a 3D space of digging energy, dig time and payload for total of 4522 cycles is built. These three parameters or a combination of them have been widely used as a measure of digging efficiency or shovel performance [4,5,7,8,16]. Then, using the K-means method [17], this space is divided into four clusters. Figure 4 shows different clusters in the 3D space.

As the results of clustering show (Figure 4), clusters have been generated based on the digging energy which means that digging energy explains the majority of variability in the data. Therefore, a classification for digging energy based on the results of clustering analysis is presented in Table 2. This table illustrates that most cycles are within the range of average to high digging energy. Among all operators, Operator C has the highest percentage of cycles in the low energy class and the lowest percentage of cycles in the extremely high energy class while Operator A has the lowest percentage of cycles in the low energy class and the highest percentage of cycles in the extremely high energy class.

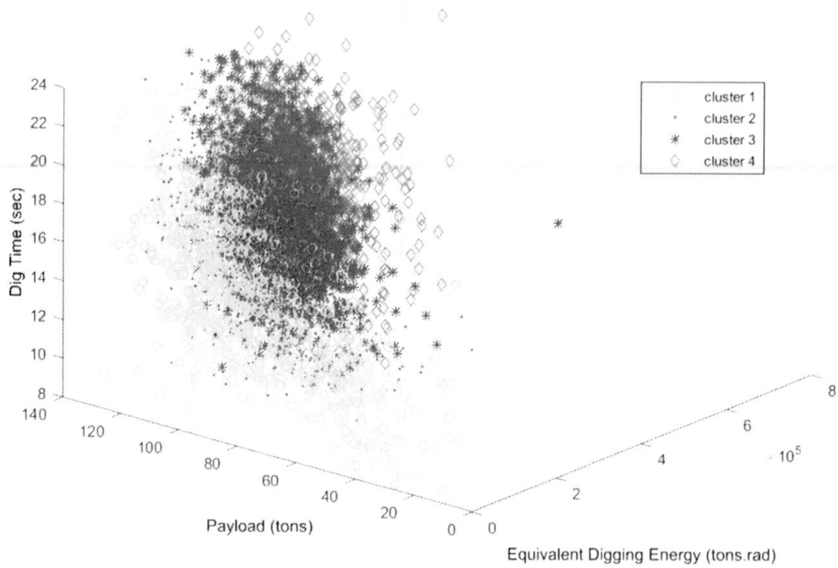

Figure 4. Generated clusters using K-means clustering method.

Table 2. Digging energy classification.

Digging Energy Class	Energy Range (× 10^5) (tons × rad)	Percentage of Cycles (All Data)	Percentage of Cycles (Operator A)	Percentage of Cycles (Operartor B)	Percentage of Cycles (Operator C)	Percentage of Cycles (Operator D)
Low Energy	<1.57	18	14.2	14.4	24	15.2
Average Energy	1.57–2.36	37	22.6	39.1	44	25.4
High Energy	2.36–3.23	33	34.2	36.2	27	37.5
Extremely High Energy	>3.23	12	29	10.3	5	21.9

In order to further investigate the relationship between digging energy, payload and dig time,Figure 5 and Figure 6 compare dig time and payload for the different classes of digging energy from Table 2.

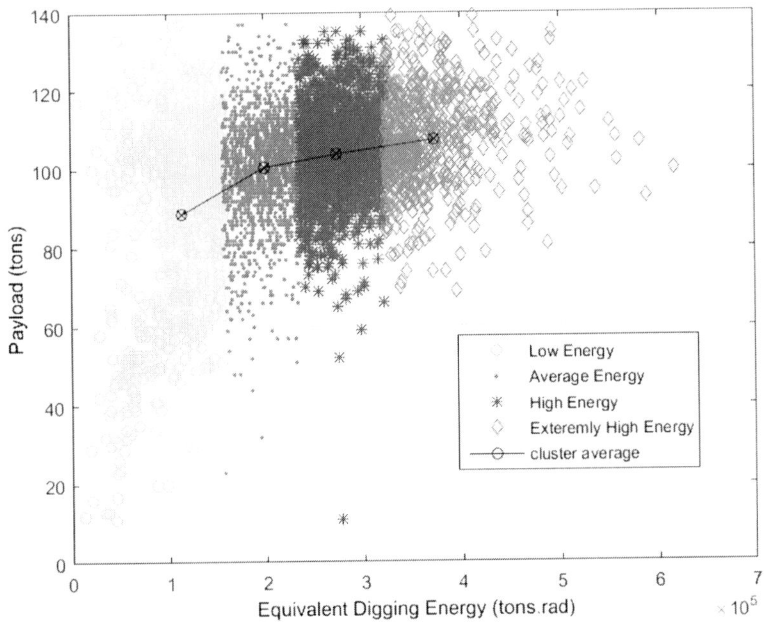

Figure 5. Result of clustering analysis for comparison of payload and equivalent digging energy.

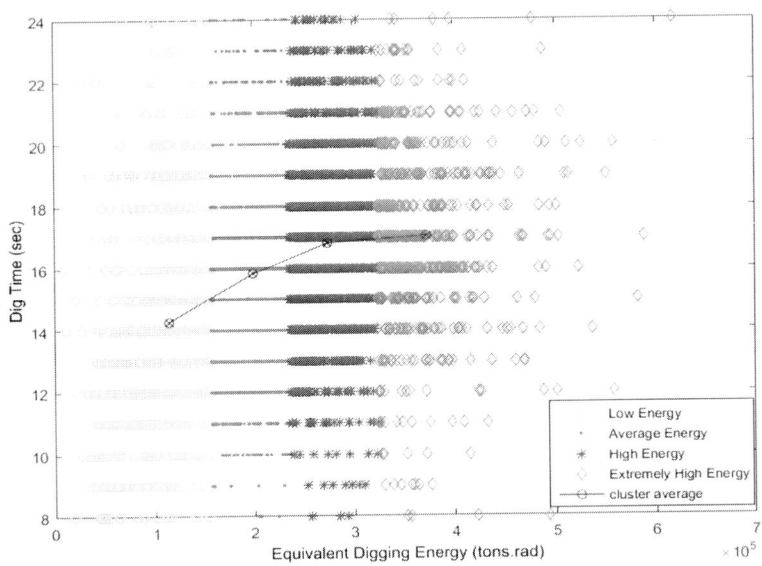

Figure 6. Result of clustering analysis for comparison of dig time and equivalent digging energy.

As the above figures show, dig time and payload vary widely across all the clusters, but Figure 5indicates that the average payload increases with increasing digging energy. Similarly, Figure 6shows that the higher energy clusters have higher average dig time. To combine the effect of dig time and payload, loading rate given by Equation (3) is analyzed.

Figure 7 shows how loading rate changes as digging energy increases. This figure indicates that there is no relationship between loading rate and digging energy (correlation coefficient = 0.0302). In addition, average loading rate is almost the same for all of the clusters (all digging energy classes). Therefore, one would conclude that variations in digging energy are caused by variations in muck-pile digging conditions or operators. However, even for one operator working in the same location (nearly same digging condition), digging energy will vary from cycle to cycle while average loading rate is almost constant. This can be explained by the effect of operator digging practice and its variability. Figure 8 compares loading rate values for different digging energy classes for Operator A. These results indicate that for this case the loading rate is independent of operator or digging condition.

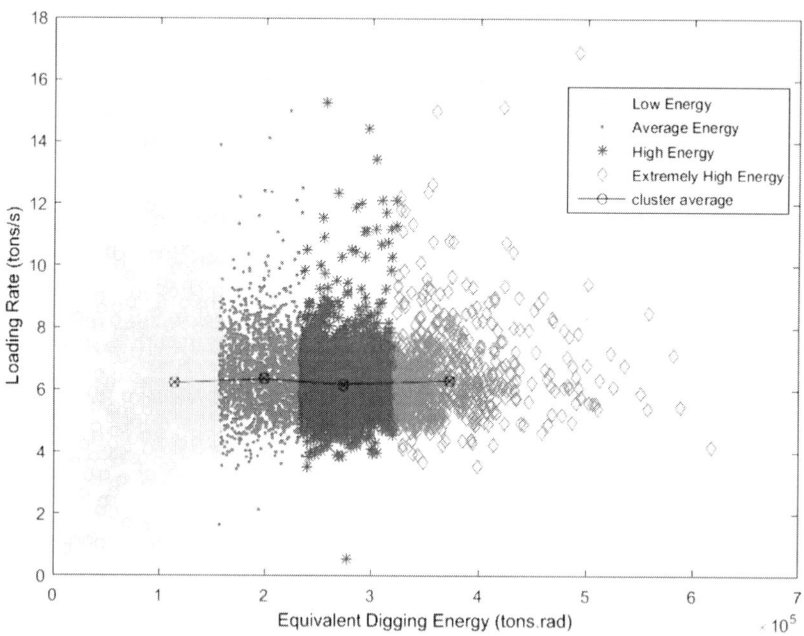

Figure 7. Loading rate *vs.* digging energy for all operators.

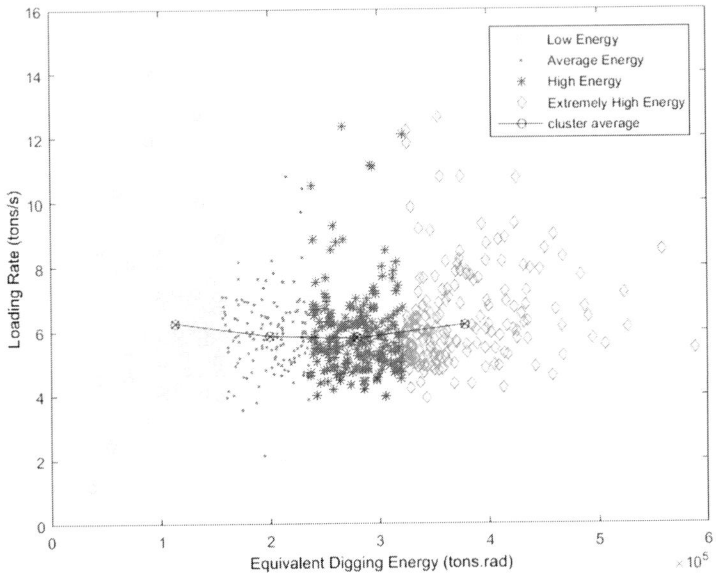

Figure 8. Loading rate *vs.* digging energy for Operator A.

In order to further understand the effect of operator digging practice on the shovel performance, the digging practice of two operators (A and B) are compared in the next section. In addition to the data presented in Section 4.2 (from the commercial monitoring system), the joysticks reference signals, as well as hoist rope retraction and crowd arm extension (from the Octagon computer), are analyzed for Operators A and B.

4.3. Operator Digging Practice

As mentioned in Section 1, digging is mainly a combination of hoist and crowd actions. There are two joysticks in the operator's cab that allow operators to control the machine. To study the effect of operator practice on digging efficiency, an evaluation of their hoist and crowd practices based on joystick signals is done. Figure 9 shows the hoist joystick reference signal for Operators A and B during a period of 250 s (5 cycles). The signals for Operators A and B have been annotated to show the start of digging (red circles), the end of digging (blue circles) and the end of the cycle (green circles). Similarly, Figure 10 shows the crowd joystick reference signals for Operators A and B for the same period of time.

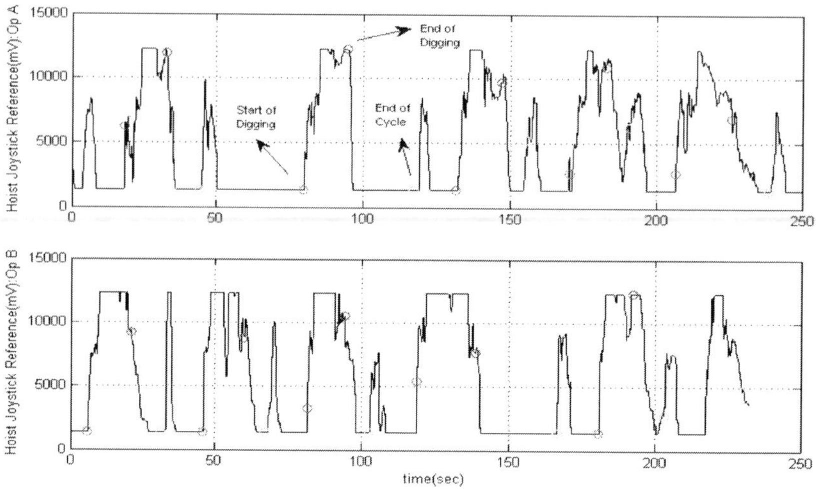

Figure 9. Hoist joystick reference.

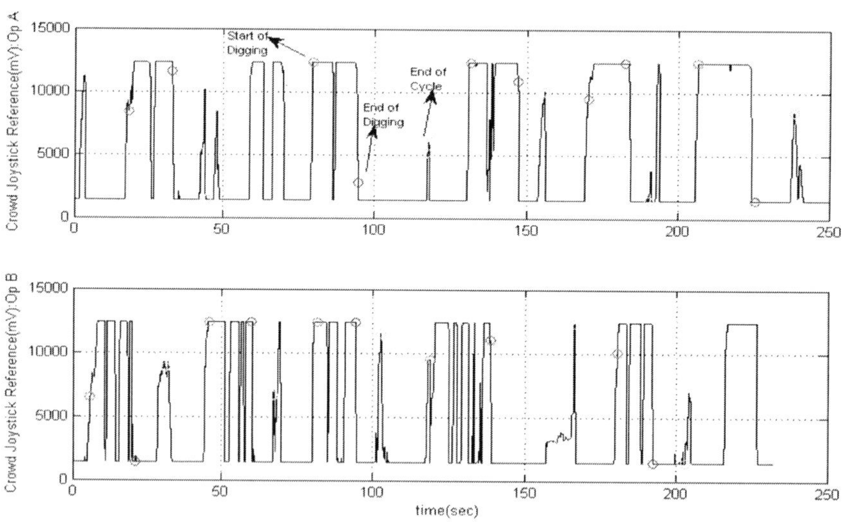

Figure 10. Crowd joystick reference.

The above figures show that each operator has a unique style in filling the dipper. Operator B has a smoother hoist action during digging while he frequently pulls the crowd joystick towards himself (retraction) which is shown as valleys in the crowd joystick reference signal. Although, the above signals exhibit some similar trends, it is clear that each operator has different digging habits/techniques. To better understand the effect of operator digging techniques on shovel performance, Figure 11 and Figure 12 compare the average dig time and payload respectively for the different digging energy classes presented in Table 2. These figures show that average dig time and payload increase with higher digging energy classes. Figure 11 shows that Operators A and B have similar average dig time values except for the low energy class where Operator A has slightly lower (~1 s) average dig time.

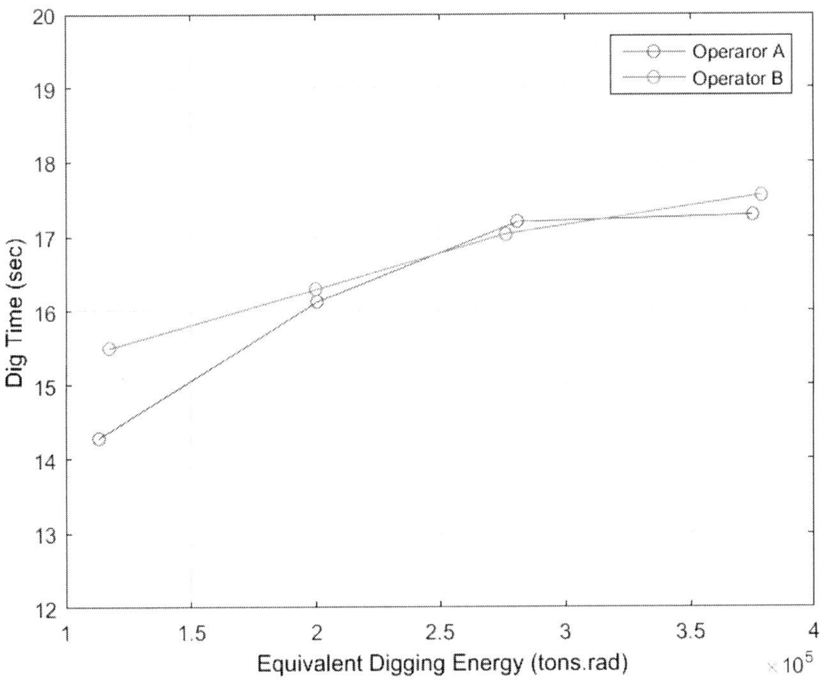

Figure 11. Average dig time for different digging energy classes.

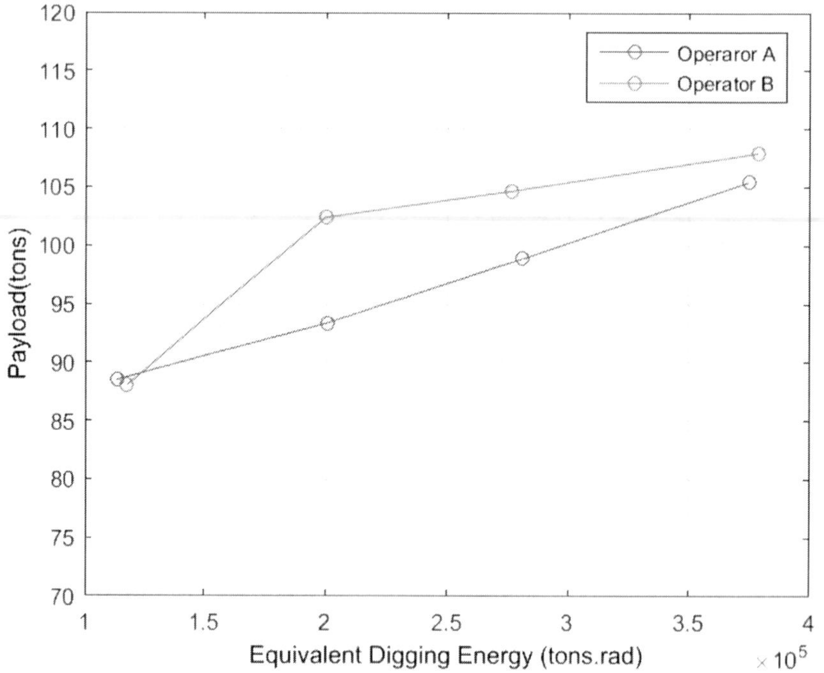

Figure 12. Average payload for different digging energy classes.

In contrast to dig time, Figure 12 shows that for average to extremely high digging energy classes, Operator B has a higher average payload. In other words, to fill the bucket to the same payload, the shovel consumed more energy when Operator A worked. One might conclude that Operator A worked in harder digging conditions since the shovel consumed more energy during digging for the same payload as Operator B (Figure 12); however, based on the authors' field observations, the blast engineer's comments and operators comments, as mentioned in Section 3.1, muck-pile digging conditions were easier for Operator A. Therefore, to understand the differences in digging energy, Figure 13 compares the digging trajectories of 5 consecutive cycles for each of the operators. This figure indicates that Operator A takes deeper cuts compared to Operator B which causes higher energy consumption.

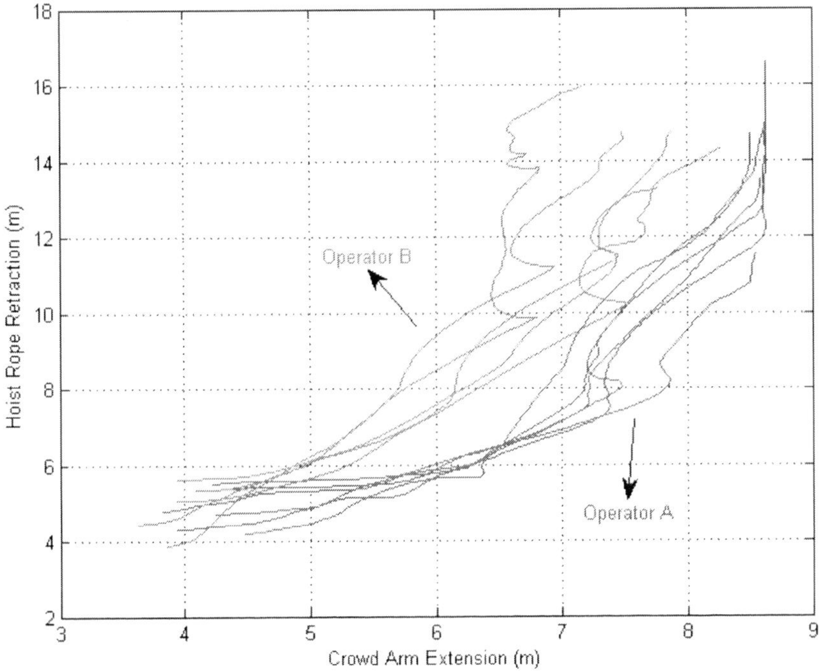

Figure 13. Digging trajectory for Operators A and B

According to past studies [16,19] the best operator digging practice is achieved by lower crowd extension speed and higher hoist rope retraction speed, which result in a decrease in the depth of cut. Such a practice should result in lower digging energy consumption per unit of loading rate known as specific digging energy. However, a limit of using specific digging energy as a measure of shovel performance is the inability to determine causes for its variations. In Section 4.1 it was shown that there is no relationship between digging energy and loading rate; therefore, based on the definition of specific digging energy, digging energy is not normalized for the effect of loading rate. Additionally, digging energy can be affected by other factors, such as muck-pile digging conditions and machine type and conditions.

To demonstrate how operators control the machine during digging, Figure 14 shows the crowd arm extension and hoist rope retraction during digging for 5 consecutive cycles for Operators A and B. In this figure, the slope of hoist rope retraction represents hoist speed and the slope of crowd arm extension represents crowd speed.

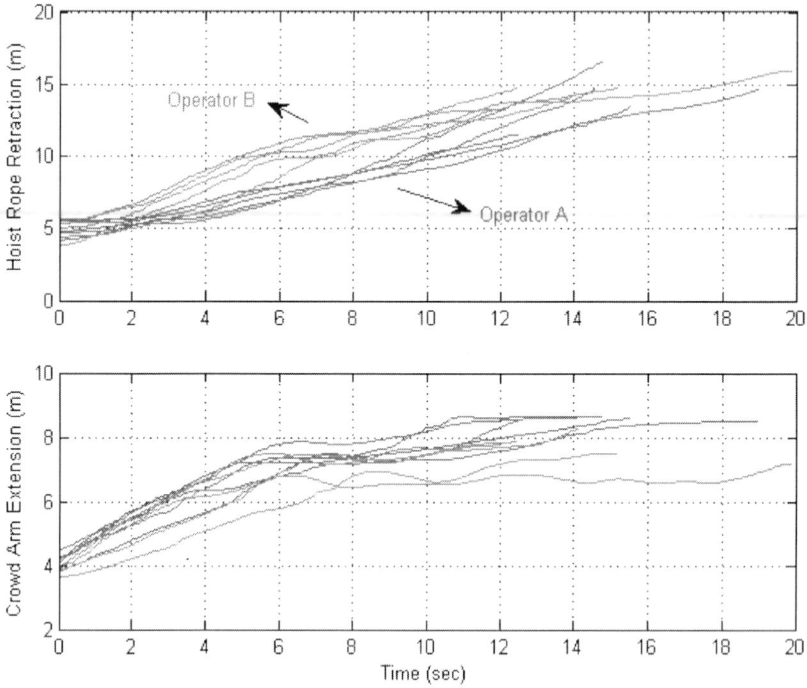

Figure 14. Hoist rope retraction and crowd arm extension for Operators A and B.

Figure 14 shows that the slope of the hoist rope retraction is almost constant for each cycle and a straight line can be fitted on the data for each cycle to estimate the hoist speed. Comparing the two operators, Operator A has a lower hoist speed. The hoist speeds and associated R-squared values for cycles in Figure 14 are presented in Table 3. In contrast to the hoist rope retraction, Figure 14shows that the crowd arm extension exhibits two different trends:

1. Constant crowd speed until the desired dipper depth of penetration is achieved (the first part of the digging);
2. Once the dipper penetrates into the bank, digging is mainly accomplished by hoist action, and the crowd speed is approximately zero.

Table 3. Crowd and hoist speed values.

Operator	Cycle #	1	2	3	4	5	Mean	Standard Deviation	Coefficient of Variation
Operator A	Crowd Speed (m/s)	0.460	0.523	0.567	0.599	0.402	0.510	0.080	0.157
	R-squared	0.994	0.986	0.986	0.989	0.959			
	Hoist Speed (m/s)	0.692	0.816	0.518	0.528	0.571	0.625	0.127	0.204
	R-squared	0.954	0.966	0.992	0.983	0.995			
Operator B	Crowd Speed (m/s)	0.379	0.654	0.553	0.462	0.493	0.508	0.103	0.202
	R-squared	0.988	0.996	0.986	0.992	0.947			
	Hoist Speed (m/s)	0.782	0.617	0.766	0.591	0.792	0.709	0.097	0.137
	R-squared	0.983	0.961	0.978	0.942	0.970			

This observation confirms that the dipper is mainly filled through the hoist action and the crowd action only helps to maintain a proper dipper depth of penetration into the bank. However, depth of penetration and crowd speed (in the first part of the digging) have effects on digging energy and shovel performance. The crowd speed values are estimated based on the slope of a straight line fitted on the crowd arm extension values in the first part of the digging. Crowd speeds and the associated R-squared values are presented in Table 3.

Table 3 shows that the average hoist speed is higher for operator B while the average crowd speeds are almost the same for both operators. The R-squared values indicate that a straight line is the best fit to describe the data. A combination of crowd and hoist speeds generates different digging trajectories, as shown in Figure 13, which directly affects the energy consumption and consequently shovel performance.

Therefore, to compare the digging performance of different operators, in contrast to the common approach of using one indicator, such as loading rate or specific digging energy, in this paper, a rating system similar to surface excavating classification systems [20,21,22] developed in the past is proposed based on the product of digging energy, loading rate, crowd speed and hoist speed:

$$N = a_1 \times a_2 \times a_3 \times a_4 \quad (5)$$

where a_1, a_2, a_3, a_4 are the numerical ratings of digging energy, loading rate, crowd speed and hoist speed, respectively. Because of high variability in digging energy, a weighted average based on the percentage of cycles within each class can be used to calculate the numerical rating of digging energy (a_1):

$$a_1 = p_1 \times 5 + p_2 \times 4 + p_3 \times 3 + p_4 \times 2 \quad (6)$$

where p_1, p_2, p_3, p_4 denote the percentage of cycles in low, average, high and extremely high energy classes, respectively.

Despite the fact that it was shown in Section 4.1 that loading rate is independent of the operator, to have a more generic equation it has been included in this approach. Additionally, digging energy is influenced by not only operator practices (crowd and hoist speeds) but also digging conditions and machine type and conditions. Therefore, to ensure that other factors such as digging conditions don't mislead the assessment, it is essential to include crowd and hoist speeds in addition to digging energy. It should be noted that the proposed formulation has operational purposes and is an experimentally derived approach that can be employed by different operations to assess digging performance of electric rope shovels operators.

The rating of each of the parameters in Equation (5) is subjective and their weights can change according to management policies. For example, if the focus of an operation is mainly on the volume produced, loading rate should have the highest weight in the rating system. Table 4suggests an example of the rating system based on the observed data during the field trial. The ratings may vary from mine to mine with different types of operations and management strategies, and can be modified by operations as more data is collected. In this study, the classification for digging energy is based on the clustering analysis performed, and the classes for loading rate, hoist speed and crowd speed have been defined based on the observed distribution of data and the discussion with mine senior engineers. To be able to have a universal classification data needs to be collected from different types of operations and from different machines. The proposed rating/classification in Table 4 can only be used as a guideline. In this table, higher loading rates, lower digging energy, higher hoist speed and lower crowd speed should have higher rating numbers. Such a rating will result in higher N values for operators with a better performance. Nvalues can be calculated for each operator per shift. To validate the proposed approach, Table 5compares the N values for Operators A and B. Although Operator B was digging harder conditions, he has a higher N value compared to Operator A which means he has a better digging performance.

Table 4. Operator rating system.

Parameters	Class / Rating			
Loading Rate (tons/s)	<5.4 4	5.4–6.9 6	6.9–8.8 8	>8.8 10
Digging Energy	Low 5	Average 4	High 3	Extremely High 2
Hoist Speed (m/s)	<0.6 1	0.6–0.7 2	0.7–0.8 3	>0.8 4
Crowd Speed (m/s)	<0.3 0.5	0.3–0.4 0.4	0.4–0.5 0.3	>0.5 0.2

Table 5. *N* values for Operators A and B.

Pataramete rs	Loading Rate (tons/s)	Digging Energy	Hoist Speed (m/s)	Crowd Speed (m/s)	N
	Average Value a_i				
Operator A	6.2 6	266,265.5 3	0.625 2	0.510 0.2	7.2
Operator B	6.5 6	231,875.7 4	0.709 3	0.508 0.2	14.4

5. CONCLUSIONS

This study presented the results of performance monitoring of an electric rope shovel operating in an open pit coal mine in Canada. The effect of operator on shovel productivity was studied. The performance of four operators was compared. It was found that among different key shovel performance indicators for each operator, payload has the lowest variability. The low variability of payload values was found to be influenced by the operator's response to the payload monitoring system.

A clustering analysis was performed to classify shovel digging effort and behaviour based on digging energy, dig time and payload. It was found that digging energy is the principal component which describes the majority of variability in the data. Therefore, based on the result of clustering analysis, a classification for digging energy was presented. It was shown that most of the cycles during the monitoring trial were in the range of average to high digging energy. The best operator practice should result in a higher percentage of cycles in the lower digging energy classes while maintaining a proper loading rate. Furthermore, average dig time and payload for different clusters were compared. It was concluded that average dig time and average payload increase toward higher digging energy classes. It was also shown that digging energy is independent of loading rate. Average loading rate was almost constant along different energy classes.

Furthermore, to demonstrate the effect of operator digging practice on shovel performance, two cases were selected and compared. It was found that the joysticks reference signals can be used to compare the operators' styles. It was shown that digging energy is not only a function of muck-pile digging conditions but also mainly is a function of digging practice. Even for the same loading rate, a good operator operating the shovel in harder digging conditions can achieve lower digging energy by adjusting the hoist and crowd speeds. It was found that the operator with the lower hoist speed and higher crowd speed takes deeper cuts in the bank and the shovel consumes more energy during digging to achieve a targeted payload.

To compare the operators' digging performance an experimentally derived rating system based on digging energy, loading rate, hoist speed and crowd speed was proposed. Given the rapid implementation of onboard shovel performance monitoring systems, the proposed rating system should be easy to implement.

It is believed that a properly determined digging energy, which has been compensated for the effect of operator digging practice, can be part of an approach to measure muck-pile diggability and indirectly blast quality. Future work will study the effect of muck-pile digging conditions to develop a diggability index based on the shovel performance and independent of operator practice.

AUTHOR CONTRIBUTIONS

Mohammad Babaei Khorzoughi and Robert Hall both contributed in all steps of preparation of the paper.

CONFLICTS OF INTEREST

The authors declare no conflict of interest.

REFERENCES

1. Hustrulid, W.A. *Blasting Principles for Open Pit Mining*; A.A. Balkema: Rotterdam, The Netherlands, 1999; Volume 1.
2. Scott, A.; McKee, D.J. The interdependence of mining and mineral beneficiation process on the performance of mining projects. In Proceeding of the Australian Institute of Mining and Metallurgy Annual Conference, Melbourne, Australia, 5–9 August 1994.
3. Osanloo, M.; Hekmat, A. Prediction of shovel productivity in the Gol-E-Gohar iron mine. *J. Min. Sci.* **2005**, *41*, 177–184.
4. Singh, S.P.; Narendrula, R. Factors affecting the productivity of loaders in surface mines. *Int. J. Min. Reclam. Environ.* **2007**, *20*, 20–32.
5. Hendricks, C. Performance Monitoring of Electric Mining Shovels. Ph.D. Thesis, McGill University, Montreal, QC, Canada, 1990.
6. Jessett, A. Tools for Performance Monitoring of Electric Rope Shovels. Master's Thesis, University of Queensland, Brisbane, Australia, 2001.
7. Onederra, I.; Brunton, I.; Battista, J.; Grace, J. Shot to shovel-understanding the impact of muck-pile conditions and operator proficiency on instantaneous shovel productivity. In Proceeding of the EXPLO 2004, Perth, WA, USA, 26–28 July 2004; pp. 205–213.
8. Patnayak, S.; Tannant, D.D.; Parsons, I.; Del Valle, V.; Wong, J. Operator and dipper tooth influence on electric shovel performance during oil sands mining. *Int. J. Min. Reclam. Environ.* **2008**, *22*, 120–145.
9. Awuah-Offei, K.; Summers, D. *Reducing Energy Consumption and Carbon Footprint through Improved Production Practices*; Technical Report; Missouri University of Science and Technology: Rolla, MO, USA, 2010.
10. Vukotic, I.M. Evaluation of Rope Shovel Operators in Surface Coal Mining Using a Multi-Attribute Decision-Making Model. Master's Thesis, West Virginia University, Morgantown, WV, USA, 2013.

11. Oskouei, M.A.; Awuah-Offei, K. Statistical methods for evaluating the effect of operators on energy efficiency of mining machines. *Min. Technol.* **2014**, *123*, 175–182.

12. Oskouei, M.A.; Awuah-Offei, K. A method for data-driven evaluation of operator impact on energy efficiency of digging machines. *Energy Effic.* **2015**, *123*, 175–182.

13. Widzyk-Capehard, E.; Lever, P. Towards rope shovel automation operation information system. In Proceeding of the CRC Mining Conference, Queensland, Australia, 15–16 June 2004.

14. Bernold, L. Quantitative assessment of backhoe operator skill. *J. Constr. Eng. Manag.* **2007**,*133*, 889–899.

15. Komljenovic, D.; Bogunovic, D.; Kecojevic, V. Dragline operator performance indicator. *Int. J. Min. Reclam. Environ.* **2010**, *24*, 34–43.

16. Awuah-Offei, K.; Frimpong, S. Cable shovel digging optimization for energy efficiency.*Mech. Mach. Theory* **2007**, *42*, 995–1006.

17. Lloyd, S. Least squares quantization in PCM. *IEEE Trans. Inform. Theory* **1982**, *28*, 129–137.

18. Arthur, D.; Vassilvitskii, S. K-means: The advantages of careful seeding. In Proceedings of the Eighteenth Annual ACM-SIAM Symposium on Discrete Algorithms, Philadelphia, PA, USA, 7 January 2007; pp. 1027–1035.

19. Awuah-Offeik, K.; Frimpong, S. Geometric simulation of material weight resistance in cable shovel loading. In Proceedings of the 15th IASTED Conference, Marina Del Rey, CA, USA, 1–3 March 2004; pp. 141–146.

20. Kirsten, H.A.D. A classification system for excavation in natural materials. *Civ. Eng. S. Afr.* **1982**, *24*, 293–308.

21. Scoble, M.J.; Muftuoglu, Y.V. Derivation of a diggability index for surface mine equipment selection. *Min. Sci. Technol.* **1984**, *1*, 305–322.

22. Hadjigeorgiou, J.; Poulin, R. Assessment of ease of excavation of surface mines. *J. Terramechanics* **1998**, *35*, 137–153.

6
CHAPTER

Bend Testing of Silicon Microcantilevers from 21°C to 770°C

David E.J. Armstrong[1], and Edmund Tarleton[1]

[1]*Department of Materials, University of Oxford, Oxford, UK*

ABSTRACT

The measurement of mechanical properties at the microscale is of interest across a wide range of engineering applications. Much recent work has demonstrated that micropillar compression can be used to measure changes in flow properties at temperatures up to 600°C. In this work, we demonstrate that an alternative microscale bend testing geometry can be used to measure elastic, plastic, and fracture behavior up to 770°C in silicon. We measure a Young's modulus value of 130 GPa at room temperature, which is seen to drop with increasing temperature to ≈125 GPa. Below 500°C, no failure is seen up to elastic strains of 3%. At 530°C, the microcantilever fractures in a brittle fashion. At temperatures of 600°C and above plastic deformation is seen before brittle fracture. The yield stresses at these temperatures are in good agreement with literature values measured using micropillar compression.

INTRODUCTION

The development of microscale mechanical testing of focused ion beam (FIB) machined elements has come a long way since the initial work by Uchic et al.1 Although the majority of reported work has focused on the compression of microscale and nanoscale pillars,2–5there has also been a substantial body of work based on the bending of microscale FIB-machined cantilevers in a wide range of materials systems.[6–10] Such microscale cantilever bend experiments have several advantages. First, as demonstrated by Gong and Wilkinson[11] and Armstrong et al.,[12] they can be used to accurately determine the elastic modulus of metals including copper and titanium with sufficient accuracy to allow the anisotropy in Young's modulus to be measured, without the need for finite-element modeling. In comparison, Young's moduli values measured from compression tests tend to significantly underestimate the elastic modulus unless the punch in effect from the pillar and any taper and misalignment is accounted for during modeling. If the elastic modulus extracted from the stress–strain curve is not accurately measured, then it stands to reason that either the stress or strain is incorrectly reported. Second, bend samples can be notched—either by FIB machining[13] or by other means such as using a sharp prenotch,[14]which allows the measurement of fracture toughness of either bulk materials[8,10,14] or of individual microstructural features such as phase interfaces[15] or grain boundaries.[13]Recent work by Jaya et al.[16] has shown that this microcantilever method produces fracture toughness values in silicon, which are in good agreement with fracture toughness values measured by other microscale means such as pillar splitting or double-edge-clamped beams as well as experimentally measured bulk fracture toughness values. Although micropillars have been tested to failure by compression in a range of brittle materials,[17,18] this does not allow the fracture toughness to be measured. In some cases, the geometry promotes the plastic flow of otherwise brittle materials,[18,19] and in others splitting is observed where cracks propagate through the sample.[20]

Although much microcantilever bending work has been carried out, it is almost exclusively reported at room temperature. In many engineering systems of interest, from aerospace to nuclear, the performance of the materials at temperature close to or above service conditions is vital to allow safe engineering design. In recent years, there has been a major effort by both equipment manufacturers and users to develop high-temperature nanoindentation and microscale deformation experiments. Most of this work has focused on either straightforward nanoindentation, for which nanohardness and indentation modulus have been reported up to 750°C.[21–25] For high-quality nanoindentation experiments to be performed, it is necessary to accurately know the area function for the indenter tip. As discussed by Wheeler et al.,[23,26] high-temperature experiments can quickly blunt indenter tips whether made from diamond or another hard material such as cubic boron nitride. Second, reactions can occur at high temperatures between samples and tips[27] (such as diamond dissolving in strong carbide formers such as tungsten or vanadium).This results in a rapid change in area functions and calibration must be performed regularly throughout the experimental procedure.

Finally at higher temperatures, pile up effects may become more pronounced, resulting in a loss of tip calibration. As such, most high-temperature work has focused on performing microscale compression tests on a range of engineering materials at temperatures up to 600°C,28where tip damage is mitigated as wear to the flat punch is negligible.

In this work, we will demonstrate the testing of microcantilevers at temperatures from 21°C to 700°C in vacuum to measure the elastic modulus, yield stress, and fracture behavior in single-crystal silicon. The values measured are in good agreement with literature data.

BEND TESTS IN SILICON

The microcantilevers were manufactured in an undoped single-crystal silicon (100) surface with the long axis of the cantilever in the ⟨001⟩ direction (Fig. 1a). A Zeiss Auriga Dual Beam FIB-scanning electron microscope (SEM) (with Ga+ ions; Carl Zeiss, Oberkochen, Germany) was used to manufacture the microcantilevers following the method described by Armstrong et al.12 The final cantilevers had a length of ≈27 μm, a width of ≈4 μm, a triangular cross section with a maximum depth of ≈5 μm (Fig. 1b). Each cantilever was imaged using the SEM column of the Aurgia FIB-SEM before testing for accurate measurement of the test geometry and afterward to study the deformation that had occurred.

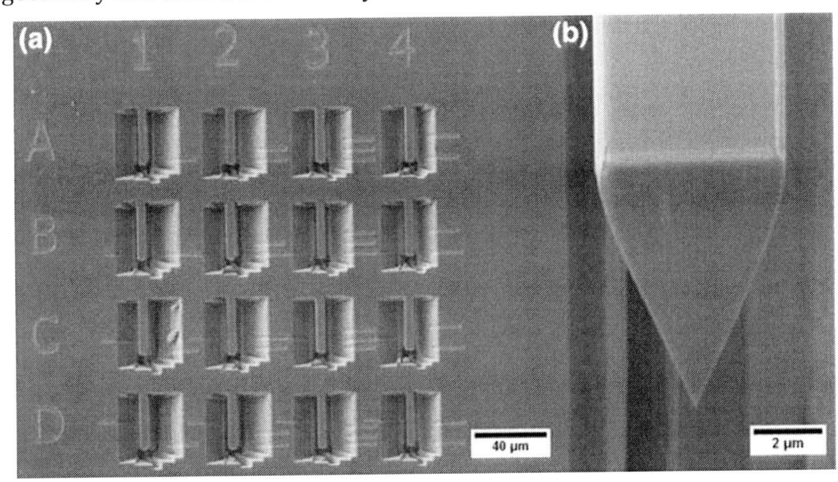

Figure 1.(a) Array of microcantilevers manufactured in single-crystal silicon prior to testing. (b) Single cantilever tilt showing end geometry (tilt corrected)

Testing was performed in a NanotestXtreme (MML Ltd., Wrexham, U.K.), nanoindenter housed in a vacuum chamber (Fig. 2a). The chamber was purged with argon prior to being evacuated to a vacuum level of 1×10^{-5} mbar prior to testing and was held at this level for all experiments.

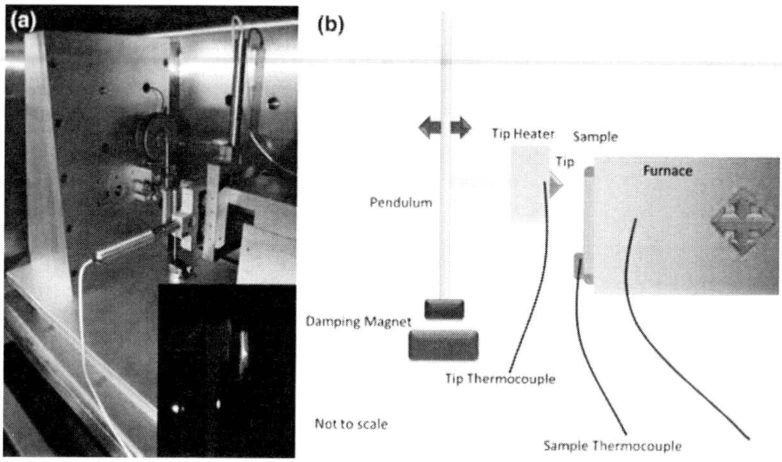

Figure 2. (a) Internal layout of nanoTestExtreme; inset shows tip in heat shield at 700°C. (b) Schematic of test layout showing thermocouple positions for matching of tip and sample temperatures

A cubic boron nitride Berkovich indenter was used, which was independently heated from the samples. The samples were mounted on a hot stage using Forta Fix Autostic FC6 high-temperature cement (Fig. 2b) with thermocouples mounted inside the furnace to monitor furnace temperature and a second thermocouple mounted using the same cement on the sample surface to allow accurate matching of the sample surface and indenter temperature. The same method used by Gibson et al.21 was used to match the sample and indenter temperature, which minimized the drift rates during the experiment. Figure 3shows displacement versus time for constant load for two tests; in the first, the temperatures of the tip and sample have been matched, and in the second, they have been intentionally unmatched. This demonstrates the importance of ensuring isothermal contact; when the sample and indenter temperatures are matched, the drift rate is minimal across a 60-s hold period, with a 10°C difference in temperature over 100 nm of drift are observed. To ensure accurate placement of the indenter tip on the microcantilever, each sample was imaged by scanning the indenter tip across the surface with a small (≈1 μN) load to produce a surface profile image (Fig. 4) with surface height data collected every 0.25 μm.

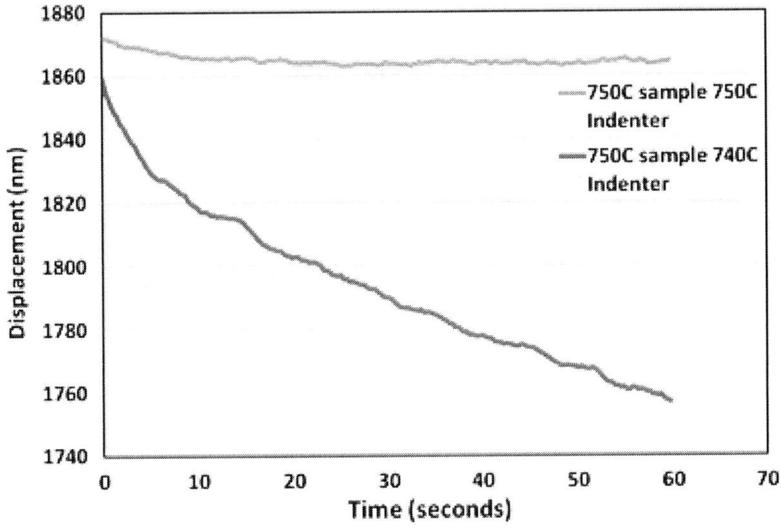

Figure 3. Time–displacement data at peak load for two tests showing the importance of isothermal contact

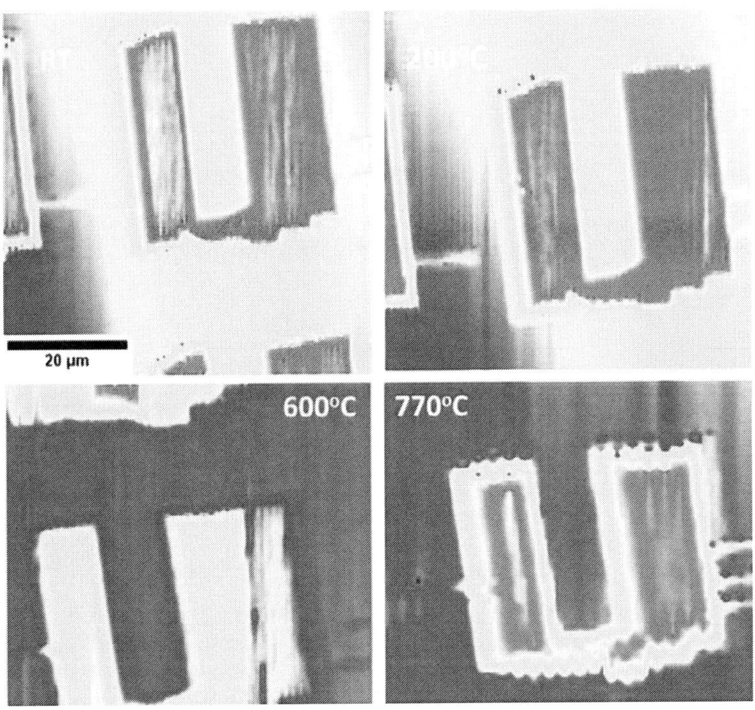

Figure 4. Surface profilometry images of microcantilevers taken from room temperature to 770°C (RT = room temperature)

The indenter tip was then placed close to the free end of the cantilever and the load applied. A constant loading rate of 0.05 mN/s was used with a maximum displacement of between 1000 nm and 3000 nm (Fig. 5).

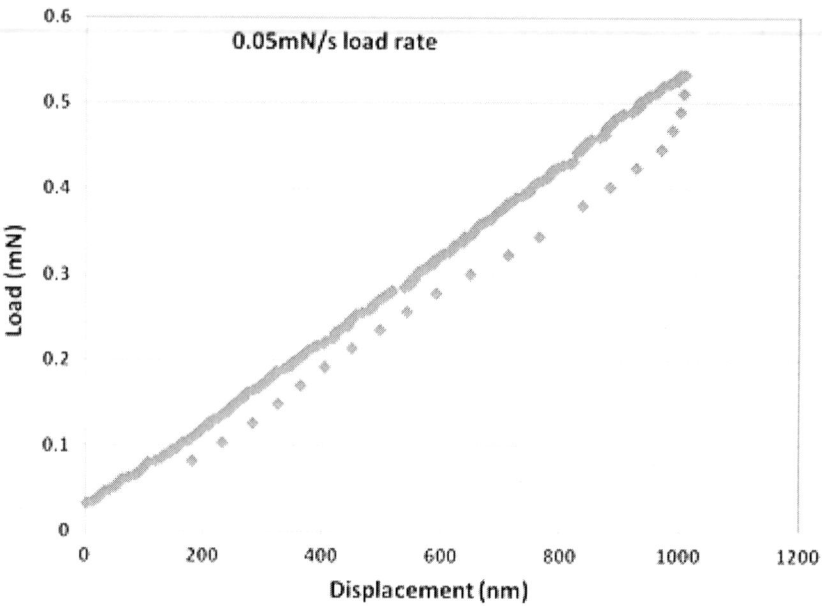

Figure 5. Load–displacement curve of microcantilever at room temperature showing elastic deformation to 1000-nm deflection

Figure 4 shows the surface scans of the microcantilevers up to 770°C, collected using the same parameters. Although some small scan artefacts can be seen at higher temperatures, the cantilever can be clearly identified for testing in each case.

Figure 6 shows the stress–strain curves calculated from the raw-load displacement data using simple beam theory (as described in Armstrong et al.12). A small amount of hysteresis is seen in the load–unload curves at lower temperatures due to plastic deformation under the sharp indenter tip as previously seen in other work.12 The Young's modulus values calculated are shown in Table I, alongside the literature values taken from Ref. 29. The Young's modulus values are seen to be consistently low compared with the literature values. As described by Gong and Armstrong,11,12 this is due to the flexure of the top surface of the fixed end of the cantilever where it is not truly encastred. To account for this, finite-element analysis (FEA) was conducted using a simple elastic model (Fig. 7) to fit to the load–displacement data, which takes into account deformation in the fixed end of the cantilever. The values from Table I are in good agreement with the literature values, which show only a small drop (≈10 GPa) across the temperature range used.

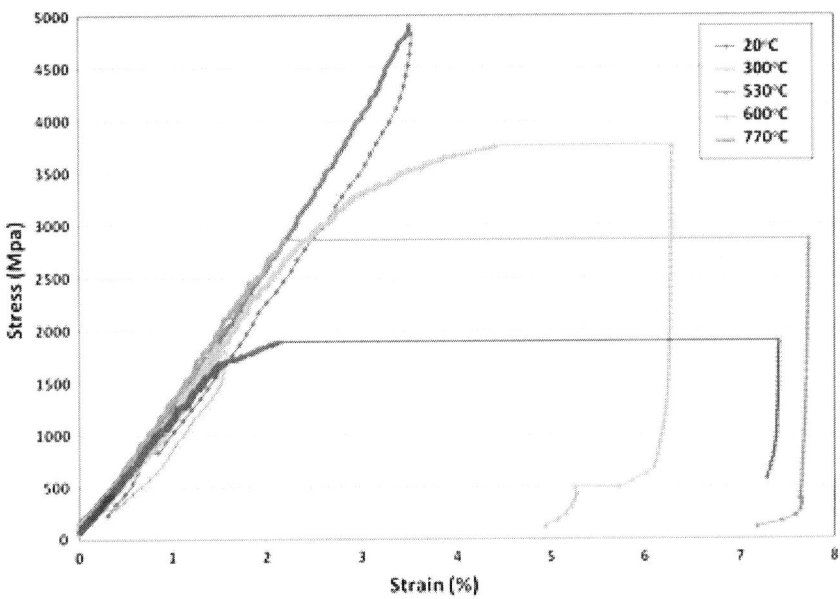

Figure 6. Stress–strain curves calculated using simple beam theory for tested microcantilevers from 20°C to 700°C

Table I. Young's' Modulus as a Function of Temperature

Temperature (°C)	Young's modulus literature value (GPa)	Young's modulus calculated using simple beam theory (GPa)	Young's modulus calculated using FEA (GPa)
21	130	110	130
200	128	105	128
300	126	109	129
530	122	105	126
600	121	110	131
770	120	105	125

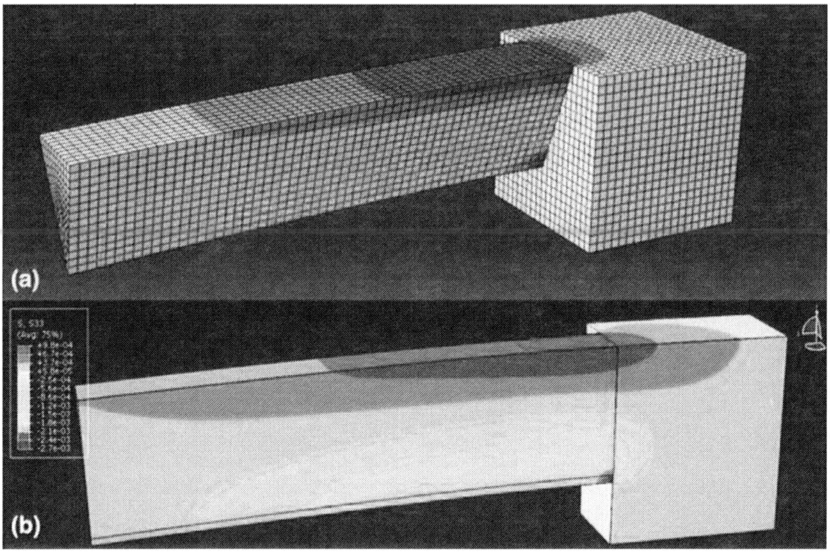

Figure 7. (a) Finite-element model of microcantilever including fixed end showing mesh (b) cut through central line of model showing peak stress distribution at maximum deflection

The cantilevers tested up to 530°C showed pure elastic deformation with no plasticity evident in the load displacement curve. Figure 8a and b shows the surface profile scans of a cantilever tested at room temperature before and after testing; no permanent deformation of the cantilever is observable. At 530°C, a change in deformation was observed. Although elastic deformation was observed initially, at a stress of 2.8 GPa the cantilever catastrophically fractured, seen by the sudden displacement burst and near vertical unloading section. At 600°C, a further change in deformation mechanism is observed; significant plasticity is observed above a stress of 2.7 GPa to a maximum load of 3.7 GPa at which load fracture again occurs. This fracture was confirmed by a surface profile scan made immediately after testing (Fig. 8c and d) and subsequently in the secondary electron image (Fig. 9a), showing that the cantilever is no longer present in the test position and through fracture has occurred. At 770°C, similar behavior was observed with a lower yield stress of 1.6 GPa followed by fracture occurring at 1.9 GPa. To confirm that plastic deformation rather than stable crack growth occurred, an interrupted test was performed at 770°C. The test was interrupted after yield had occurred but before failure of the cantilever by fracture. This can be seen in Fig. 9b. Slip bands on the sample are clearly seen in the inset of the figure running across the width of the microcantilever, clearly showing that plastic deformation is occurring in silicon at these temperatures on the expected ⟨111⟩ planes.

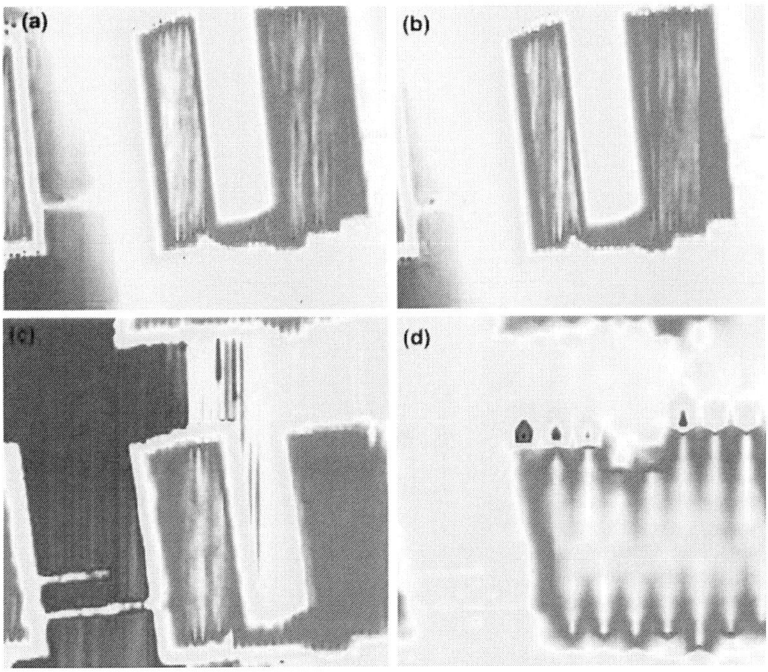

Figure 8. Surface profile scans (a) room temperature before testing, (b) room temperature after testing, (c) 600°C before testing, and (d) 600°C after testing. No permanent deflection of the cantilever is seen after testing at room temperature, indicating that the deformation is purely elastic, whereas fracture has clearly occurred in the 600°C test

These data can be looked at in relation to other data in the literature regarding the high-temperature deformation of silicon. Pearson et al. studied the fracture behavior of 20-μm diameter high-purity silicon whiskers from room temperature to 800°C. Below ≈600°C, the samples deformed elastically followed by brittle fracture, and above 600°C the samples underwent a small amount of plastic deformation before brittle failure and above 650°C significant plastic deformation was observed. This transition in behavior is in good agreement with the work of Hirsch and Roberts,30 who studied the brittle-to-ductile transition in notched silicon samples of ≈1 mm2 cross-sectional area, tested in a four-point bend. They observed a transition that was logarithmically dependent on the strain rate of testing, with brittle-to-ductile transitions occurring between 550°C and 700°C.

Figure 9. (a) Microcantilever tested at 600°C to fracture. (b) Microcantilever tested at 770°C to yield and stopped before fracture

More recently, several high-temperature microcompression experiments have been performed on silicon pillars, with which the data in this study can be compared. Korte et al.31 used an experimental setup similar to that used in this study to compress FIB prepared micropillars from room temperature to 500°C with a ⟨100⟩ pillar normal direction. They found that there was a pronounced effect of both temperature and size on the failure behaver and yield stress. Although large diameter (typically

4000 nm) failed in a brittle manner by reducing the diameter to less than 1000 nm, plastic deformation occurred at room temperature. When the temperature of compression of 2000-nm diameter pillars was increased from 25°C to 500°C, a transition was seen from a fracture-based failure mechanism (at a ≈ 6 GPa) to a splitting mechanism. This mechanism had limited plasticity at 200°C at similar stress levels. Finally, full plastic deformation was observed at 500°C, with a yield stress of ≈3 GPa. Wheeler et al.27 observed a transition in lithographically prepared pillars (also with a ⟨100⟩ orientation), which failed catastrophically up to 100°C, to a splitting mechanism between 200°C and 300°C followed by plastic flow at ≈400°C. Wheeler's and Korte's values of yield stress of between 3 GPa and 4 GPa are higher than the values measured in this work (1.6–1.9 GPa). This discrepancy this could be for several reasons, including the differences in size between the pillars and cantilevers, manufacturing methods, and the difference in loading modes between pure compression and bending. In particular, the bending deformation will promote the propagation of any cracks due to defects in the tension side of the cantilever induced in the manufacture process. Rabier et al.32 looked at several different sized micropillars manufactured by FIB machining, oriented in a ⟨123⟩ direction to ensure single-slip system activation. Again, larger pillars (2000-nm diameter) showed cracking and splitting, whereas smaller pillars (980 nm diameter) showed plastic flow even at room temperature. In these smallest pillars, the yield stress dropped from ≈7 GPa at 25°C to ≈3 GPa at 400°C.

The trend in the data for yield stress and critical temperature for change in deformation behavior measured using both macroscopic methods and microcompression testing are in good agreement with the yield stress behavior with increasing temperature and transition temperature values measured in this study. These microscale bend tests have advantages in probing true brittle behavior as the tensile side promotes true crack growth rather than a complex pillar splitting mechanism seen in compression testing. However, the differences in yield stresses measured at elevated temperatures need further study as to their controlling mechanisms.

CONCLUSION

This study has shown that microcantilever experiments can be performed at temperatures up to 770°C; the measured values for the elastic, plastic, and fracture behavior of single-crystal silicon are in good agreement with data in the literature from both bulk and microscale testing methods. Further work will focus on applying these high-temperature microcantilever techniques on notched samples to allow fracture toughness values to be measured and understanding differences in behavior between compression and bend tests.

ACKNOWLEDGEMENTS

DEJA acknowledges the Royal Academy of Engineering for a Research Fellowship. Some of the work was supported financially by the EPSRC under program Grant EP/H018921/.

REFERENCES

1. M.D. Uchic, D.M. Dimiduk, J.N. Florando, and W.D. Nix, *Science* 305, 986 (2004).
2. J.R. Greer and J.T.M. De Hosson, *Prog. Mater. Sci.* 56, 654 (2011).
3. E.M. Grieveson, D.E.J. Armstrong, S. Xu, and S.G. Roberts, *J. Nucl. Mater.* 430, 119 (2012).
4. D. Kiener, P. Hosemann, S.A. Maloy, and A.M. Minor, *Nat. Mater.* 10, 608 (2011).
5. M.D. Uchic, P.A. Shade, and D.M. Dimiduk, *JOM* 61, 36 (2009).
6. D.E.J. Armstrong, A.S.M.A. Haseeb, S.G. Roberts, A.J. Wilkinson, and K. Bade, *Thin Solid Films* 520, 4369 (2012).
7. D.E.J. Armstrong, M.E. Rogers, and S.G. Roberts, *Scr. Mater.* 61, 741 (2009)
8. D. Di Maio and S.G. Roberts, *J. Mater. Res.* 20, 299 (2005).
9. J.C. Gong and A.J. Wilkinson, *Acta Mater.* 57, 5693 (2009).
10. S. Wurster, C. Motz, M. Jenko, and R. Pippan, *Adv. Eng. Mater.* 12, 61 (2010).
11. J.C. Gong and A. Wilkinson, *Philos. Mag. Lett.* 90, 503 (2010).
12. D.E.J. Armstrong, A.J. Wilkinson, and S.G. Roberts, *J. Mater. Res.* 24, 3268 (2009).
13. D.E.J. Armstrong, A.J. Wilkinson, and S.G. Roberts, *Philos. Mag. Lett.* 91, 394 (2011).

14. S. Wurster, C. Motz, and R. Pippan, *Philos. Mag.* 92, 1803 (2012).

15. F. Iqbal, J. Ast, M. Goken, and K. Durst, *Acta Mater.* 60, 1193 (2012).

16. B.N. Jaya, C. Kirchlechner, and G. Dehm, *J. Mater. Res.* 30, 686 (2015).

17. P.R. Howie, S. Korte, and W.J. Clegg, *J. Mater. Res.* 27, 141 (2012).

18. S. Korte and W.J. Clegg, *Scr. Mater.* 60, 807 (2009).

19. S. Korte and W.J. Clegg, *Philos. Mag.* 91, 1150 (2011).

20. F. Ostlund, P.R. Howie, R. Ghisleni, S. Korte, K. Leifer, W.J. Clegg, and J. Michler,*Philos. Mag.* 91, 1190 (2011).

21. J.S.K.L. Gibson, S.G. Roberts, and D.E.J. Armstrong, *Mater. Sci. Eng. A* 625, 380 (2015).

22. J.C. Trenkle, C.E. Packard, and C.A. Schuh, *Rev. Sci. Instrum.* 81, 073901 (2010).

23. J.M. Wheeler, D.E.J. Armstrong, W. Heinz, and R. Schwaiger, *Curr. Opin. Solid State Mater. Sci.* (2015).

24. C.A. Schuh, C.E. Packard, and A.C. Lund, *J. Mater. Res.* 21, 725 (2006).

25. A. Sawant, S. Tin, and J.C. Zhao, *Superalloys* (New York: Wiley, 2008), pp. 863–871.

26. J.M. Wheeler, R.A. Oliver, and T.W. Clyne, *Diam. Relat. Mater.* 19, 1348 (2010).

27. J.M. Wheeler and J. Michler, *Rev. Sci. Instrum.* 84, 045103 (2013).

28. C. Walter, J.M. Wheeler, J.S. Barnard, R. Raghavan, S. Korte-Kerzel, P. Gille, J. Michler, and W.J. Clegg, *Acta Mater.* 61, 7189 (2013).

29. N. Ono, K. Kitamura, K. Nakajima, and Y. Shimanuki, *Jpn. J. Appl. Phys.* 1, 368 (2000).

30. P.B. Hirsch and S.G. Roberts, *Philos. Mag. A* 64, 55 (1991).

31. S. Korte, J.S. Barnard, R.J. Steam, and W.J. Clegg, *Int. J. Plast* 27, 1853 (2011).

32. J. Rabier, A. Montagne, J.M. Wheeler, J.L. Demenet, J. Michler, and R. Ghisleni,*Phys. Status Solidi C* 10, 11 (2013).

7

CHAPTER

The History and Future Challenges of Calcined Petroleum Coke Production and Use in Aluminum Smelting

Les Edwards

[1]Rain CII Carbon, Covington, LA 70433, USA.

ABSTRACT

Calcined petroleum coke is used for the production of carbon anodes in the Hall-Héroult aluminum smelting process due to a combination of low impurity levels, ready availability, and relatively low cost. This article provides a review of the history and use of calcined petroleum coke for anode production and describes the different calcining technologies used by the industry. The article discusses the impact of changes in crude oil quality and refining economics over the last 10 years as well as the impact on green petroleum coke quality and availability. The industry has adapted well to quality changes in recent times, and the blending of different quality cokes by smelters is becoming increasingly important. The world has a plentiful supply of green petroleum coke, but the next wave of aluminum smelting capacity growth will put further pressure on the supply of the higher quality cokes traditionally favored by the industry.

INTRODUCTION

Calcined petroleum coke (CPC) has been used for more than 120 years to produce the carbon anodes used in the Hall-Héroult aluminum electrolysis process. Several other forms of carbon were tested in the early stages, but none proved to have the right combination of low impurity levels, relatively low cost, and ready availability. Prebaked anodes are produced with 55–65% CPC, 13–15% coal tar pitch binder, and 20–30% recycled anode butts. The anodes are consumed at a net consumption rate of approximately 400 kg carbon/ton aluminum for modern smelting cells according to reaction 1. They provide an important source of heat for the electrolysis process, reducing the cell voltage and overall energy requirement relative to cells operating with nonconsumable anodes.1

$$2Al_2O_3 + 3C = 4Al + 3CO_2 \quad (1)$$

CPC is produced by heating or calcining green petroleum coke (GPC) at temperatures greater than 1200°C. The production of GPC has remained essentially the same since 1929 when the modern delayed coking process was born. This was followed in 1935 by the development of the rotary kiln calcining process, which is the most commonly used technology in the Western world. CPC production has increased significantly in China over the last 10 years, and shaft calcining is the dominant technology used in this region. The aluminum industry has had a ready supply of good-quality GPC and CPC for many years, but the situation has become more challenging over the last 10 years due to a general trend toward higher impurity levels resulting from changes in crude and refining economics.

This article will review the history of GPC and CPC production and its use in aluminum smelting and provide comments on current and future challenges.

Early History

An excellent review of the history of aluminum production is provided in the Hall-Héroult Centennial book,2 which also includes a chapter on anode production. It was fortuitous that petroleum coke had been discovered about 30 years before Hall and Héroult made their independent discoveries on producing aluminum. Work had already been done on developing coke as a filler material in carbon products by Charles Brush and Washington Lawrence in 1877.3 Brush discovered that calcining coke prior to forming pitch-bonded shapes produced better quality carbons, and Hall used electrodes supplied by Brush in his early work. In Europe, purified anthracite was believed to have been used as an alternative to CPC during these first years of production.

The commercial production of aluminum by the Pittsburgh Reduction Company (later Alcoa) used carbon anodes made from CPC produced nearby by the Faraday Carbon Company. An early generation anode from the Alcoa Tennessee smelter, circa 1926, is shown in Fig. 1. The anode has a fine grain texture suggesting that the CPC was pulverized first. The slightly bent shape and presence of longitudinal score marks on the external surface is consistent with an extrusion process for forming.

Figure 1. Early generation Alcoa Tennessee anode (5″ diam. × 18″ long) in foreground with potroom photo in background (courtesy Steve Lindsay, Alcoa)

Green Petroleum Coke Production

GPC production began as early as the 1860s in the United States and went through many process changes until 1929, when Standard Oil developed the delayed coking process. A good review of the history and production of GPC can be found in the tutorial by Ellis and Paul.4 Delayed coking is a "carbon rejection" process used to upgrade the heavy residuum or resid from the bottom of vacuum distillation towers and other refinery process units. Delayed coking significantly increases the recovery of valuable gas and liquid products such as gas, gasoline, gas oils, and so on, as a result of the higher temperatures (approximately 500°C) and more severe thermal treatment. This causes some of the larger, higher molecular weight hydrocarbons to crack into smaller molecules that are then recovered in a fractionator. Polymerization and cross-linking reactions lay down a solid residue of carbon referred to as GPC.

A delayed coker uses vertical drums in a semicontinuous process, as shown in Fig. 2. While one drum is being filled with hot resid and forming coke, solid coke is cut out of the other drum using a high pressure water jet. Not all refineries have delayed cokers, and some produce asphalt or sell resid into the fuel oil market. A typical refinery flow sheet with a delayed coker is shown in Fig. 3. The economic value of GPC is relatively low (2–3%) compared with the value of the liquid and gaseous products (97–98%), so GPC is considered a byproduct at best, for most refineries.

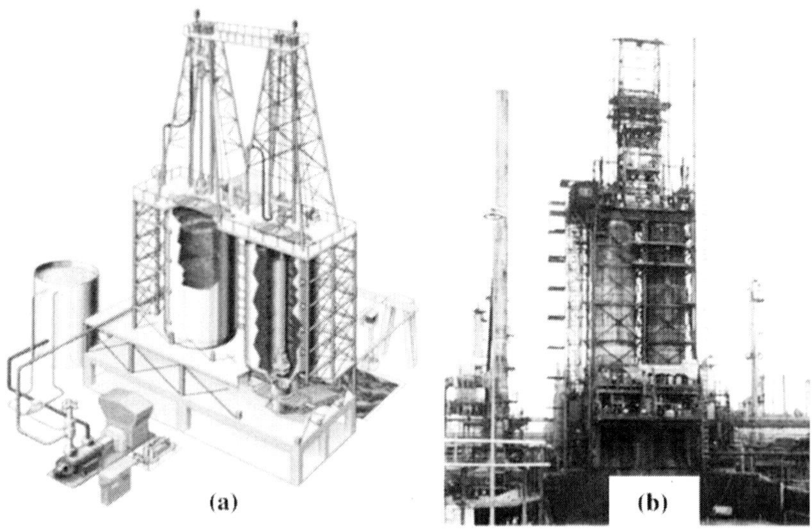

Figure 2. Delayed coker operation: schematic (a) and delayed coker (b)

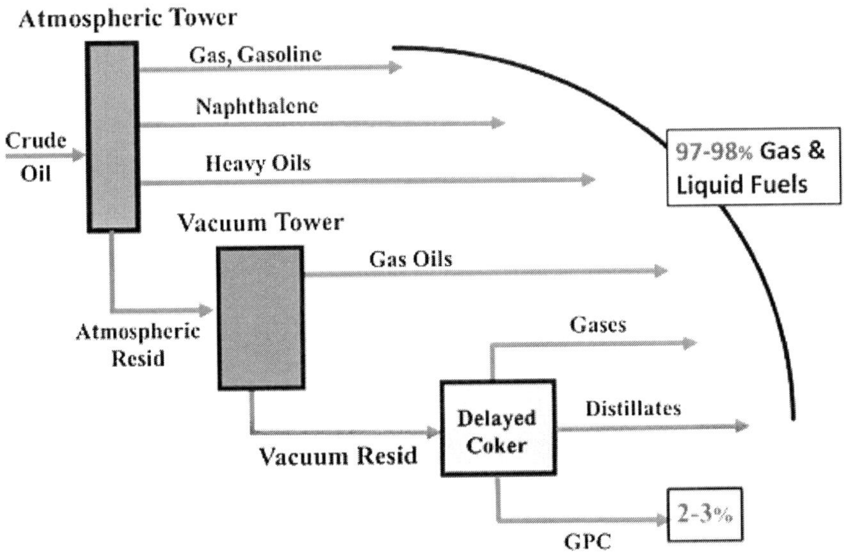

Figure 3. Simplified refinery flow sheet with delayed coker showing relative product values

Oil refineries that process heavy crude oils (those with a high specific gravity or low API gravity) produce more resid, and for this reason, delayed cokers are common in these refineries. They produce a greater volume of GPC relative to those running a lighter crude blend. Most GPC (approximately 75%) is sold as a fuel for power generation and cement production, but the higher quality GPC (lower sulfur and trace metals) is sold to the calcining industry. The global production of GPC was around 115 million DMT (dry metric tons) in 2013 with around 54 million DMT produced in the United States.

GPC Properties

GPC has several different structural forms commonly referred to as needle coke, sponge coke, and shot coke. Crude oil quality plays a major role in determining which of these is produced, although the coker operation can also play a role.5 Figure 4 compares the typical structures of calcined needle, sponge coke, and shot coke with their optical textures, which can be viewed when looking at polished cross-sections under polarized light with an optical microscope.6,7

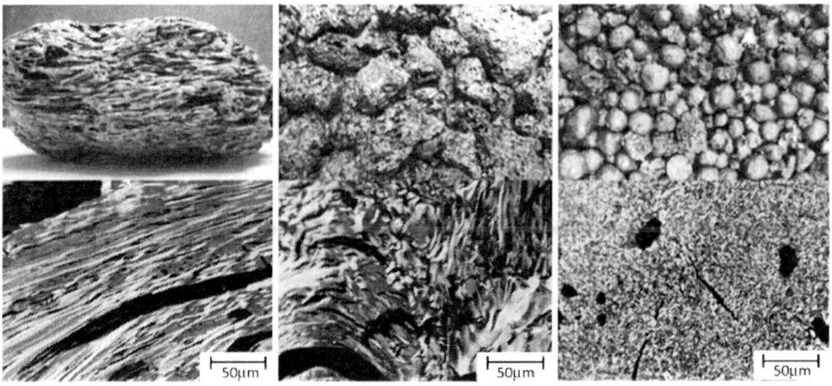

Needle **Anisotropic** | Sponge **Mixed** | Shot **Isotropic**

Figure 4. Delayed coke types and optical textures

Needle coke is a premium product with very low S and impurity levels and a highly layered or anisotropic structure. It has a low coefficient of thermal expansion (CTE < 2.0×10^{-6}/K), making it the material of choice for the production of the graphite electrodes used in steel-producing electric arc furnaces. Needle coke is produced from highly aromatic feedstocks such as decant or slurry oil produced in a fluid catalytic cracker.

Sponge coke is the preferred structure for anode production and CTEs are typically in the range of $3.5–4.8 \times 10^{-6}$/K. It has a mixed optical texture with a wide range of domain sizes. The open porosity in sponge coke allows good pitch penetration during mixing, and a mechanically strong, interlocking structure develops after anode baking. Shot coke has a characteristic spherical particle shape and a dense, highly isotropic texture sometimes referred to as a granular texture. Shot coke has a high CTE (>5.5) and typically higher levels of sulfur and trace metal impurities, particularly V and Ni. It is formed from crudes with high levels of resins and asphaltenes, which are large molecular weight precursors.

All GPC has residual volatile matter (VM) when it is cut from the drum. The VM level is dependent on the coking severity, but typical ranges are 9–14%. Table I gives an example of three sponge cokes used by calciners for anode applications and a typical fuel-grade shot coke.

Table I. Sponge GPC properties

	Anode grade			Fuel/TiO_2 Shot
	Sponge A	Sponge B	Sponge C	
Sulfur, %	0.5	2.0	4.6	6.5
Vanadium, ppm	15	180	460	1930
Nickel, ppm	65	170	190	410
Iron, ppm	190	190	110	25
Calcium, ppm	180	50	60	65
Silicon, ppm	185	50	90	150
Sodium, ppm	45	50	90	100
VM content, %	11.2	10.4	11.8	11.2
HGI	105	95	80	35

The Hardgrove Grindability Index (HGI) is a measure of hardness (lower value = harder), and the data highlight the hardness of shot coke relative to sponge coke. Independent or merchant calciners (those not integrated with refineries) typically blend different quality GPCs together to produce CPC that meets anode specifications for S, V, Ni, and other trace metals. The CPC product can also be blended by calciners prior to shipment.

The S, V, and Ni levels are determined by the crude blend used at the refinery with heavy, sour crudes containing higher levels of each. These three impurities are intimately bound in the carbon structure and cannot be removed. A more detailed description of the sources of impurities in GPC can be found elsewhere.8 Sponge GPC used by the calcining industry is typically classified as low sulfur (<2.5%, e.g., cokes A and B in Table I) or high sulfur (>3.5%, e.g., coke C in Table I).

Coke Calcining

GPC must be calcined or heat-treated at temperatures above 1200°C before it can be used to make anodes. It is necessary to drive out the VM during calcining so that the coke shrinks and densifies. Figure 5 shows the change in a coke's real density and bulk density during calcining. Anodes produced from GPC would result in uncontrolled shrinkage and cracking during baking, producing unacceptable quality.9 GPC is an electrical nonconductor and calcining transforms the structure into a more ordered, electrically conductive form of carbon with an acceptably low reactivity to air (oxygen) and CO_2. The CPC used for anode production has low ash levels (<0.3%) and a high carbon content (typically >97% depending on the S level).

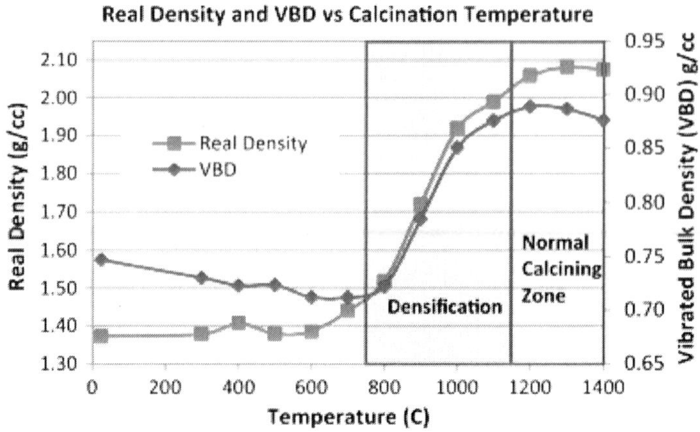

Figure 5. Real density and vibrated bulk density changes during calcining

Calciners always prefer GPC with a lower VM content because it results in a calcined product with lower porosity and higher bulk density. Figure 6 shows the relationship between the VM content of a single GPC source and the vibrated bulk density (VBD) of the CPC product. The VBD test is used widely in the industry as a simple, indirect measure of coke porosity. For an anode producer, higher VBD CPC can help in the production of higher density anodes, which is desirable to maximize the life of the anode in the cell before the anode butt is removed for recycling.

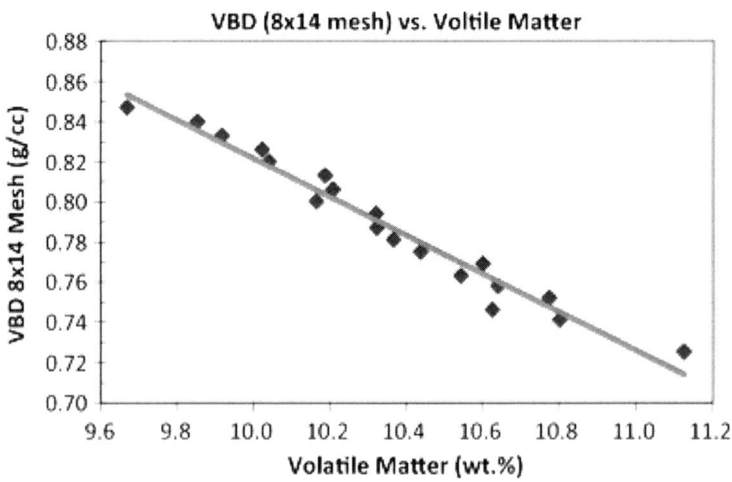

Figure 6. Relationship between VM content and CPC VBD for a single-source GPC

Table II shows the properties of CPC most commonly specified by aluminum smelters.

Table II. Typical specifications for CPC

Property	Maximum or minimum	Typical specification	Broader specification
Moisture, %	Max	0.3	0.5
Ash, %	Max	0.3	0.5
Sulfur, %	Max	3.0	3.5
Vanadium, ppm	Max	350	500
Nickel, ppm	Max	250	300
Iron, ppm	Max	300	400
Calcium, ppm	Max	200	250
Silicon, ppm	Max	250	300
Sodium, ppm	Max	150	200
Real density, g/cc	Min	2.05	2.03
VBD (28 × 48 mesh = 0.3–0.6 mm)	Min	0.85	0.82
Particle size (+4 mesh = 4.75 mm)	Min	25	25
Dedust oil, %	Max	0.3	0.5

The real density is used as a measure of the degree of calcination, but other measurements such as Lc and specific electrical resistivity can also be used. There are at least four different bulk density tests in common use by the industry, and Table II shows the ASTM D4292 VBD test. Many of these tests are strongly influenced by minor sample preparation differences,10 which can make the comparison of results between laboratories difficult. A small amount of dedust oil (approximately 0.3 wt.%) is almost always added to CPC to reduce dusting. One other property requested by some smelters is CO_2 reactivity. This can be measured analytically using an ISO method11 or can be predicted quite reliably using an equation developed by R&D Carbon based on the CPC S, Ca, and Na levels.12

Coke Calcining Technology

Early calcining of petroleum coke used both electric and gas fired vertical calciners,2 the latter being a likely predecessor of the shaft calcining technology still used today in some parts of Russia and extensively in China. Electric calciners were used first, followed by the gas-fired Woodall-Duckam and Glover-West vertical types.3

Rotary Kiln

The first rotary kiln for calcining GPC was used at Great Lakes Carbon in 1935,2 although rotary kilns had been used in the cement industry prior to that. Rotary kilns for CPC production are steel shells lined with high-temperature refractory brick; kilns can measure 30–85 m in length and

2.4–4.4 m in diameter. The kiln is sloped and turns slowly to move the coke through the kiln with a residence time of 45–60 min. The calcined coke exits the kiln at a temperature of 1200°C to 1350°C and is typically cooled by direct quenching with water in a rotary cooler.

Most kilns have a natural gas- or oil-fired burner at the discharge end that is used to heat the kiln from a cold start. It can also be used to provide additional heat for calcining, but most of the heat comes from the combustion of VM inside the kiln. Although rotary kiln technology has not changed significantly since its introduction, kilns have grown larger in size and most modern kilns are fitted with tertiary air fans as shown in Fig. 7. These are mounted on the kiln shell and blow air into the kiln in the zone where the volatile matter is combusted. This increases the efficiency of VM combustion and productivity from the kiln and reduces the need for supplemental fuel.13

Figure 7. Rotary kiln schematic and kilns equipped with tertiary air fans

Another modification in kilns that require supplemental fuel has been the introduction of oxygen near the coke discharge end. This reduces the flue gas flow (compared with burning natural gas) and improves the product yield. It can also reduce operating costs depending on the relative costs of natural gas/fuel oil versus oxygen.14 Large, modern kilns are also fitted with internal refractory "lifters" to improve heat transfer to the coke bed and improve the bulk density of the calcined product.15

Many rotary kilns are built with waste heat energy recovery systems to capture heat from the hot flue gases exiting the kiln countercurrent to the coke bed. The heat is recovered from the combustion of VM in the GPC and the fine coke particles carried over from the kiln. Complete combustion takes place in a pyroscrubber and the hot flue gas passes through a heat recovery steam generator. The steam can be sold to nearby plants or used to generate power with a steam turbine generator. Waste heat-recovery systems have become pretty much mandatory for new calciner projects to improve the project economics.

Rotary kilns are still the predominant technology used to calcine coke outside China and they offer excellent economies of scale with a high level of automation. A single, modern kiln can produce up to 350,000 t/year of CPC.

Rotary Hearth Calciner

The rotary hearth calciner shown in Fig. 8 is another calciner used around the world, although it is much less common than a rotary kiln. The design concept and calcining process of rotary hearth calciners have been described previously.16 The coke moves slowly from the outside perimeter of the rotary hearth kiln guided by rabbles toward the inner soak zone before exiting the hearth and being cooled with water in a rotating unit similar to the one described for the rotary kiln.

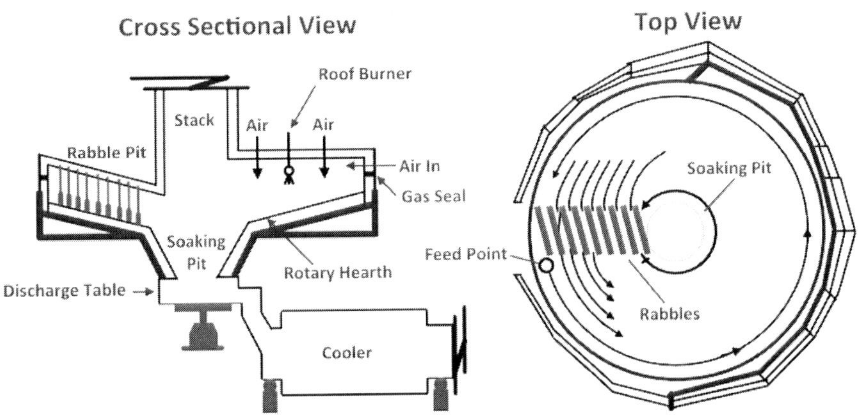

Figure 8. Schematic of rotary hearth calciner

Rotary hearth calciners can also be configured to recover waste heat energy similar to rotary kilns. Compared with rotary kilns, rotary hearth calciners typically have a faster initial heat up rate for the coke, which can lead to lower CPC bulk densities when higher VM cokes are used. Rotary hearth calciners result in less fines loss from the coke bed, resulting in a higher recovery of calcined coke per ton of GPC. This is an economic advantage when operating the calciner but leads to a product with a higher fines content (<0.5 mm), which is not always desirable for an anode paste plant.

Shaft Calcining

Shaft calcining technology was initially transferred to China from Russia, and early shaft calciners were developed in Europe. A small shaft calciner designed and built by the Riedhammer company in the 1930s is still operating in Germany and used for the production of calcined anthracite. It is similar to some of the older shaft calciners operating in China but has been highly automated. Today, shaft calcining is the dominant technology in China, although rotary kiln calciners are also used.

In a shaft calciner, the VM from the GPC is combusted inside flue walls, meaning that there is no contact between the flue gas stream and coke bed as there is in rotary kilns and rotary hearths. The coke moves very slowly down the shafts with a residence time in the range of 24–48 h. The rate is controlled by the removal of CPC from the bottom of the shafts using a slide gate or rotary valve. The coke is cooled near the bottom of the shaft by a cooling water jacket. Some older shaft calciners require a great deal of manpower to operate, but modern ones have automated feed and discharge systems, as shown in Fig. 9. A lot of the newer shaft calciners are also being built with waste heat-recovery systems.

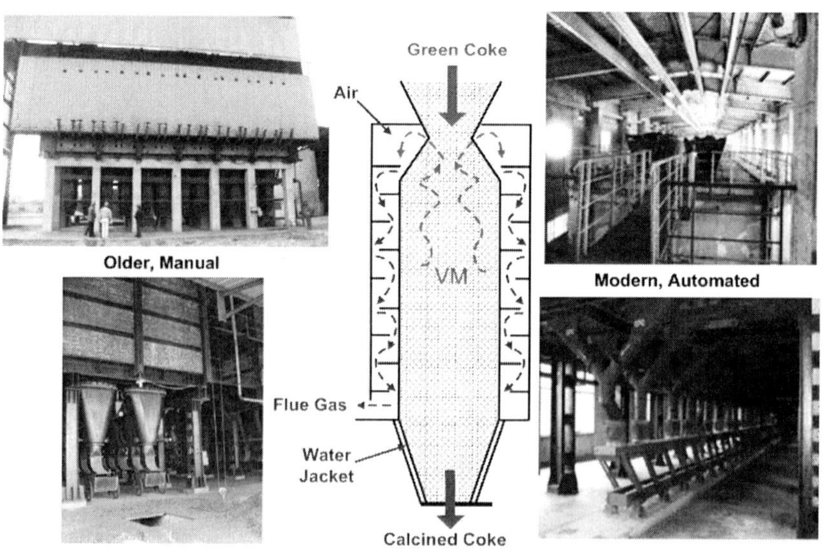

Figure 9. Shaft calciner showing principal of operation in schematic

Differences in the production and quality of shaft calcined coke versus rotary kiln and rotary hearth CPC have been well documented.16,17 Shaft CPC has a lower porosity and higher bulk density due to the slower heat-up rate and is coarser in particle size. Shaft CPC produced in older calciners can be dustier due to the agglomeration of fine particles on the surface of coarser particles. Newer shaft calciners with automated discharge systems and good dust collection systems do a better job of removing some of this dust. Figure 10shows scanning electron microscopy (SEM) images of CPC produced in a shaft calciner (Fig. 10a) versus a rotary kiln (Fig. 10b), highlighting the agglomeration problem with shaft CPC.

(a) Shaft Calciner (b) Rotary Kiln

Figure 10. SEM images of shaft (a) and rotary kiln CPC (b)

Shaft CPC can sometimes cause problems during unloading due to its higher packing density, higher angle of repose, and poorer flow characteristics relative to rotary kiln calciners. Some anode plants have also experienced problems with silo discharge and dead storage. Some trace metal impurities like calcium can also be higher in shaft CPC. High Ca levels are quite common in Chinese GPC and CPC due to the frequent use of high total acid number crudes.8 This is easy to monitor, but it is also important to examine differences in the impurity levels between coarse and fine particles. Elevated Ca levels in the fines fraction is undesirable because it increases anode CO_2 reactivities. Table III shows an example of three different low S, low V calcined cokes from northeast China with elevated Ca levels.

Table III. Chemical analysis results of several Low S Chinese CPC sources

S, %	V, ppm	Ni, ppm	Ca, ppm	Fe, ppm	Si, ppm	Na, ppm	RD, g/cc
0.47	22	281	256	157	23	49	2.12
0.43	20	378	394	168	100	30	2.13
0.44	19	316	231	100	16	44	2.12

RD real density

Changes in Petroleum Coke Quality: Current and Future Challenges

Availability of Low-Sulfur GPC and Quality Changes

The greatest change to have impacted the CPC industry in recent times has been the availability of suitable quality GPC. Over the last 10 years, gaining access to a reliable supply of low S, low metals GPC has become more challenging, and the industry no longer enjoys a ready supply relative to the demand from the calcining and aluminum industries. The reasons for this have been well documented18,19 and are driven by changes in crude oil quality and refining economics.

Most of the world's newly constructed refineries are configured to process heavy, sour crude oils that sell at a significant discount compared with light, sweet crudes. Many existing refineries, particularly in the United States, have made capital investments to allow the processing of more heavy, sour crudes like those from Canada. These changes have directly impacted the quality and volume of GPC produced. The general trend has been an increase in trace metals like V and Ni and an increase in S levels. The production of low S sponge coke (<2.5%) in the United States has decreased by approximately 50% over the last 10 years. In many other parts of the of the world, including China, most of the growth in new GPC production is coke with higher S and metals levels and more isotropic textures.

The limited supply of low S GPC has driven prices higher. Figure 11 shows GPC price data published by the Jacobs Consultancy group in their quarterly PACE Blend report. The different categories (A–D) reflect different GPC sulfur levels. These can vary somewhat quarter to quarter, but approximate averages are shown. The record high prices in 2011 reflect the relative tightness of these cokes in the global supply. What is apparent from Fig. 11 is the sustained high differential in price between the low S category A cokes and the other higher S cokes since 2010.

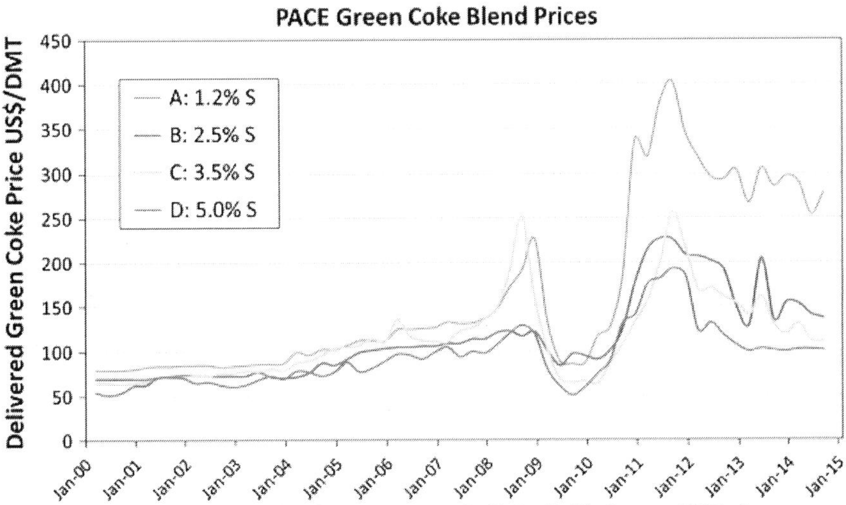

Figure 11. Prices for freely negotiated, U.S. Gulf Coast GPC (courtesy Jacobs Consultancy)

The CPC S level specification is the most critical for a smelter because it must operate within SO_2 emission limits. SO_2 limits vary by region but CPC S limits typically fall in the range of 1.8–3.0%. Trace metal impurities in CPC are the next most important consideration, particularly V and Ni because they impact aluminum metal purity. It is desirable for trace metal impurities like V, Ni, Ca, Fe, and Si to be as low as possible. Historically, typical smelter CPC maximum specifications may have been 2.5% S, 250 ppm V, 180 ppm Ni, 150 ppm Ca, 250 ppm Fe, and 150 ppm Si.

The relaxation of CPC specifications for V, Ni, and S has been underway for some time, but the number of smelters willing to relax specifications has increased since 2011 as a direct consequence of higher CPC prices. Calciners with the flexibility to source different GPC qualities can reduce raw material costs and offer lower priced CPC when one or more specifications are relaxed. The two specifications offering the greatest potential for cost savings are V and S, particularly V.

Historically, these two impurities have tended to follow each other in typical anode sponge cokes, but as shown in Fig. 12, these relationships are not as clear-cut today. The data are based on the analysis of a wide range of GPC sources used for anode CPC applications over the last 15 years. The relationships tend to be more scattered for some of the higher S and V GPC. The relationships can also vary by region as a result of crude oil differences. GPC produced with Marlim crude from Brazil, for example, shows a higher V/S ratio relative to most other GPC. Today, meeting historical V specifications of 200–300 ppm requires a higher proportion of more expensive, low V/S cokes like the category A cokes shown in Fig. 11.

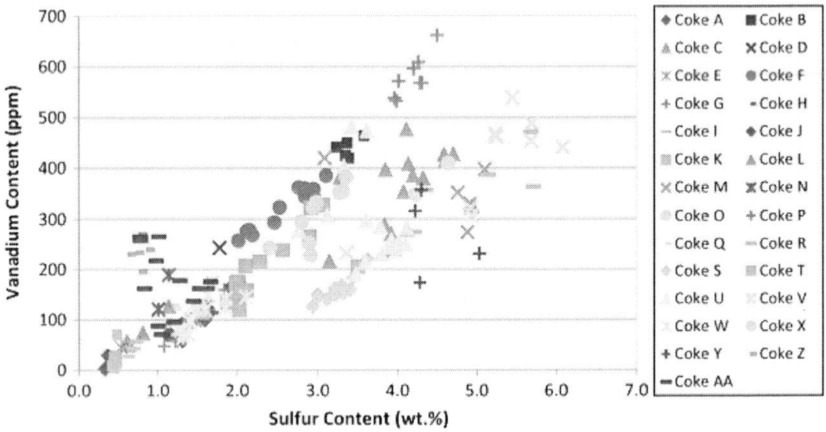

Figure 12. V and S levels in GPC used in anode blends

The trend graph in Fig. 13 shows how much average V levels have increased in anode grade coke shipped from Rain CII Carbon's US calciners to aluminum smelting customers over the last 20 years. Prior to 2003, average V levels were less than 200 ppm and well below specification maximums, reflecting the ready supply of low V/S GPC. V levels have increased since that time, particularly over the last 4 years, and are shipped much closer to specification maximums today.

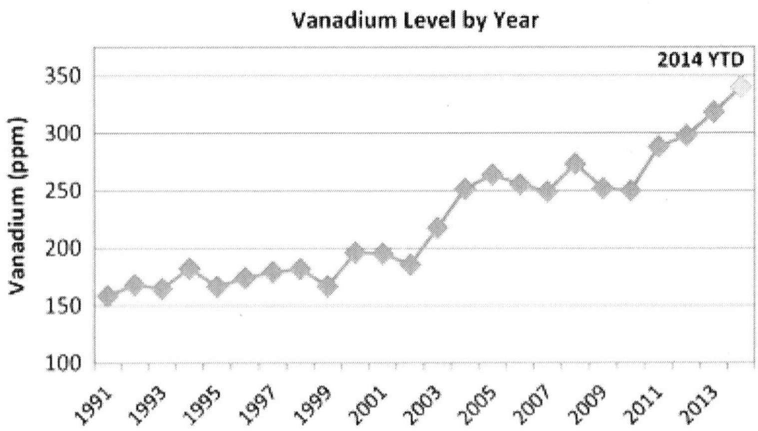

Figure 13. Average V level of anode-grade CPC shipments

In addition to looking at V average data, it is also useful to examine the change in the distribution of V maximum specifications over the last 10 years from 2004 to 2014. Figure 14 shows the change in specifications for smelters supplied from the same U.S. calciners mentioned above. Today, several smelters are taking CPC with a V maximum specification of up to 500 ppm. It is important to note that some of these smelters blend the CPC with other lower V/S CPC, which is covered in the section below on blending.

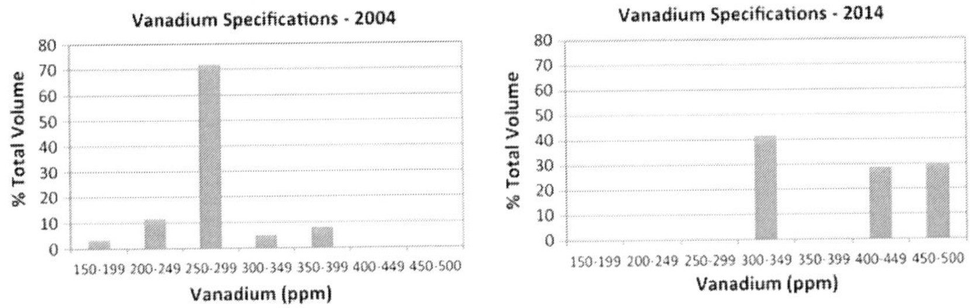

Figure 14. Change in distribution of vanadium-level specifications

Concerns about increasing V levels have been well documented in previous papers and a 2013 paper20 contains a useful reference list. That paper provides a good case study example of a smelter that switched to a significantly higher V level coke due to a sudden and unexpected change in GPC quality. A relatively minor crude change at a refinery in close proximity to the calciner that supplied CPC to the smelter caused an increase of approximately 150 ppm the GPC V level. The crude change provided a net benefit of $10–12 million per month for the refinery in the form of reduced crude costs. This illustrates the compelling economics for refineries to optimize crude blends, and more examples are given in a refinery paper on this subject.21

The V specification for the above smelter was increased from 330 ppm to 450 ppm and the V level in the CPC shipped to the smelter increased rapidly, as shown in Fig. 15. The paper cited above20 described some of the work done at the smelter to improve operational practices, such as anode covering. Today, the smelter is operating as well as it ever has historically in terms of anode and smelter performance, and the higher V CPC continues to provide significant cost savings in the form of lower CPC costs.

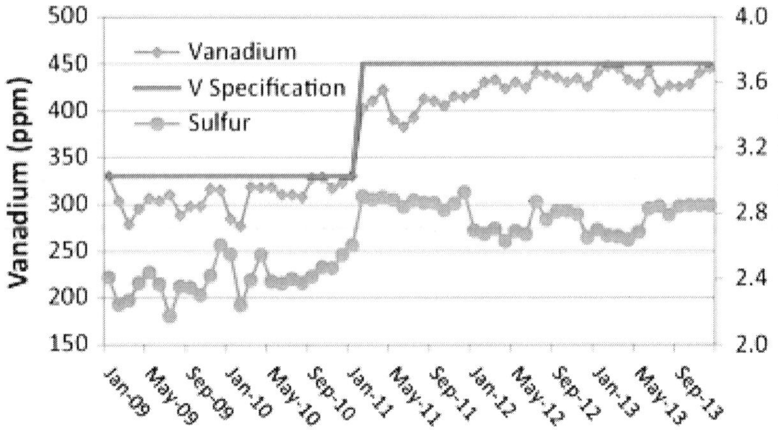

Figure 15. V and S increase in CPC supplied to smelter and V specification

For smelters producing P1020 primary aluminum product, higher V levels in CPC do not present any particular problems for metal purity. Higher V levels can be more problematic for smelters producing higher purity aluminum products depending on the end-use application. A significant body of work has already been undertaken on this issue,22–24and more is underway. For some alloys, higher V levels may even be advantageous.25,26

The situation with CPC S increases is similar to V, but less dramatic. Most smelters have a hard limit on SO_2 emissions, which typically sets an upper limit on the CPC S specification unless the smelter has the ability to scrub SO_2 from the baking furnace and potrooms. One trend that has emerged over the last 10 years is the request from smelters for both minimum and maximum S specifications. S in CPC and anodes has a beneficial effect on minimizing anode air and CO_2 reactivity, as well as excess carbon consumption.27 These smelters prefer to operate with anodes at the highest S level allowed by the smelter's SO_2emission limit. The higher S can be particularly beneficial when the smelter experiences Na and Ca contamination from poorly cleaned butts. The S in the anode helps to mitigate the catalytic effect of these two impurities on air and CO_2 reactivity.

One frequent question is whether there are any negative consequences of increasing anode S levels. Little data on this subject have been published, but one recent paper28 raised concerns about the potentially negative impact on cell current efficiencies, whereas earlier papers reported no impact.29 Thorne et al., on the other hand,30,31 reported correlations between anode reaction overpotentials and the level of impurities in CPC like S and V. Anodes made with higher S and V level CPC resulted in lower overpotentials, but it is not clear whether this is related to the impurity content or structure (anisotropic versus isotropic). More work is needed in this area.

The relaxation of smelter CPC specifications combined with the tight supply of low S GPC has resulted in a much wider range of GPC qualities being used in anode blends today. A good summary of the range of cokes being used can be found in a 2012 paper.[19] One coke with a highly isotropic texture (non-shot with irregularly shaped particles) is now being used routinely as a blend component by several calciners at levels of up to 10–15%. Previous concerns about thermal shock cracking with the use of isotropic cokes have not materialized at these relatively low levels, but other technology changes, such as the widespread use of anode slots, have significantly reduced, and in most cases eliminated, cracking. The use of isotropic cokes can provide both a raw material cost saving and an improvement in anode density, and testing with shot coke[32] is continuing.

Historical Perspective on Quality Changes

At this point, it is worth reflecting on historical concerns about the lack of suitable quality GPC for anode production. This is not a new problem or concern and many papers can be found in the literature from the mid-1970s through the late 1980s that discuss increasing S, V, and Ni levels, as well as the threat of imminent shortages of GPC. Two examples are cited in the reference list, one from 1977[33] and one from 1989.[34] The level of concern was high enough to initiate work on alternative carbons like solvent-refined coal.[35]

Despite these concerns, the industry did not experience any sustained shortages until around 2007, just before the global financial crisis (GFC). Both GPC and CPC were in short supply at that time and prices increased rapidly. The run-up in GPC prices is shown in Fig. 11. The GFC quickly changed that due to smelter closures and curtailments, which reduced the demand for CPC. The market tightened again in the 2010–2012 period, but continued smelter curtailments in the West in 2013 and 2014 have moderated the demand for CPC outside China. This situation will likely change again once aluminum prices recover, curtailed capacity is brought back online, and more new smelters are built. The sentiment for the long-term demand growth of aluminum remains very positive from this perspective.

Impact of Shale Oil

A more recent development beginning to impact GPC quality in the United States is the growth of shale oil production (also known as "tight oil"). Figure 16 shows the current and projected growth in shale oil production and illustrates why it is having a dramatic and positive impact on American energy independence, particularly when combined with a similar growth in shale gas production. This energy revolution is being driven by the development of two technologies now used in combination, namely horizontal drilling and hydraulic fracturing. This serves as a good example of how quickly technology breakthroughs can change an industry.

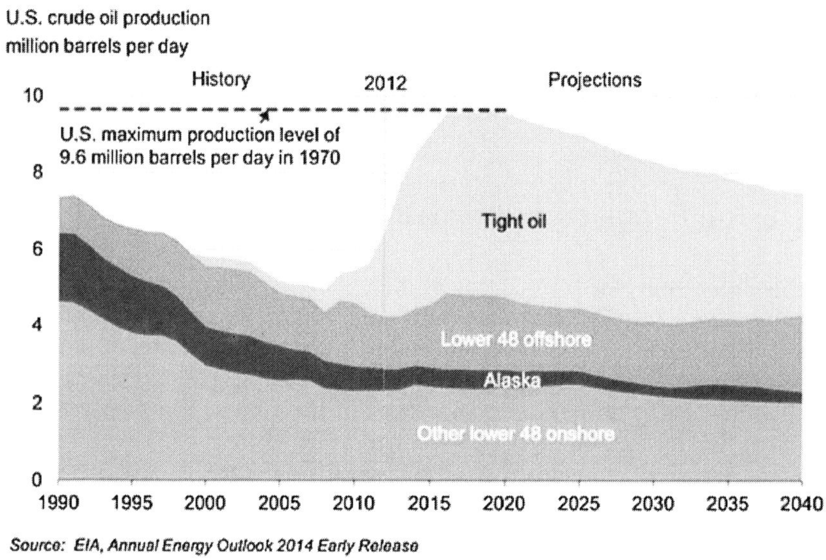

Figure 16. Growth in shale oil (tight oil) production in the United States

Shale oil is a light, sweet crude oil with a low specific gravity and low sulfur level. Given the declining supply of low S, low metals GPC in the United States, shale oil would initially seem to be good news for the production of low S GPC. The experience thus far has been the opposite and the reasons for this are well covered in Bartholomew's paper.18 In brief, shale oil contains very little "bottoms" (higher molecular weight hydrocarbons) and therefore makes very little coke.

Refineries using shale oil are using it as part of a crude blend to take advantage of its lower S and lower specific gravity by blending it with heavier, higher S crude oils. The impurities in the other crudes concentrate in the GPC product, which can increase S and V levels. For the refinery, this strategy can help reduce overall crude costs. Table IV shows the modeled impact on GPC quality and volume for a refinery switching out some of its low S imported crude in 2010 and replacing a portion of it with shale oil in 2012.

Table IV. Modeled impact of shale oil addition on GPC quality

Crude slate	2010	2012
Percentage imported	52	17
Percentage domestic	48	46
Percentage shale	0	37
GPC volume and quality		
MT/day	951	934
S%	2.87	3.75
Vanadium, ppm	272	393
Nickel, ppm	134	154

MT metric ton

The refinery reduced the volume of expensive, imported light sweet crude from 52% to 17% by blending in 37% shale oil in 2012. With the much lower S level and specific gravity of the shale oil, the refinery imported a cheaper, higher S, heavy crude in 2012 and maintained the overall refinery S and API gravity target. The volume of coke produced remained the same because the higher proportion of imported crude made more GPC but the S, V and Ni all increased. Although this is a model-generated example, the industry has seen an increase in the S level of some low S cokes in the United States by as much as 30–40% as a result of shale oil use. The impact on V and Ni levels has not been as significant.

The long-term impact of shale oil use on anode grade GPC production and quality in North America is not yet clear. Large quantities of shale oil are currently being moved by railcars and barges in the United States, but extensive pipeline networks are currently under construction. Once these pipelines are completed, shale oil may end up being a natural crude source for some refineries but not others. Some low S refineries may end up running a higher percentage of more traditional low S crudes, which could improve GPC quality and volumes. At least one American refinery is considering a capital investment to run 100% shale crude. This refinery has a delayed coker and no one is yet certain as to the quality of coke that would result, but it may produce a new source of good quality, low S GPC.

Environmental Impact of Increasing Sulfur Levels

All new calciners built in the world outside China in the last 20 years have been built with SO_2 scrubbers. Retrofitting scrubbers to existing calciners is also becoming more common as environmental regulations tighten. Using the United States as an example, a new National Ambient Air Quality Standard (NAAQS) was proposed in 2010 and will come into full effect in mid-2018. The current NAAQS has an allowable maximum concentration of SO_2 in the ambient air of 140 ppb (parts per billion) averaged over a 24-h period. The new NAAQS allows a concentration of 75 ppb averaged over a 1-h period. All states in the United States are now working on compliance plans to meet the new NAAQS. Rain CII Carbon has recently added a fluidized-bed SO_2 scrubber to its Lake Charles calciner in Louisiana (shown in Fig. 17) and is in the process of adding a similar scrubber to its Chalmette calciner in readiness for the new NAAQS.

As average CPC S levels continue to trend upward, the need for SO_2 scrubbing at smelters will also increase. Long term, we can probably expect new smelters to require the addition of SO_2 scrubbing given that scrubbing technology is mature and readily available even if it does add to capital and operating costs. Alcoa recently reported on a new SO_2 scrubbing technology it is developing as a potential retrofit for smelters to reduce SO_2 emissions.36

Blending CPC at Smelters

Coke blending at smelters is not a new idea,37 but more and more smelters are either being built with blending systems or are adding them as retrofit projects. A primary goal is to allow the blending of CPC from different suppliers, which gives the smelter the flexibility to source different quality CPC from the most geographically favorable regions. A common practice is to source low S and low metals CPC from one region for blending with lower cost, higher S and V CPC from another region.

Figure 17. Baghouse, SO_2 scrubber, and lime silo (L to R) at Lake Charles calciner

For smelters using a combination of shaft CPC and rotary kiln/rotary hearth CPC, it is important that the smelter has the ability to blend these cokes thoroughly. The porosities and bulk densities of the cokes are very different and will result in large pitch level and anode density variations if they are processed on a campaign basis. Differences in the fines levels and hardness levels of the cokes can also lead to greater variation in fraction preparation, particularly with respect to the ball mill operation. Blending the cokes at a more or less constant ratio can help minimize these problems and improve anode quality and consistency. Recent papers38,39 provide some good smelter data on the benefits of blending CPC.

Blending systems at most smelters are usually relatively simple weigh belt systems. Some smelters have adopted more sophisticated blending strategies by concentrating different coke types in different parts of the aggregate recipe. In one example,[40] the highest VBD coke is concentrated in the coarser fractions of the recipe to maximize anode density. In another example,[41] higher S, lower CO_2 reactivity cokes are used in the ball mill fines fraction to reduce the reactivity of the binder matrix, which is usually the highest reactivity phase in the anode.

As smelters get larger in size, the need to use multiple coke suppliers to mitigate supply risks and optimize economics will continue to increase. The capital investment required to add a CPC blending system is not high relative to the total smelter investment cost, and it has a big payback in terms of consistency if the smelter uses a blend of low and high S coke and/or a blend of rotary/shaft calcined coke.

SUMMARY AND CONCLUSION

GPC and CPC production have continued to evolve over the last century of aluminum production. During the last 10 years, the industry has started using a wider range of GPC qualities than it has historically, and the availability of good-quality GPC remains challenging as refiners optimize their economics by sourcing lower cost crudes. In the United States, shale oil is beginning to impact both the volume and quality of GPC produced, but the long-term impact remains unclear.

More smelters are blending CPC from different suppliers and this trend is expected to continue along with an increase in the sophistication of blending practices. The need to blend becomes even more important for smelters using a combination of rotary and shaft CPC as a result of the different porosity and bulk density profiles of these cokes.

The trend toward increasing CPC sulfur levels in combination with tightening SO_2emissions limits is driving more calciners to add SO_2 scrubbing, and it is now a mandatory technology for new calcining plants. In the future, we are likely to see more smelters built with SO_2 scrubbing, and work is underway on developing alternative scrubbing technologies.

The aluminum industry continues to adapt well to changes in CPC quality such as higher V and S levels and more isotropic textures, and many smelters have pursued a strategy of relaxing specifications to procure lower cost CPC. In 2013, the world produced approximately 115 million DMT of GPC, and only about 25% of this was used by calciners for production of CPC for aluminum and other industries. There is no shortage of GPC; rather, it is a matter of the industry continuing to use what is available.

Demand for CPC outside China has moderated over the last couple of years with the curtailment of smelting capacity in the West, but this looks set to change in the near future. Demand for aluminum exceeded primary aluminum production for the first time in many years in 2014. Once existing capacity is restarted and/or new capacity is built to satisfy the forecasted aluminum demand, we should expect the supply and demand balance for CPC and GPC to tighten once again with the usual market impact.

REFERENCES

1. J. Keniry, *Proc. Seventh Australasian Aluminium Smelting Technology Conference and Workshop*, ed. B.J. Welch and M. Skyllas-Kazacos (Sydney, Australia: University of New South Wales, 2001), pp. 584–597.
2. W.S. Petersen and R.E. Miller, eds., *Hall Héroult Centennial: First Century of Aluminum Process Technology 1886–1986* (Warrendale, PA: TMS, 1986), pp. 130–143.
3. C.L. Mantell, *Carbon and Graphite Handbook* (New York, NY: Wiley, 1968), Chap. 15.
4. P.J. Ellis and C.A. Paul (Paper presented at the AIChE 1998 Spring National Meeting, New Orleans LA, 8–12 March 1998).
5. P. Rhedey and D. DuTremblay, *Light Metals 1977*, ed. K.B. Higbie (New York, NY: The Metallurgical Society of AIME, 1977), pp. 301–315.
6. Innus, A. Jomphe, and H. Darmstadt, *Light Metals 2013*, ed. B. Sadler (Warrendale, PA: TMS, 2013), pp. 1069–1073.
7. K. Neyrey, L. Edwards, A. Ross, and F. Vogt, *Light Metals 2005*, ed. H. Kvande (Warrendale, PA: TMS, 2005), pp. 607–612.
8. L. Edwards, *Light Metals 2014*, ed. J. Grandfield (Warrendale, PA: TMS, 2014), pp. 1093–1098.
9. Q. Nguyen, *Light Metals 1985*, ed. H.O. Bohner (Warrendale, PA: TMS, 1985), pp. 903–913.
10. M. Lubin, L. Edwards, and L.-P. Lossius, *Light Metals 2013*, ed. B. Sadler (Warrendale, PA: TMS, 2013), pp. 1063–1068.

11. ISO 12981-1, Carbonaceous Materials Used in the Production of Aluminium—Calcined Coke—Determination of the Reactivity to Carbon Dioxide. Part 1—Loss in Mass Method, 2011.

12. S. Hume, W.K. Fischer, R.C. Perruchoud, and B.J. Welch, *Light Metals 1993*, ed. S.K. Das (Warrendale, PA: TMS, 1993), pp. 525–531.

13. F.J. Farago and R.R. Sood, *Light Metals 1976,* ed. S.R. Leavitt (New York, NY: The Metallurgical Society of AIME, 1976), pp. 351–362.

14. D.R. Mason and H.C. Rolseth, *Light Metals 1985*, ed. H.O. Bohner (Warrendale, PA: TMS, 1985), pp. 863–869.

15. M.F. Vogt, G.R. Jones, and G.A. Tyler, *Light Metals 1984,* ed. J.P. McGeer (Warrendale, PA: TMS, 1984), pp. 1697–1714.

16. K. Ries, *Light Metals 2009,* ed. G. Bearne (Warrendale, PA: TMS, 2009), pp. 945–950.

17. L. Edwards, *Light Metals 2011,* ed. S.J. Lindsay (Warrendale, PA: TMS, 2011), pp. 895–900.

18. K. Bartholomew, *Light Metals 2013*, ed. B. Sadler (Warrendale, PA: TMS, 2013), pp. 15–20.

19. L. Edwards, N. Backhouse, H. Darmstadt, and M-J Dion, *Light Metals 2012*, ed. C.E. Suarez (Warrendale, PA: TMS, 2012), pp. 1207–1212.

20. Weber, J. Gavin, C. Coney, W. Marcrum, L. Crabtree, and L. Edwards, *Light Metals 2013*, ed. B. Sadler (Warrendale, PA: TMS, 2013), pp. 21–25.

21. T.W. Dixon, *Light Metals 2009,* ed. G. Bearne (Warrendale, PA: TMS, 2009), pp. 941–944.

22. J. Grandfield, L. Sweet, A. Beer, S. Zhu, X. Chen, and M. Easton, *Light Metals 2014*, ed. J. Grandfield (Warrendale, PA: TMS, 2014), pp. 969–974.

23. J. Grandfield, L. Sweet, C. Davidson, J. Mitchell, A. Beer, S. Zhu, X. Chen, and M. Easton, *Light Metals 2013*, ed. B. Sadler (Warrendale, PA: TMS, 2013), pp. 39–45.

24. S. Zhu, J.-Y. Yao, L. Sweet, M. Easton, J. Taylor, P. Robinson, and N. Parson, *JOM*, 65, 584 (2013).

25. G. Jha, S. Ningileri, X. Liu, and R. Bowers, *Light Metals 2013*, ed. B. Sadler (Warrendale, PA: TMS, 2013), pp. 929–934.

26. G. Jha, F. Cannova, and B. Sadler, *Light Metals 2012*, ed. C.E. Suarez (Warrendale, PA: TMS, 2012), pp. 1303–1306.

27. Z. Kuang and J. Thonstad, *Carbon* 33, 1479 (1995).

28. S. Pietrzyk and J. Thonstad, *Light Metals 2012*, ed. C.E. Suarez (Warrendale, PA: TMS, 2012), pp. 659–664.

29. E. Barrillon and J. Pinoir, *Light Metals 1977*, ed. K.B. Higbie (New York: The Metallurgical Society of AIME, 1977), pp. 289–299.

30. R.J. Thorne, C. Sommerseth, A.M. Svensson, E. Sandnes, L.P. Lossius, H. Linga, and A.P. Ratvik, *Light Metals 2014*, ed. J. Grandfield (Warrendale, PA: TMS, 2014), pp. 1213–1217.

31. R.J. Thorne, C. Sommerseth, E. Sandnes, O. Kjos, T.A. Aarhaug, L.P. Lossius, H. Linga, and A.P. Ratvik, *Light Metals 2013*, ed. B. Sadler (Warrendale, PA: TMS, 2013), pp. 1207–1211.

32. L. Edwards, F. Vogt, M. Robinette, R. Love, A. Ross, M. McClung, R.J. Roush, and W. Morgan, *Light Metals 2009*, ed. G. Bearne (Warrendale, PA: TMS, 2009), pp. 985–990.

33. H. Brandt, *Light Metals 1977*, ed. K.B. Higbie (New York, NY: The Metallurgical Society of AIME, 1977), pp. 333–349.

34. U. Mannweiler, W. Schnidt-Hatting, D. Rodriguez, and A. Maitland, *Light Metals 1989*, ed. P.G. Campbell (Warrendale, PA: TMS, 1989), pp. 449–454.

35. E. Hollingshead and P. Rhedey, *Light Metals 1985*, ed. H.O. Bohner (Warrendale, PA: TMS, 1985), pp. 589–605.

36. "Alcoa pilots new in-duct scrubber technology." *Light Metal Age* 71 (5) (2013).

37. L. Edwards, F. Vogt, and J. Wilson, *Light Metals 2001*, ed. J. Anjier (Warrendale, PA: TMS, 2001), pp. 689–694.

38. Ndjom, M.S. Malik, A. Al Marzouqi, T.K. Sahu, and S.A. Rabba, *Light Metals 2013*, ed. B. Sadler (Warrendale, PA: TMS, 2013), pp. 1105–1110.

39. E.G.M. Mofor, S. Akhmetov, T.K. Sahu, J. Blasques, D. Whitfield, G. Meintjes, K.S. Rajwinder, A. Jassim, S.A. Rabba, H. Devadiga, and K. Al Aswad, *Light Metals 2014*, ed. J. Grandfield (Warrendale, PA: TMS, 2014), pp. 1199–1202.

40. Kato, H. Truci, F. Figueiredo, A. Nascimento, and V. Piffer, *Light Metals 2007*, ed. M. Sørlie (Warrendale, PA: TMS, 2007), pp. 891–894.

41. M. Gendron, S. Whelan, and K. Cantin, *Light Metals 2008*, ed. D.H. DeYoung (Warrendale, PA: TMS, 2008), pp. 861–864.

8

CHAPTER

Oxidation of a Commercial Nickel-Based Superalloy under Static Loading

B. J. Foss,[1] M. C. Hardy,[1] D. J. Child,[1] D. S. Mcphail,[2] and B. A. Shollock[3]

[1]*Rolls-Royce plc, Derby DE24 8BJ, UK.*

[2]*Department of Materials, Imperial College London, London SW7 2AZ, UK.*

[3]*WMG, University of Warwick, Coventry CV4 7AL, UK*

ABSTRACT

The current demands of the aviation industry for increased gas-turbine efficiency necessitate higher turbine entry temperatures, requiring that alloys exhibit superior oxidation resistance. The synergistic effects of oxidation and mechanical stresses pose a complex issue. The purpose of the current research was to examine the effects of stress on the oxidation and oxygen transport in a commercial nickel-based superalloy. Fine grain RR1000 in both polished and shot-peened conditions was studied for classic (zero load) and statically loaded conditions using integrated two-stage isotopic tracing combined with focused-ion-beam secondary ion mass spectrometry (FIB-SIMS). Cr_2O_3 external oxide formed with semicontinuous TiO_2 above and below. Preferential grain boundary Al_2O_3internal oxide formation, γ'-dissolution, and recrystallization occurred subsurface. Oxidation mechanisms were dominated by anionic/cationic growth in the external oxide with inward oxygen transport, initially through the partially unprotective external oxide, then along internal oxide/alloy interfaces. Loading did not influence the oxidation products formed but did bring about expedited oxidation kinetics and changes to the oxide morphology. The oxygen diffusivity $D_O{}^*$ ($\times 10^{-13}$ cm^2s^{-1}) ranged from 0.39 for the polished alloy to 3.7 for the shot-peened condition under compressive stress. Arguably, the most significant effects took place in the subsurface regions. Increased oxidation kinetics were attributed to the development of fast cation diffusion paths as the alloy deformed by creep.

INTRODUCTION

Increased turbine entry temperatures (TETs) for improved turbine efficiency push Ni-based superalloys to their operational limits. Higher temperatures require excellent alloy degradation resistance in oxidizing environments to meet the operational service life of the component. Failure to meet this requirement results in excessive oxidation, which reduces the component's ability to maintain its mechanical integrity. The oxidation resistance will depend on the ability of the alloys to form, and maintain, a compact and adherent scale that will protect the alloy from further oxidation attack.1 Loss of the protective oxide scale becomes problematic as it exposes the parent alloy to further oxidation attack. Stresses in the oxide scale due to intrinsic growth stresses, thermal stresses, or externally applied stresses all increase the likelihood of scale failure.2,3 The influence of environmental factors on alloy stability at elevated temperatures is now of great importance, in both new alloy design, and those currently in service. The distinct characteristics of the oxide formed and its kinetics depend on several variables, including composition, microstructure, environment, temperature, and surface condition. The surface condition of the alloy plays an important role in its resistance against oxidation-induced damage.4,5 Shot-peening is a widely used surface treatment primarily sought for its ability to offset high mechanical loads at the surface, but shot-peening effects could be both beneficial and detrimental to an alloy's long-term integrity because the process itself imposes high levels of surface and subsurface deformation.

For disk applications, the composition of the class of nickel-based alloys normally permits the formation of a continuous chromia healing layer beneath the transient oxides that rapidly form from the less noble alloy components. The formation of this layer should prevent further oxidation attack as it acts as a barrier between the metal and the exposure environment; however, further attack can occur at either the oxide/gas or the metal/oxide interfaces, if either the metal cations or oxygen anions penetrate the layer. The latter case is known as internal oxidation and is often the main cause of oxidation failures.6 In addition to the above, and of particular importance to nickel-based alloys, is the depletion of Ni_3Al precipitate-forming elements from the alloy to form oxidation products themselves. Consequently, a subsurface depleted zone develops that ultimately weakens the near surface of the alloy and reduces the load bearing cross section of the component. Further consequences of compositional changes in the subsurface include void formation, volume changes, alloy–scale interface stability and phase transformations. Each of these relate to the different rates at which the various components in the alloy oxidize. In addition to this microstructural instability, as the oxide continues to grow, the stress required to crack the oxide decreases,2,3 and the likelihood of further exposure of the base alloy to the environment increases.

This investigation reveals the oxidation behavior of the fine-grain (FG) polycrystalline nickel superalloy fine grain RR1000 in both the polished shot-peened surface conditions at temperatures in the range 700–800°C both without applied load, termed classic oxidation conditions, and under a static bending load. The oxidation and oxidation-induced microstructural changes have been investigated using several analytical techniques and are discussed using the exposure at 800°C as an illustration in this article. The different methodologies have allowed for quantification and characterization of the governing mechanisms pertaining to the surface and subsurface oxidation processes.

EXPERIMENTAL

Alloy

The FG RR1000 material studied was produced via a powder metallurgy route followed by extrusion, forging, and a two-step heat treatment to achieve the desired fine-grained microstructure, γ' size and volume fraction. Table I gives the nominal alloy composition (in wt.%) and reveals the various components added to maintain a combination of strength, damage tolerance, and creep resistance.7

Table I. Nominal composition of FG RR1000 in wt.%

Al	Co	Cr	Hf	Mo	Ta	Ti	C	B	Zr	Ni
3.0	18.5	15	0.5	5.0	2.0	3.6	0.027	0.015	0.06	Bal.

Hardy et al.7 provided a detailed rationale for each of the elements added and their limitations in the alloy design.

Oxidation Experiments

Exposures made use of a customized isotopic exchange facility, which permitted careful control of the exposure environment. This particular setup facilitated the use of the $^{18}O_2$isotope in two-stage isotopic exposure experiments. The samples were first exposed in research-grade oxygen (referred to as $^{16}O_2$) at pO_2 = 200 mbar. The exchange furnace was reevacuated before introducing the $^{18}O_2$-enriched oxygen also at pO_2 = 200 mbar. The samples were then quench-cooled in air by rolling off the exchange furnace. Bending oxidation under combined isotopic oxidation conditions used a bespoke bending rig based on a simple bent beam test where the load is applied to the ends of the test sample. All specimens were loaded to a nominal stress of 850 MPa, which is below the yield stress at 800°C. To achieve the desired stress required application of the following relationship:8

$$H = \frac{\sin\left(\frac{L\sigma}{ktE(T)}\right)\mathrm{k}tE(T)}{\sigma} \tag{1}$$

where H is the distance between two ends of the holder required to achieve the desired stress σ; L and t are the length and thickness of the specimen, respectively; $E(T)$ is the Young's modulus at temperature T; and k is an empirical constant equal to 1.28. This relationship then states that achieving the desired load requires a measurement of H only. Measurement of H was accomplished using a calibrated digital Vernier caliper with an accuracy of ±0.01 mm. Table II provides the sample dimensions and holder separation as well as the elastic strain as calculated for the 850 MPa applied stress and the Young's modulus for the 800°C exposures.

Table II. Material and loading parameters used in bending oxidation tests at 800°C

Surface finish	L (mm)	t (mm)	H (mm)	Strain, ε (%)
Polished	59.95	1.21	59.58	0.50
Shot-peened	59.16	1.10	58.73	0.50

Please refer to text for definition of the parameters

Please refer to text for definition of the parameters

Characterization

Oxide cross-section preparation, imaging, and focused-ion-beam secondary ion mass spectrometry (FIB-SIMS) used an FEI FIB-200 SIMS workstation (FEI Company, Hillsboro, OR) equipped with a quadrupole detector as described by McPhail.9 Cross-sectioning applied various milling angles, ranging from 15° to 30° to the surface to produce ramped cross sections of the external oxide (EO) and oxidation-affected zone (OAZ). Note that the OAZ may contain internal oxidation (IO), γ'-dissolution (GPD), and recrystallization (RX). Imaging used a 50-pA gallium primary ion beam current in both ion-induced secondary electron (IISEI) and total positive ion (TPII) modes with at least two cross sections analyzed at each temperature and time. The SIMS maps were collected for $^{16}O^-$ and $^{18}O^-$ isotopes. Mapping used beam currents of 100–300 pA depending on the sample analyzed; at these currents, the beam has a diameter 20–30 nm. A raster of 256 × 221 pixels achieved suitable lateral resolution for all specimens while limiting the beam-induced damage. Samples were also examined in a Zeiss Auriga (Carl Zeiss, Oberkochen, Germany) fitted with an Oxford Instruments SMax 80 EDX detector (Oxford Instruments, Oxfordshire, U.K.) to evaluate the oxide compositions.

RESULTS

Classic Oxidation

Ion imaging provided useful, complementary information from the sectioned samples. Figure 1 shows the typical oxide cross-sectional morphology of polished (Fig. 1a) and shot-peened (Fig. 1b) samples following exposure at 800°C for 120 h (72 h $^{16}O_2$ + 48 h $^{18}O_2$), whereas Fig. 2 presents a schematic diagram showing the features observed in the oxidized samples of FG RR1000. The micrographs demonstrate the similarity in the oxide morphologies produced between the two surface finishes; however, the severity of the oxidation damage incurred after shot-peening increases particularly in the subsurface region.

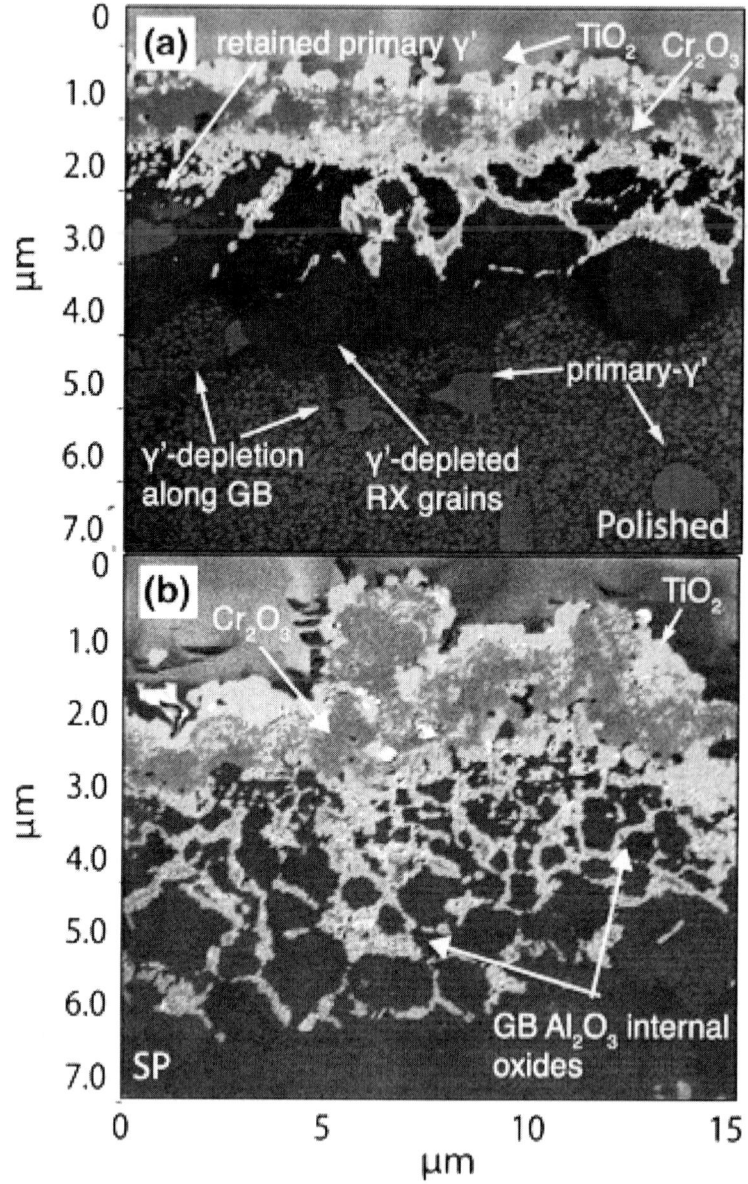

Figure 1. Oxide cross sections of FGRR1000 following classic oxidation: (a) TPII polished and (b) TPII shot-peened after exposure 800°C for 120 h

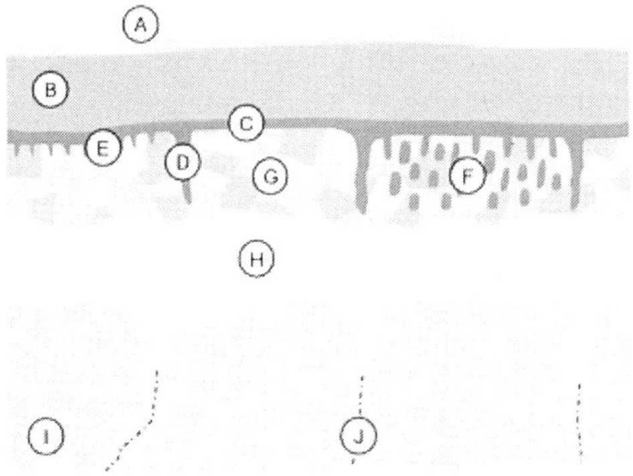

A. Rutile (TiO_2) outer scale/particles
B. Chromia scale
C. Alumina layer
D. Intergranular alumina fingers
E. Intragranular alumina fingers
F. Internal alumina islands
G. Recrystallised zone
H. Gamma-prime depleted zone
I. Gamma-prime particles
J. Grain boundary carbides

Figure . 2. Schematic diagram highlighting the features observed in oxidized FG RR1000

The external oxide growth observed in both the polished and shot-peened alloys follows that observed for other superalloys of similar compositions tested at equivalent temperatures.10–14 Energy-dispersive x-ray analyses showed that the overall oxide morphology from the external scale to internal oxides of Cr_2O_3/TiO_2 in the outer scale to Al_2O_3 internally agrees with the expected thermodynamic behavior; however, and in contrast to the thermodynamic predictions, the alloy formed an outer layer of TiO_2, believed to be at the expense of its oxidation resistance.14,15 Cationic diffusion is inferred from the known transport mechanisms for Cr_2O_3 growth 16,17 and from the fact that titanium must diffuse outwards to form above the outer scale. This requires titanium diffusion through the chromia and an adequate solubility of the chromium sublattice. In support of this, Naoumidis et al.18 previously reported a solubility of 18% on the chromium sublattice at 1000°C. Although the temperatures are considerably higher than those examined in this article, titanium is still expected to have a large enough solubility to facilitate titanium diffusion to the surface.

FIB-SIMS mapping shows the relative distributions of the ^{16}O and ^{18}O species in the oxide cross sections for the image presented in Fig. 1a. The pixel intensities have been normalized to the maximum counts in each array. The $^{16}O^-$ map following exposure at 800°C for 120 h represents the oxide growth that occurs in the first part of the two-stage exposure (72 h). The maps confirm the existence of the external scale with the internal oxides directly beneath in the base alloy. Note that the relative intensity of the $^{16}O^-$ signals in internal oxide regions, compared with the external oxide, does not necessarily imply higher oxygen contents. Likewise, the $^{18}O^-$ map (Fig. 3b) signifies the oxygen transport and oxide growth during the second half of the two-stage exposure (48 h). The proficiency of the FIB-SIMS mapping technique reveals the new oxide growth as well as the identifying short-circuit oxygen diffusion paths such as boundaries.19,20 FIBS-SIMS mapping showed that inward anionic diffusion took place to form the internal oxides during the oxide growth for both surface conditions. A comparison of the $^{16}O^-$ and $^{18}O^-$ion distributions at all temperatures studied indicated that the external scale contains ^{18}O-rich regions at the surface and within the bulk external oxide. In the region of the external scale closest to the oxide/gas interface, the oxide is composed of mostly ^{18}O, whereas in the region closest to the oxide-metal interface, the oxides contain a mixture of ^{16}O and ^{18}O species. This illustrates oxygen transport taking place through the external oxide to form new oxide at the alloy/oxide interface, but at the same time new oxide forms at the gas/oxide interface, which relies on outward cation diffusion. Indeed, these findings agree with other studies, which previously revealed countercurrent mechanisms in growing chromia scales,16,17,21,22 which is consistent with the observations that the Cr_2O_3normally grows through outward cationic growth (i.e., *p*-type behavior).16 Therefore, the oxide grows through simultaneous countercurrent transport of inwardly diffusing oxygen anions and outwardly diffusing titanium and chromium cations.

Oxidation Under Bending Load

Bending loads do not influence the nature of the oxidation products formed but do cause significant morphological changes in the shot-peened alloy derivative at 800°C. The external bending load induced changes to the oxide morphology in both the polished (Fig. 4) and shot-peened (Fig. 5) samples following exposure at 800°C for 120 h. In the case of the polished sample, for example, Fig. 4a outlines how the external oxides now form subsurface protrusions, consisting largely of Cr_2O_3.

As with the samples tested under classic oxidation conditions, shot-peened samples exhibited clear differences in the oxide and near-surface morphology under the influence of an external load compared to the polished samples, with the most notable morphological difference observed in the subsurface of the compressively loaded side of the bending specimen. Concerning the external oxide, Fig. 4 shows that the alloy undergoes more extensive and continuous TiO_2 formation under compressive loads. In the case of the shot-peened alloy (Fig. 5), a continuous oxide layer forms beneath the external oxide but separated from the external oxide by an intermediate internal oxidation/recrystallized layer. This continuous oxide layer forms under both loading conditions at 800°C, but is more extensive under compression (Fig. 5a). The presence of such a layer does not prevent grain boundary internal oxide precipitation deeper into the alloy. Under compressive loads, isolated precipitates connect to the continuous oxide layer by fine ligaments. The tensile-loaded specimen exhibits relatively more extensive internal oxidation below this layer.

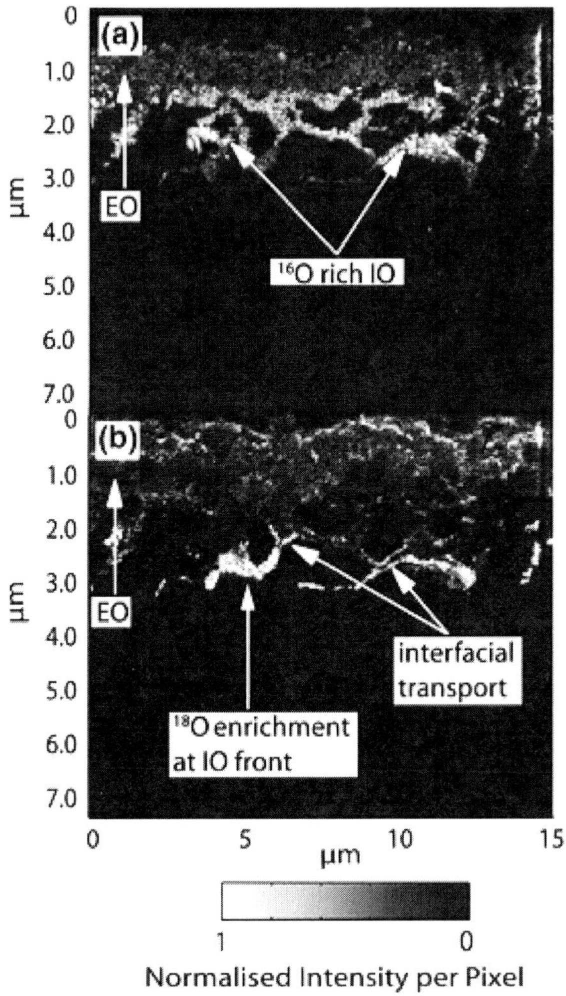

Figure 3.FIB-SIMS elemental mapping: (a) $^{16}O^-$ and (b) $^{18}O^-$ of the oxide cross section given in Fig. 1a for the polished condition after exposure 800°C for 120 h

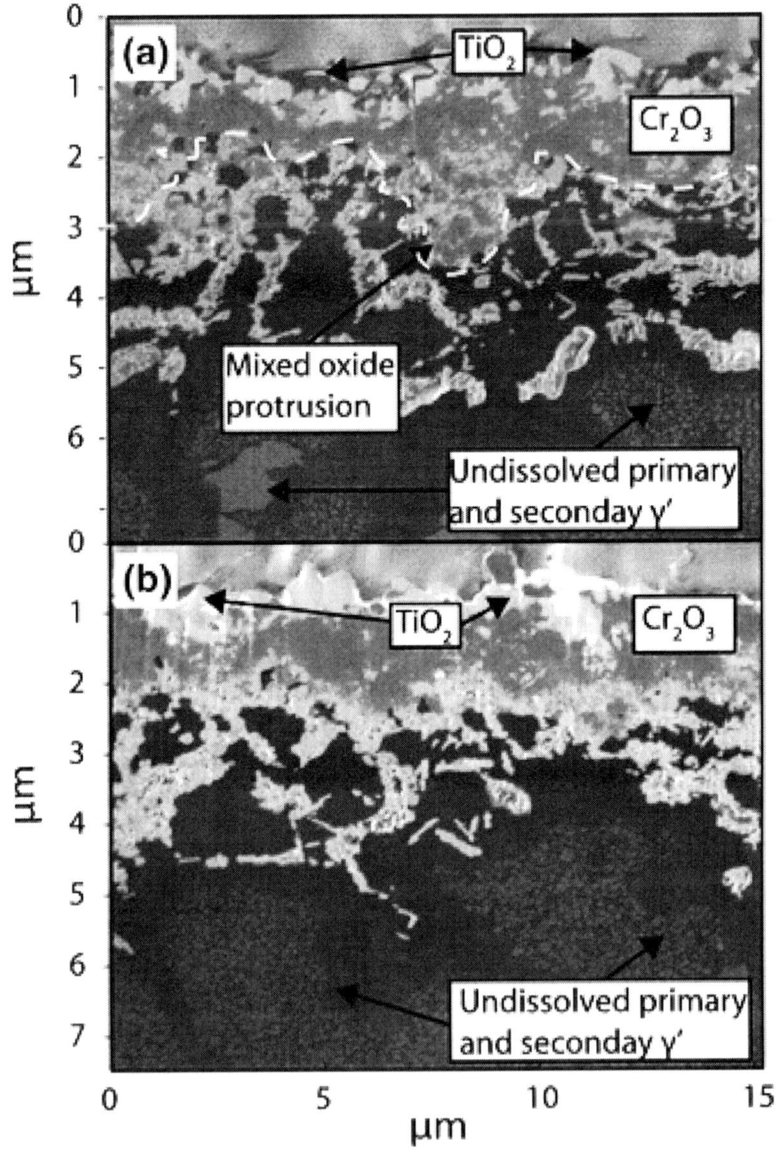

Figure 4. Oxide cross sections of polished specimens under bending loads: (a) TPII compression and (b) TPII tension after exposure 800°C for 120 h

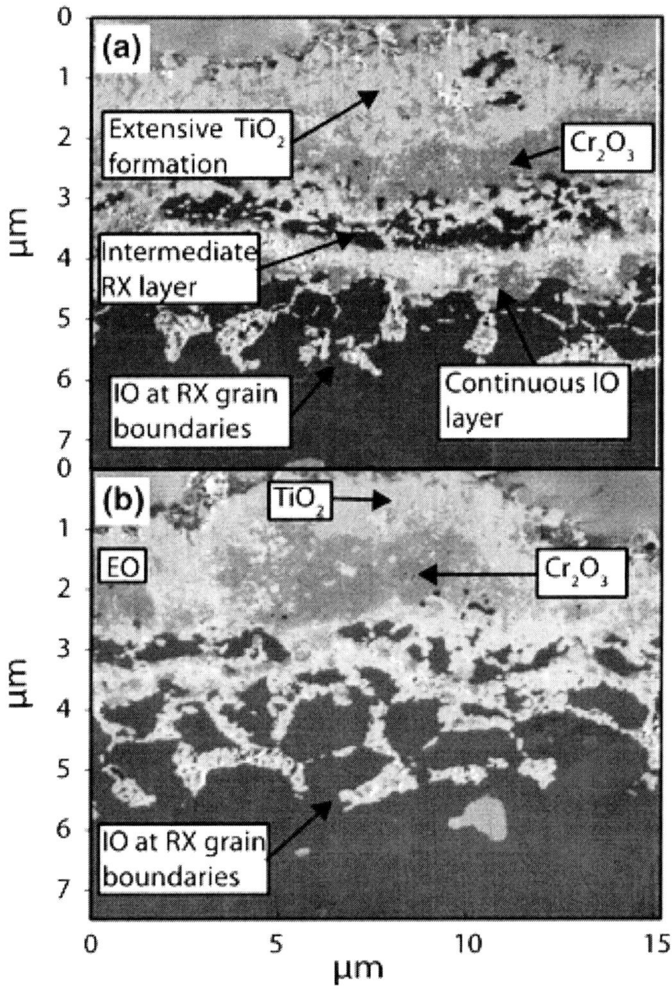

Figure 5. Oxide cross sections of shot-peened FG RR1000 under bending loads: (a) TPII compression and (b) TPII tension after exposure 800°C for 120 h

The $^{16}O^-$ and $^{18}O^-$ maps for each surface and loading condition confirm the countercurrent transport mechanism previously described for classic oxidation, with the example of the shot-peened sample given in Fig. 6 (compression) and Fig. 7 (tension). The $^{18}O^-$ distribution throughout the entirety of the external oxide corroborates the inward oxygen (anionic) diffusion and outward cation diffusion to form the external TiO_2 and Cr_2O_3 external scale. Anionic diffusion through the bulk alloy itself takes place initially along the grain boundaries of the parent alloy, where the internal oxides nucleate, with the interface of the grain boundary oxides and the parent/recrystallized grains providing a fast diffusion path for oxygen transport in the alloy. As with the unloaded specimens, the enhanced oxygen diffusivity along these interfaces expedites the grain boundary internal oxidation rate.23–25

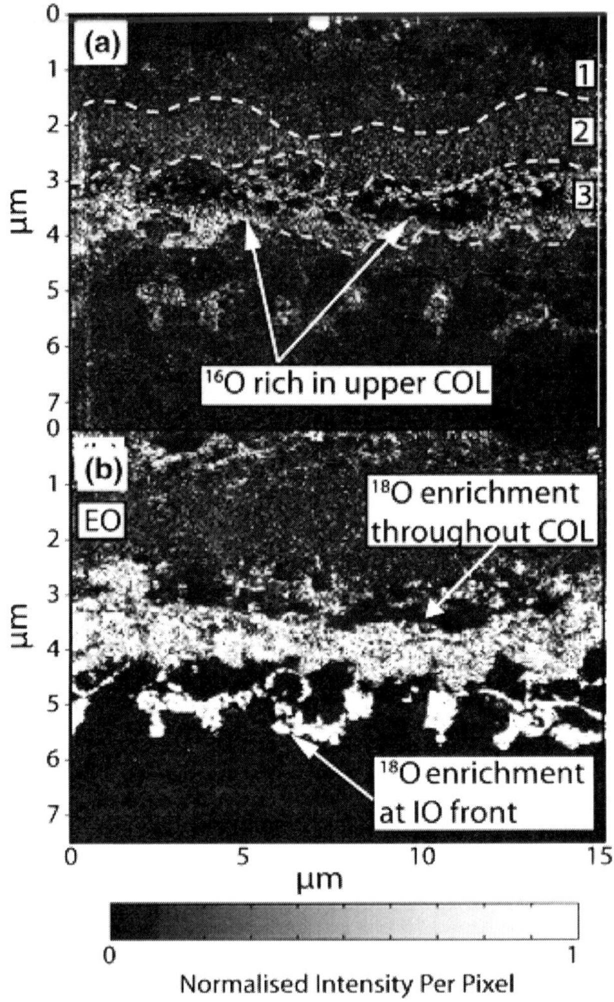

Figure 6. FIB-SIMS elemental mapping: (a) $^{16}O^-$ and (b) $^{18}O^-$ of the compression oxide cross sections for the shot-peened sample given in Fig. 5a after exposure 800°C for 120 h

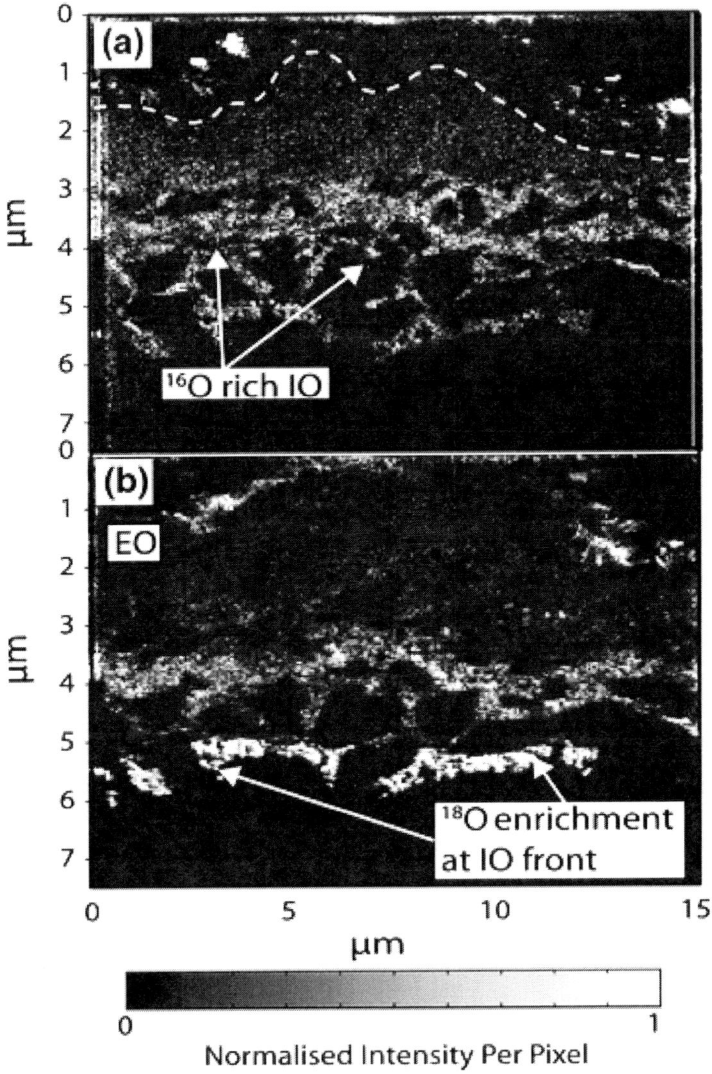

Figure 7. FIB-SIMS elemental mapping: (a) $^{16}O^-$ and (b) $^{18}O^-$ of the tension oxide cross sections for the shot-peened sample given in Fig. 5b after exposure 800°C for 120 h

The overall depth of the scale and internal oxide region did not significantly differ between compression and tension for the shot-peened condition. The $^{16}O^-$ map for the shot-peened compressive specimen (Fig. 6a), however, reveals a subtle difference in the normalized intensity across the external oxide. This is visualized as a layer of low $^{16}O^-$ intensity (labelled 1) below a layer of higher $^{16}O^-$ intensity (labeled 2); the dashed white line defines the interface of these two layers. This boundary coincides with the interface of the rutile and chromia layers seen in Fig. 5b and, therefore, relates to the different secondary ion yields of the chromia and rutile. The $^{18}O^-$ map (Fig. 6b), however, does not demonstrate this layering. Below the external scale, the internal oxides that reside among the intermediate recrystallized layer demonstrate both ^{16}O and ^{18}O enrichment. The continuous layer is also separable into two layers of contrasting $^{16}O^-$ intensity, where the dashed green line (Fig. 6a) defines the interface of these two regions; however, the region having the higher $^{16}O^-$ signal (labeled 3) sits above that of lower intensity area. As with the external oxide, the $^{18}O^-$ distribution (Fig. 6b) does not show this separation in the continuous layer. The sustained $^{18}O^-$ signal in the grain boundary internal oxides illustrates the continued oxygen diffusion through the continuous layer. The $^{16}O^-$ map corresponding to oxidation under tension (Fig. 7a) demonstrates a similar layering in the external oxide (see the dashed white line) but not in the continuous oxide layer. For both surface conditions, this layered structure indicates the temporal evolution of the structure, with the intermediate chromia-rich and TiO_2 layers forming early in the exposure in both cases, with some internal oxides, particularly in the tension side. Internal oxides (Fig. 7b) form predominately in the second stage of the oxidation exposure, with the continuous layer in the compressive region (Fig. 7b).

Oxygen Tracer Diffusion Coefficient in External Oxide

The oxygen tracer diffusion coefficient D∗OD0∗ in the external oxide was evaluated for all oxidation conditions using the $^{16}O^-$ and $^{18}O^-$ FIB-SIMS maps. An analysis used the isotopic fraction *v* of the oxygen isotopes distribution, as this mitigates the spurious variations in signal intensity due to sample topography, localized charging, and instrumental factors. Neglecting the contribution from $^{17}O^-$, the isotopic fraction was calculated as:

$$v = \frac{^{18}O^-}{[^{16}O^- + ^{18}O^-]} \tag{2}$$

where $^{16}O^-$ and 18O−18O− are the $^{16}O^-$ and $^{18}O^-$ intensities obtained from the FIB-SIMS maps. The application of Eq. 2 on a pointwise basis produced isotopic fraction (IF) maps, which indicated regions of ^{18}O enrichment. Because the maps are also *m* × *n* elemental arrays containing the spatial distribution and the intensities of the $^{16}O^-$ and $^{18}O^-$ species, it was possible to compute isotopic enrichment depth profiles (IEDPs).

IEDPs were computed using an integrated line scan (ILS) method in MATLAB. This method computes the average of the integrated secondary ion intensity for a line scan width of N pixels at each depth increment x, to give a column vector $C(x)$. Mathematically, this may be represented as:

$$C(x) = \frac{1}{N} \frac{\sum_n^N {}^{18}O^-}{\left[\sum_n^N {}^{16}O^- + \sum_n^N {}^{18}O^-\right]} \tag{3}$$

Finally, D_O^* was estimated by fitting the profiles (using a nonlinear least-squares method) according to Fick's second law of diffusion for a semi-infinite media with a constant source:26

$$C'(x) = \frac{C(x) - C_{bg}}{C_g - C_{bg}} = \text{erfc}\left(\frac{x}{2\sqrt{D_O^* t_{18}}}\right) \tag{4}$$

where $C'(x)$ is the isotopic fraction normalized to the tracer natural isotopic abundance C_{bg} (0.24%), and the tracer gas phase concentration C_g (97.0%), x is the distance from the surface, and t_{18} is the $^{18}O_2$ exposure time. Figure 8 shows this method applied to the polished alloy oxidized under classical conditions (i.e., Figure 1a and b), giving both the IF map and the fitted isotopic enrichment profile for the external oxide.

Table III. Comparison of various kinetic parameters for classical, tensile loading and compressive loading

Parameter	Specimen	Polished	Shot-Peened
D_O^* ($\times 10^{-13}$ cm^{2}s^{-1})	Classical	0.39	1.4
	Compression	0.62	3.7
	Tension	0.41	3.4
R_{comp} or Δ_{comp} (μm)	EO	1.48 (0.58)	1.01 (0.02)
	IO	1.51 (1.48)	1.09 (0.17)
	GPD/RX	1.50 (2.00)	1.09 (0.65)
R_{ten} (%) or Δ_{ten} (μm)	EO	1.29 (0.35)	1.03 (0.06)
	IO	1.38 (1.13)	1.06 (0.28)
	GPD/RX	1.40 (1.58)	1.07 (0.51)
R_{IOA}	Compression	1.76	1.34
	Tension	1.47	1.08

The subscripts comp. and ten.—applied to R and Δ—indicate values for compression or tension, respectively. For the relative depth increase (R), the values in parenthesis are the absolute increase (i.e., Δ_{ten} or Δ_{comp})

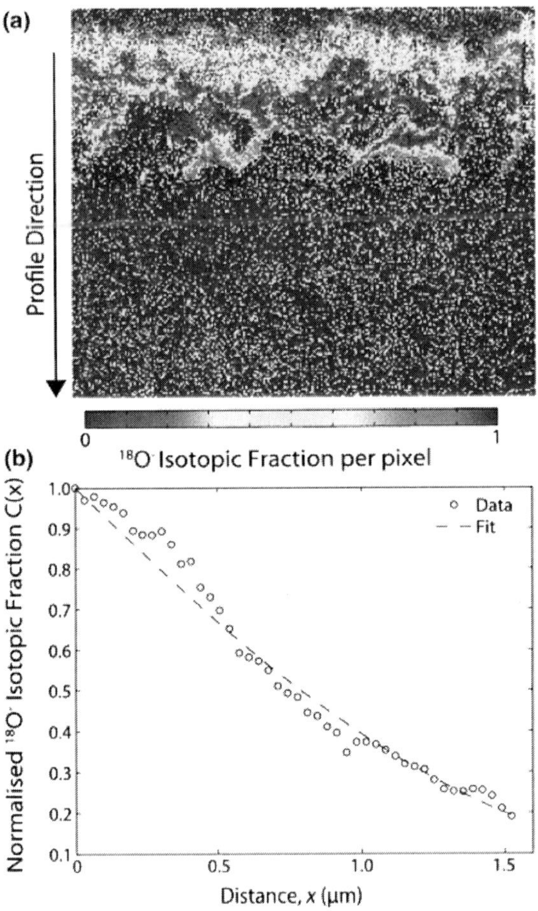

Figure 8. Example determination of oxygen tracer diffusion in external oxide using FIB-SIMS data for polished FG RR1000 oxidized under classic conditions: (a) $^{18}O^-$ isotopic fraction map and (b) fitted isotopic enrichment profile

Table III lists the tracer diffusion coefficients calculated for classical and bending load specimens in both the polished and shot-peened conditions.

The values should be treated as effective diffusion coefficients as they account for both the bulk and grain boundary diffusivities. As shown, the shot-peened specimens consistently show higher diffusivities than the polished specimens. The bending loads appear to have only a small influence on the diffusivities, which increase by less than an order of magnitude for all examples. Importantly, the coefficients here agree with the apparent oxygen diffusion coefficients measured for Cr_2O_3 scales grown on Ni-30Cr alloys at the same temperature.[17]

Oxidation Damage

Depth measurements of the external scale, internal oxide, and γ'-depleted/recrystallized region clearly highlighted the stress-related effects during oxidation under compressive and tensile loading. These stress effects do not portray straightforward relationships, and therefore, Table III interprets the load effects in terms of depth enhancement ratios *R*, the absolute depth increase Δ, and relative internal oxide area increase R_{IOA} for both polished and shot-peened specimens.

The polished specimens are most affected by the application of either compressive or tensile loads. This applies to all the parameters measured with the GPD/RX exhibiting both the greatest relative (~50% in compression) and absolute depth increases. Compressive loads have the greatest influence on the oxidation damage with relative increases of ~50% for all damage parameters with the exception of the IO area where a 76% increase occurs. In contrast, the shot-peened specimens appear to be less influenced by the applied loads. Nearly all of the damage parameters demonstrate an increase less than 10%. Under compression, the internal oxide depth increases only slightly; however, the internal oxide area increases by 34%. Note that this specimen also revealed significant changes in the internal oxide morphology.

DISCUSSION

Induced internal stresses from shot-peening under classic oxidation conditions and external loading of polished samples modify the nature of the external and internal oxides and/or accelerates the oxidation kinetics. The difference in oxide morphologies comes as a surprise given that most of the current literature only reports expedited kinetics.[27–32]The isotopic tracer experiments indicated that shot-peening and applied stresses have very little influence on the inward oxygen diffusion, thus eliminating enhancement due to a greater contribution from inward anionic growth. This suggests that the shot-peening and mechanical loads applied to polished samples act to influence the diffusivity of the oxidized solute species (i.e., chromium and titanium) in the parent and recrystallized alloy,[33–35] and this is proposed to result from defects induced from creep deformation. This could explain the increased internal oxidation kinetics realized in the polished specimens under load. This appears to contradict much of the literature available (see for example,

Refs.27,36,37), which infers the greater penetration depths from stress-assisted grain boundary diffusion. The premise of this argument states that the hydrostatic stress increases the oxygen diffusivity. For example, modeling conducted by Karabela et al.27,28described accelerated internal oxidation kinetics by first computing the hydrostatic stress gradient through crystal plasticity modeling and then the oxygen diffusion. The authors successfully rationalized the amplified oxygen penetration depths through this method; however, this analysis did not account for flux of the metallic species and the oxidation processes themselves, which inherently modify the media in which the oxygen diffuses through. Furthermore, the approach itself implies that a compressive stress gradient should have a reverse effect and decrease the inward oxygen flux. Clearly, this is not consistent with the current study as compressive stresses generated increased internal oxidation penetration depths. The increased penetration depths presumably result from enhanced aluminum diffusion in the alloy, like the chromium and titanium to form the surface oxides. In support of this, both compression and tension sides demonstrate relatively higher γ'-dissolution/recrystallization depths under load, indicative of an increased aluminum flux toward the internal oxidation front. To maintain internal oxide growth, the oxygen flux would have to equal that of the solute at the precipitation front to maintain the mass balance. SIMS mapping demonstrated a buildup of ^{18}O at an already established ^{16}O internal oxide front due to rapid surface diffusion along the internal oxide/alloy interface. Then, it is reasonable to assume that the internal oxidation front is already saturated with oxygen, so the oxygen potential is high enough for continued oxide precipitation. Hence, although the stress may influence the oxygen diffusion, the stresses could have a negligible effect with an already rapid surface diffusion taking place.

In the shot-peened specimens, a very different response occurs under external loads, which reflects the number of competing microstructural processes. The applied stresses did not increase the oxidation kinetics in the polished alloy above those of the shot-peened alloy. Even with defect annihilation in the shot-peened structure, it appears that the overall number defects reaches equilibrium. As such, the shot-peened alloy realizes no further enhancement of the oxidation kinetics at 800°C, but significant morphological changes to the oxide microstructure occur.

ACKNOWLEDGMENTS

B.J.F. would like to acknowledge the support of the U.K. Engineering and Physical Sciences Research Council (EPSRC) and Rolls-Royce plc for financial support of his PhD research. BAS would like to thank the Royal Society and Rolls-Royce plc for her Industry Fellowship.

REFERENCES

1. M. Brady, I. Wright, and B. Gleeson, *JOM* 52 (1), 16 (2000).
2. H. Evans, *Int. Mater. Rev.* 40, 1 (1995).
3. P. Hancock and J.R. Nicholls, *Mater. High Temp.* 12, 209 (1994).
4. J. Evans, *J. Mater. Eng. Perform.* 19, 1001 (2010).
5. S. Cruchley, M. Taylor, H. Evans, P. Bowen, M. Hardy, and S. Stekovic, *Superalloys 2012,* eds., E.S. Huron, R.C. Reed. M.C. Hardy, M.J. Mills, R.E. Montero, P.D. Portella, and J. Telesman (Warrendale, PA: TMS, 2012), p. 751.
6. D.J. Young, *High Temperature Oxidation and Corrosion of Metals*, 1st ed. (Amsterdam: Elsevier Science Ltd., 2008), p. 247.
7. M. Hardy, B. Zirbel, G. Shen, and R. Shankar, *Superalloys 2004,* eds., K.A. Green, H. Harada, T.E. Howson, T.M. Pollock, R.C. Reed, J.J. Shirra, and S. Walston (Warrendale, PA: TMS, 2004), p. 83.
8. A.J. Sedriks, *Corrosion Testing Made Easy: Stress Corrosion Cracking Test Methods*, Vol. 1 (Houston: National Association of Corrosion Engineers, 1990).
9. D. McPhail, R.J. Chater, and L. Li, *Microchim. Acta* 161, 387 (2008).
10. A. Encinas-Oropesa, G. Drew, M. Hardy, A. Leggett, J. Nicholls, and N.J. Simm, *Superalloys 2008,* eds., R.C. Reed, K.A. Green, P. Caron, T. Gabb, M.G. Fahrmann, E.S. Huron, and S.A. Woodard (Warrendale, PA: TMS, 2008), p. 609.
11. C.K. Sudbrack, S.L. Draper, T.T. Gorman, J. Telesman, T.P. Gabb, and D.R. Hull, *Superalloys 2012,* eds., E.S. Huron, R.C. Reed. M.C. Hardy, M.J. Mills, R.E. Montero, P.D. Portella, and J. Telesman (Warrendale, PA: TMS, 2012), p. 863.
12. M. Taylor, H. Evans, S. Stekovic, and M. Hardy, *Mater. High Temp.* 29, 145 (2012).
13. L. Zheng, M. Zhang, and J. Dong, *Appl. Surf. Sci.* 256, 7510 (2010).
14. J. Chen, P. Rogers, and J. Little, *Oxid. Met.* 47, 381 (1997).

15. S. Cruchley, H. Evans, M. Taylor, M. Hardy, and S. Stekovic, *Corros. Sci.* 75, 58 (2013).

16. S. Tsai, A. Huntz, and C. Dolin, *Oxid. Met.* 43, 581 (1995).

17. S.C. Tsai, A.M. Huntz, and C. Dolin, *Mater. Sci. Eng., A* 212, 6 (1996).

18. A. Naoumidis, H. Schulze, W. Jungen, and P. Lersch, *J. Eur. Ceram. Soc.* 7, 55 (1991).

19. A. Alibhai, R. Chater, D. McPhail, and B.A. Shollock, *Appl. Surf. Sci.* 203, 630 (2003).

20. T. Ooi, D. McPhail, R. Chater, and B.A. Shollock, *Surf. Coat. Technol.* 201, 3885 (2006).

21. Huntz and S. Tsai, *J. Mater. Sci. Lett.* 13, 821 (1994).

22. A. Huntz, J. Balmain, S. Tsai, K. Messaoudi, M. Loudjani, B. Lesage, and J. Li, *Scripta Mater.* 37, 651 (1997).

23. Y. Shida, F. Stott, B. Bastow, D. Whittle, and G. Wood, *Oxid. Met.* 18, 93 (1982).

24. D. Whittle, Y. Shida, G. Wood, F. Stott, and B. Bastow, *Philos. Mag. A* 46, 931 (1982).

25. F. Stott, Y. Shida, D. Whittle, G. Wood, and B. Bastow, *Oxid. Met.* 18, 127 (1982)

26. J. Crank, *The Mathematics of Diffusion*, 2nd ed. (London: Oxford University Press, 1980), p. 20.

27. A. Karabela, L. Zhao, B. Lin, J. Tong, and M. Hardy, *Mater. Sci. Eng., A* 567, 46 (2013).

28. A. Karabela, L.G. Zhao, J. Tong, N.J. Simms, J.R. Nicholls, and M.C. Hardy, *Mater. Sci. Eng., A* 528, 6194 (2011).

29. G. Calvarin Amiri, A. Huntz, and R. Molins, *Mater. High Temp.* 18, 91 (2001).

30. G. Calvarin-Amiri, A. Gourgues, A. Huntz, and R. Molins, *Mater. Sci. Eng., A* 298, 200 (2001).

31. G. Calvarin-Amiri, R. Molins, and A. Huntz, *Oxid. Met.* 53, 399 (2000).

32. A.M. Limarga and D.S. Wilkinson, *Acta Mater.* 55, 251 (2007).

33. D. Caplan and M. Cohen, *Corros. Sci.* 6, 321 (1996).

34. F. Wang, *Oxid. Met.* 48, 215 (1997).

35. R. Naraparaju, *Corros. Eng., Sci. Technol.* 44, 211 (2009).

36. L.G. Zhao, *J. Eng. Mater. Technol.* 133, 031002 (2011).

37. U. Krupp, *Int. Mater. Rev.* 50, 83 (2005).

9

CHAPTER

Geochemistry and Mineralogy of Tuff in Zhongliangshan Mine, Chongqing, Southwestern China

Jianhua Zou [1,2,*], Heming Tian [2] and Tian Li [2]

[1]*College of Geosciences and Survey Engineering, China University of Mining and Technology (Beijing), Beijing 100083, China*

[2]*Chongqing Key Laboratory of Exogenic Mineralization and Mine Environment, Chongqing Institute of Geology and Mineral Resources, Chongqing 400042, China*

ABSTRACT

Coal-bearing strata that host rare metal deposits are currently a hot issue in the field of coal geology. The purpose of this paper is to illustrate the mineralogy, geochemistry, and potential economic significance of rare metals in the late Permian tuff in Zhongliangshan mine, Chongqing, southwestern China. The methods applied in this study are X-ray fluorescence spectrometry (XRF), inductively coupled mass spectrometry (ICP-MS), X-ray diffraction analysis (XRD) plus Siroquant, and scanning electron microscopy in conjunction with an energy-dispersive X-ray spectrometry (SEM-EDX). The results indicate that some trace elements including Li, Be, Sc, V, Cr, Co, Ni, Cu, Zn, Ga, Zr, Nb, Cd, Sb, REE, Hf, Ta, Re, Th, and U are enriched in the tuff from Zhongliangshan mine. The minerals in the tuff mainly include kaolinite, illite, pyrite, anatase, calcite, gypsum, quartz, and traces of minerals such as zircon, florencite, jarosite, and barite. The tuff is of mafic volcanic origin with features of alkali basalt. Some minerals including florencite, gypsum, barite and a portion of anatase and zircon have been derived from hydrothermal solutions. It is suggested that Zhongliangshan tuff is a potential polymetallic ore and the recovery of these valuable elements needs to be further investigated.

Keywords: geochemistry; mineralogy; origin; tuff; Chongqing

1. INTRODUCTION

With the depletion of traditional rare metal deposits, coal deposits as promising alternative sources for rare metals have attracted much attention in recent years [1,2,3,4,5,6,7,8,9]. At present, germanium is the most successful rare metal element that has been extracted from coal ash [4,8,9]. The three well-known coal-bearing strata hosted Ge deposits include Lincang (Yunnan Province) and Wulantuga (Inner Mongolia) of China, and Spetzugli of Russia, are the main sources for the industrial Ge at present and for the foreseeable future [1,4,8,9]. The super-large coal-bearing strata hosted gallium deposit in the Jungar Coalfield (Inner Mongolia), China, is another typical example discovered in 2006 [10], which was considered as the third and the most outstanding discovery after the coal-bearing strata hosted uranium and germanium deposits [2,10]. Moreover, aluminum is also enriched in Jungar coalfield [1,10]. In 2010, another new type of coal-bearing strata hosted Nb (Ta)-Zr (Hf)-REE-Ga polymetallic deposit of volcanic origin was discovered in the late Permian coal-bearing strata of eastern Yunnan, southwestern China [11]. Similar polymetallic deposits have since been discovered in some coalfields from southern China [1,3]. Similar to most typical areas enriched in rare metals in coal-bearing strata, the tectonic controls on the localization of the metalliferous coal deposits and the mechanisms of rare-metal mineralization in south China and south Primorye of Russia have been studied comparatively in detail [3]. The possible recovery of rare earth elements from coal and its combustion products such as fly ash is an exciting new research area [2,12,13,14,15,16], because coal and its combustion derivation (fly ash) may have elevated concentrations of these rare metals.

The purpose of this paper is to discuss the mineralogical and geochemical compositions of tuff layer in late Permian coal-bearing strata of Zhongliangshan mine, Chongqing, southwestern China. It also contributes to the discussion on the origin and potential prospects of rare metals mineralization of the tuff.

2. GEOLOGICAL SETTING

The Zhongliangshan mine is located in the urban area of Chongqing, southwestern China (Figure 1). The coal-bearing sequence is the late Permian Longtan Formation (P_3l), which is composed of the light gray, gray, dark gray mudstone, sandy mudstone, siltstone, sandstone and coal seams. This formation is enriched in brachiopods, fern, cephalopods, bivalves, trilobite and other fossils. The Longtan Formation was deposited in a continental–marine transitional environment and has a thickness varying from 26.5 to 105.02 m, with an average of 71.08 m. It contains 10 coal seams, which are identified as K1 to K10 from top to bottom. The Changxin Formation conformably overlies the Longtan Formation and is mainly composed of thick layers of brown-gray, dark gray limestone intercalated with thin layers of mudstone and flint nodules. Some fossils including brachiopods, spindle dragonflies, sponges, corals, and trilobites are enriched in the Changxin Formation. The Maokou Formation disconformably underlies the Longtan Formation, which consists of thick layers of light gray to dark gray bioclastic limestone.

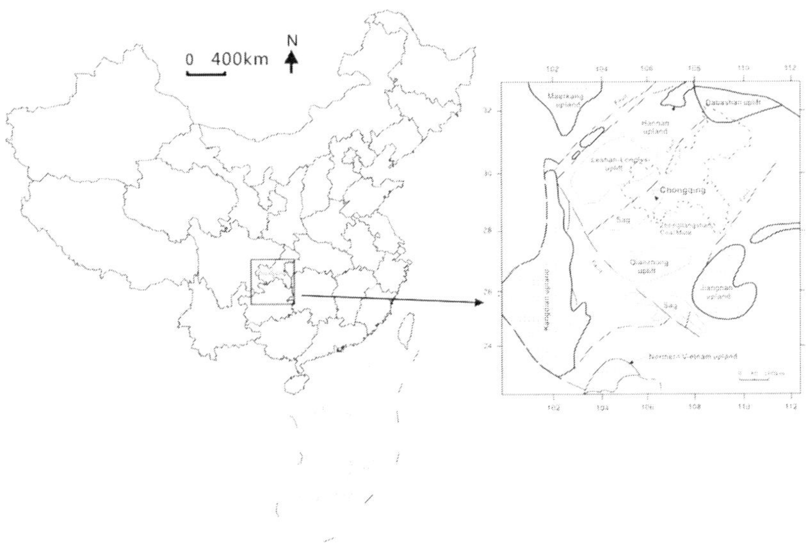

Figure 1. Location of the Zhongliangshan Mine, Chongqing, southwestern China.

The tuff layer, with a thickness mostly of 2–5 m, light-gray or light-gray-white in color, and a conchoidal fracture and a soapy feel, is located at the lowermost Longtan Formation. The K10 coal seam conformably overlies the tuff layer, which has a disconformable contact with the underlying Maokou Formation (middle Permian) (Figure 2). The tuff is enriched in pyrite and shows massive bedding structure. The tuff was derived from the basalt eruption and deposited directly on the weathered surface of the Maokou Formation limestone, and then was subjected to weathering, leaching, and eluviation [17,18]. It is usually described as bauxite or bauxitic mudstone during core sample identification or field lithological description [17,18].

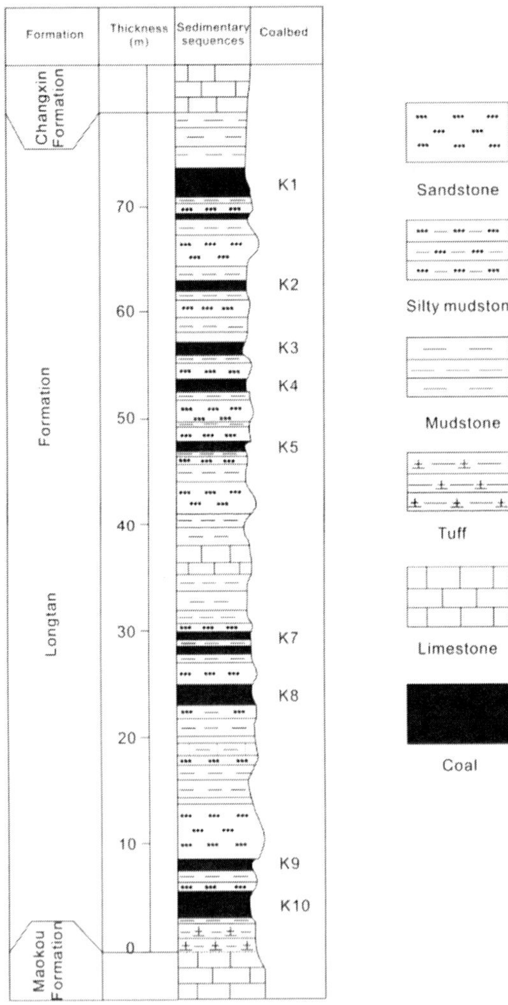

Figure 2. Generalized sedimentary sequence at the Zhongliangshan Mine, Chongqing.

3. SAMPLES AND ANALYTICAL PROCEDURES

A total of 21 bench samples were taken from the tuff layer in the Zhongliangshan mine, following the Chinese Standard GB/T 482-2008 [19]. Each tuff bench sample was cut over an area 10-cm wide, 10-cm deep and 10-cm thick. All collected samples were immediately stored in plastic bags to minimize contamination and oxidation. Large chips were selected at random from each sample for preparation of polished sections and also kept for later reference if required. The remainder of each sample was crushed and ground to pass through the 200-mesh sieve for analysis.

The loss of ignition (LOI) of each sample was determined according to ASTM standard D3174 [20]. All samples were analyzed by X-ray diffraction (XRD) using a D8 advance powder diffractmeter with Ni-filtered Cu-Kα radiation and a scintillation detector. The XRD pattern was recorded over a 2θ interval of 2.6°–70°, with a step size of 0.02°. X-ray diffractograms of the tuff samples were subjected to quantitative mineralogical analysis using Siroquant™ of China University of Mining and Technology (Beijing), a commercial interpretation software developed by Taylor [21] based on the principles for diffractogram profiling set out by Rietveld [22]. Further details indicating the use of this technique for coal-related materials are given by Ward *et al.* [23,24] and Ruan and Ward [25]. A Scanning Electron Microscope in conjunction with an energy-dispersive X-ray spectrometer (SEM-EDX, JEOL JSM-6610LV+OXFORD X-max, Tokyo, Japan), with an accelerating voltage of 20 kV, was used to study morphology and microstructure of minerals, and also to determine the distribution of some elements in tuff samples under a high vacuum mode in Chongqing Institute of Geology and Mineral Resources.

Percentages of major element oxides including SiO_2, TiO_2, Al_2O_3, Fe_2O_3, MnO, MgO, CaO, Na_2O, K_2O, and P_2O_5 in the tuff samples were determined by X-ray fluorescence spectrometry (XRF) in Chongqing Institute of Geology and Mineral Resources. The contents of trace elements were determined by inductively coupled mass spectrometry (Thermo X series II ICP-MS, Thermo Fisher Scientific, Waltham, MA, USA) in Chongqing Institute of Geology and Mineral Resources. The procedures of ICP-MS were: weigh 0.25 g sample in a 50 mL Teflon beaker; add 20 mL HNO_3-$HClO_4$-HF (volume ratio of 4:1:5) and 2 mL H_2SO_4; place on a temperature controlled heating plate and heat to 230 °C until like wet salt; then heat to 280 °C and evaporate to dryness; turn off the heating plate to cool the sample for 3 min; add 8 mL concentrated aqua regia; incubate for 10 min; transfer the solution to a 25 mL plastic flask; mix and volume; take 5 mL of solution to a 25 mL volumetric flask and dilute to the mark; and study using high resolution inductively coupled plasma mass spectrometry.

4. RESULTS

4.1. Minerals

The proportion of each crystalline phase of the tuff identified by X-ray diffractometry plus Siroquant is given in Table 1. The minerals in Zhongliangshan tuff mainly include kaolinite, illite, pyrite, anatase, calcite, gypsum, and quartz. Some trace minerals such as zircon, florencite, jarosite, and barite, are observed under SEM-EDX.

Table 1. Mineral compositions of Tuff by XRD analysis and Siroquant (%).

Sample	Kaolinite	Illite	Pyrite	Anatase	Calcite	Quartz	Gypsum
S140SE7-1	83.2	-	13.6	2.7	-	0.5	-
S140SE7-2	85.1	-	9.3	3.5	-	2.1	-
S140SE7-3	65.4	-	11.6	8.2	-	14.8	-
S140SE7-4	69.9	-	20.2	6.7	-	3.3	-
S140SE7-5	53.7	-	38.2	2.8	-	-	5.3
S140SE7-6	61.4	-	30.8	3.2	-	-	4.6
S140SE7-7	61.9	-	19.8	4.1	-	-	14.2
S140SE7-8	78.7	-	14.8	3.3	-	0.1	3.1
S140SE7-9	80	-	10.6	7.5	-	0.3	1.7
S140SE7-10	65.8	-	26.5	4.7	-	0.3	2.7
S140SE7-11	72.4	-	16.1	8.8	-	0.3	2.5
S140SE7-12	73.4	6.7	2.8	14.3	-	0.4	2.4
S140SE7-13	49.9	18.6	10.6	9.7	6.1	0.3	4.8
S140SE7-14	11.4	3.8	28.8	0.8	45	-	10.3
S140SE7-15	52.9	21	3.1	8.4	10.7	0.5	3.5
S140SE7-16	81.5	4.4	1.4	6.9	3.8	0.1	1.8
S140SE7-17	86.4	4.4	1.1	6.9	-	-	1.2
S140SE7-18	80.7	6	2.4	7.2	0.4	0.4	2.8
S140SE7-19	34.5	31.3	1.9	3.6	24.7	0.3	3.7
S140SE7-20	67.2	18.6	1	10.5	0.1	0.5	2.2
S140SE7-21	56	35.3	0.4	7.2	0.3	-	0.7
Average	65.30	15.01	12.62	6.24	11.39	1.61	3.97

4.1.1. Kaolinite and Illite

Kaolinite is the dominant mineral of the tuff in Zhongliangshan (Figure 3). The average content of kaolinite is up to 65.3%, and all the samples are richer than 50% except for samples S140SE7-13, S140SE7-14, and S140SE7-19. Kaolinite occurs mainly as matrix material (Figure 4A), and to a lesser extent, as vermicular (Figure 4B) and individual massive (Figure 4C). Illite occurs at the lower part of the profile (Figure 3).

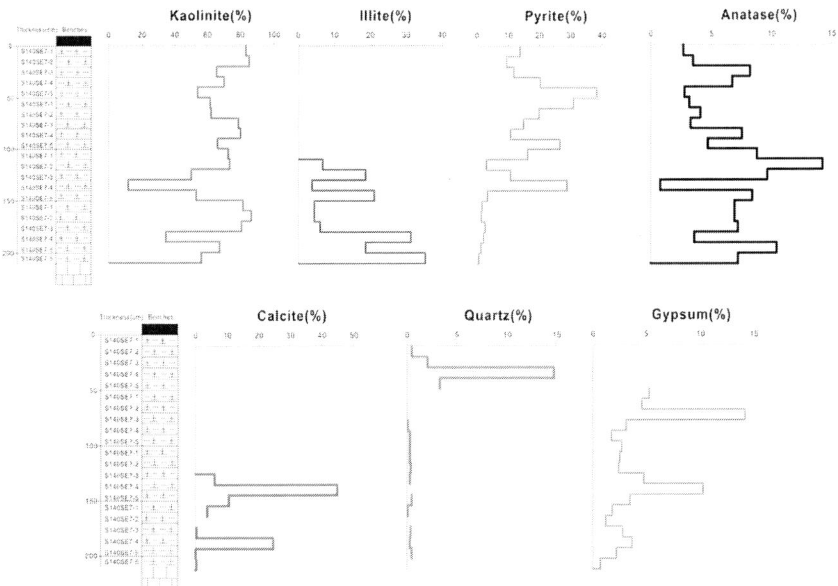

Figure 3. Vertical variations of minerals from the tuff in the Zhongliangshan mine.

4.1.2. Pyrite, Jarosite and Barite

Pyrite distributes widely in the tuff samples and its content varies from 0.4% to 38.2% (12.6% on average). Its content is gradually decreasing from top to bottom (Figure 3), suggesting that the upper portion has been more subjected to seawater. Pyrite mainly occurs as discrete particles (Figure 4A,D), lumps (Figure 4D), and in some cases, as cubic crystal and pentagonal dodecahedron (Figure 4E). Jarosite occurs as fracture-fillings (Figure 4F), indicating a weathering product of pyrite. Barite is located on the edge of jarosite (Figure 4F), which may be formed by the reaction of jarosite with the hydrothermal solution containing Ba.

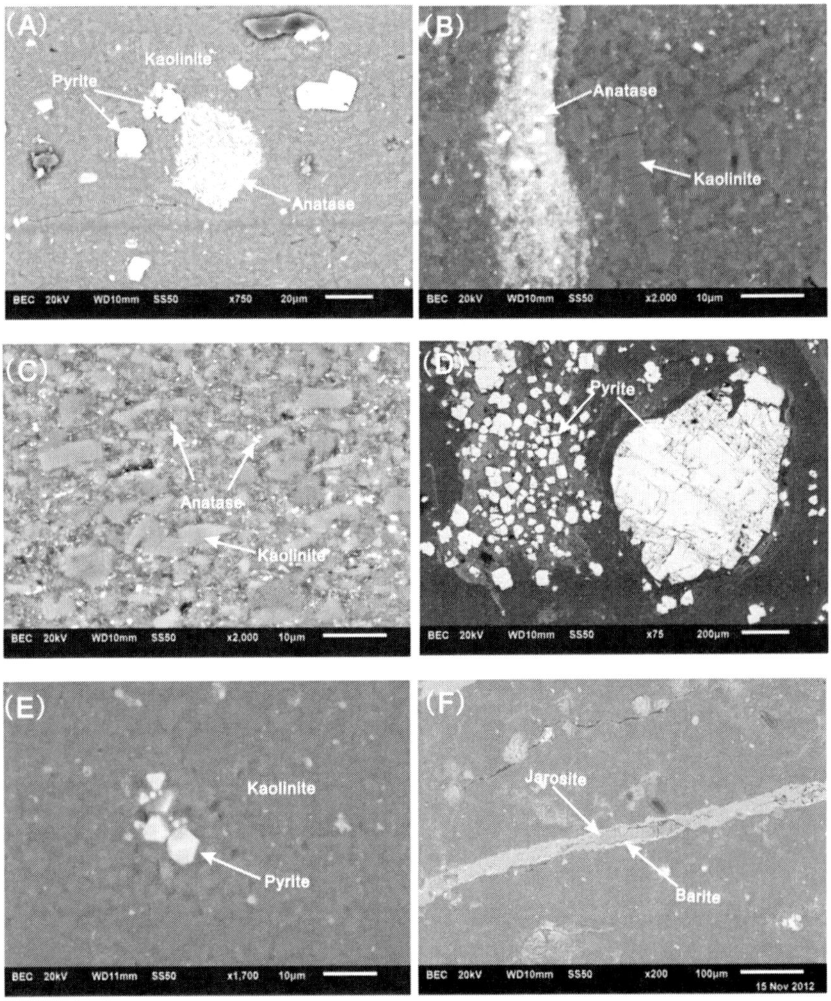

Figure 4. Back scattered images of minerals in the Zhongliangshan tuff: (**A**) kaolinite, pyrite and anatase in sample S140SE7-1; (**B**) kaolinite and anatase in sample S140SE7-8; (**C**) kaolinite and anatase in sample S140SE7-1; (**D**) pyrite in sample S140SE7-4; (**E**) pyrite and kaolinite in sample S140SE7-6; and (**F**) jarosite and barite in sample S140SE7-18.

4.1.3. Anatase

Anatase is present evenly in the tuff samples and varies from 0.8% to 14.3% with an average of 6.2%. The content of Nb in the anatase is up to 0.18% determined by SEM-EDX. Anatase occurs mainly as irregular fine particles (Figure 4B,C) or as colloidal (Figure 4A and Figure 5A) in the kaolinite matrix.

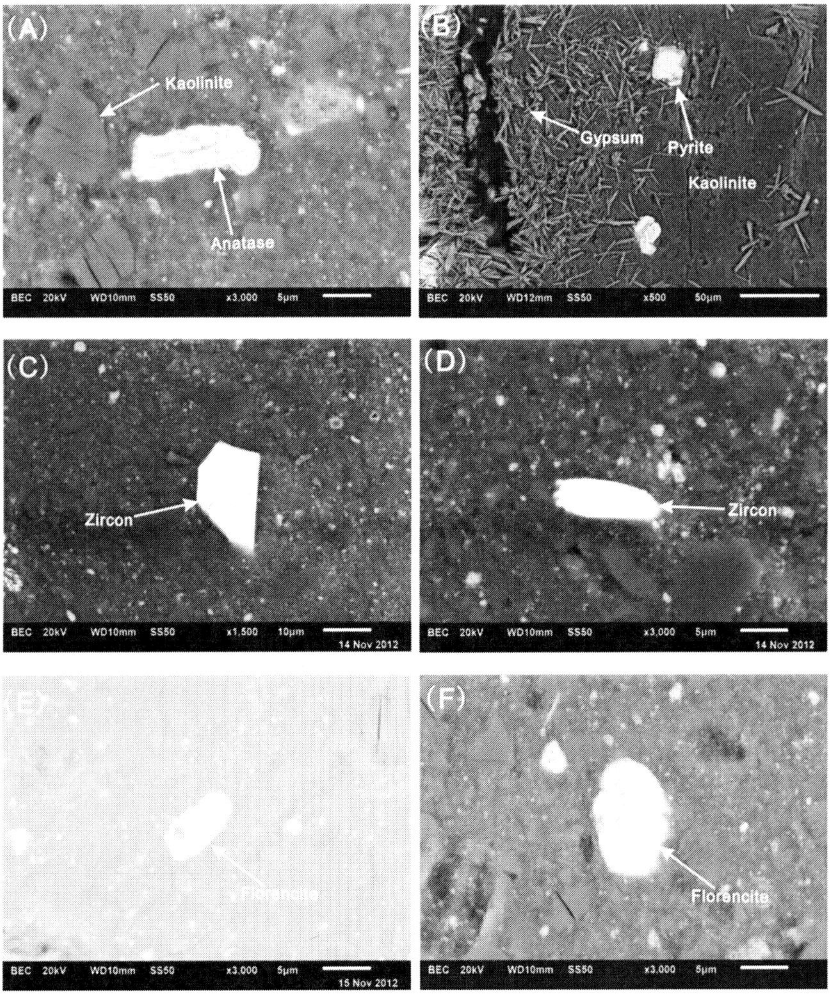

Figure 5. Back scattered images of minerals in the Zhongliangshan tuff: (**A**) kaolinite and anatase in sample S140SE7-4; (**B**) gypsum, pyrite, and kaolinite in sample S140SE7-10; (**C**) zircon in sample S140SE7-15; (**D**) zircon in the sample S140SE7-15; (**E**) florencite in sample S140SE7-21; and (**F**) florencite in sample S140SE7-4.

4.1.4. Calcite and Gypsum

Calcite distributes at the lower portion of the profile (Table 1, Figure 3), similar to that of illite. Gypsum occurs as radiating forms in the tuff and is present on the edge of fractures (Figure 5B), indicating an epigenetic origin.

4.1.5. Zircon and Florencite

Although zircon and florencite are at concentration below the detection limit of the XRD and Siroquant analysis, they have been observed under SEM-EDX in the tuff samples of the present study. Zircon occurs as subhedral (Figure 5C) and long axis (Figure 5D) in the kaolinite matrix. Florencite occurs as ellipsoidal form in kaolinite; however, minerals containing medium (M-REE) and heavy-rare earth elements (H-REE) have not been observed (M-REE include Eu, Gd, Tb, Dy, and Y; and H-REE include Ho, Er, Tm, Yb, and Lu [7]).

4.2. Major Elements

The loss of ignition of the tuff samples varies from 13.94% to 23.56%, with an average of 17.7%. The major element oxides are mainly represented by SiO_2 (35.3% on average) and Al_2O_3 (29.23%), followed by Fe_2O_3 (10.95%) and TiO_2 (3.82%) (Table 2). The ratio of SiO_2/Al_2O_3 is from 1.16 to 1.26 and averages 1.21, higher than the theoretical value of kaolinite (1.18). The ratio of TiO_2/Al_2O_3 is from 0.09 to 0.15, with an average of 0.13.

4.3. Trace Elements

Compared with the average concentration of the Upper Continental Crust (UCC) [26], some trace elements are enriched in the tuff samples from Zhongliangshan mine (Table 2). The concentration coefficients (CC, the ratio of the trace-element concentrations in investigated samples *vs.* UCC) of trace elements higher than 10 include Li, Cr, Cu, Cd, Sb and Re; whereas the elements with CC between 5 and 10 include V, Ni, Zr, Hf, and U. Elements Be, Sc, Co, Zn, Ga, Nb, REE, Ta, and Th, have a CC between 2 and 5. Elements Rb, Sr, Ba, and Tl are depleted, with a CC < 0.5. Other trace elements have concentrations close to the UCC, with CC between 0.5 and 2.

Table 2. Elemental concentrations in Tuff samples from the Zhongliangshan Mine (elements in μg/g, Oxides in %).

Sample	LOI	SiO_2	TiO_2	Al_2O_3	Fe_2O_3	MnO	MgO	CaO	Na_2O	K_2O	P_2O_5	FeO	SiO_2/Al_2O_3	TiO_2/Al_2O_3	Li	Be	Sc	V	Cr	Co	Ni	Cu
S140SE7-1	19.87	36.25	2.77	31.23	9.1	0.012	0.26	0.12	0.14	0.1	0.075	0.6	1.16	0.09	454	9.46	33.4	371	317	36.8	176	247
S140SE7-2	17.87	38.9	3.21	33.6	5.52	0.009	0.26	0.1	0.31	0.1	0.04	0.55	1.16	0.1	445	9.91	38	507	379	30.8	124	243
S140SE7-3	18.15	36.84	3.82	30.98	9.52	0.02	0.2	0.11	0.22	0.1	0.051	0.5	1.19	0.12	349	7.55	31.8	621	467	30.3	110	245
S140SE7-4	19.66	32.69	3.63	27.44	15.46	0.21	0.19	0.14	0.17	0.1	0.051	0.8	1.19	0.13	367	6.33	29.1	648	469	32.8	96.2	259
S140SE7-5	23.56	24.78	2.57	20.01	28.09	0.015	0.18	0.36	0.13	0.085	0.037	0.75	1.24	0.13	337	5.66	25.6	500	375	42.3	120	258
S140SE7-6	21.65	28.05	2.92	23.17	23	0.015	0.22	0.31	0.15	0.094	0.046	0.6	1.21	0.13	299	5.42	24	453	325	32.7	114	235
S140SE7-7	19.52	32.7	3.28	27.62	16.08	0.012	0.27	0.12	0.15	0.11	0.043	0.4	1.18	0.12	421	6.95	28.2	489	386	39.7	132	286
S140SE7-8	19.06	33.86	3.35	28.75	14.17	0.013	0.31	0.12	0.17	0.13	0.048	0.4	1.18	0.12	401	6.61	32.4	530	344	45	138	278
S140SE7-9	16.45	38.23	3.87	32.89	7.81	0.008	0.25	0.11	0.16	0.11	0.049	0.35	1.16	0.12	442	7.11	33.3	592	388	41.4	104	251
S140SE7-10	20.62	29.97	2.89	24.88	20.9	0.011	0.23	0.17	0.11	0.13	0.03	0.55	1.2	0.12	311	4.87	24.7	408	308	65.4	161	282
S140SE7-11	17.82	35.72	4.38	29.64	11.54	0.004	0.15	0.26	0.21	0.17	0.066	0.5	1.21	0.15	321	7.34	33.2	669	419	40.7	119	316
S140SE7-12	15.42	40.12	4.65	33.07	5.18	0.007	0.2	0.26	0.49	0.43	0.086	0.25	1.21	0.14	248	7.59	36.5	743	547	39.4	78.8	330
S140SE7-13	17.93	35.93	4.39	29.44	9.57	0.008	0.2	1.56	0.3	0.43	0.067	0.6	1.22	0.15	219	6.48	30.7	639	422	39.1	104	266
S140SE7-14	18.09	15.54	1.55	11.23	31.32	0.039	0.43	20.38	0.16	0.65	0.017	1.75	1.38	0.14	31.5	2.16	12.6	253	114	65.6	156	146
S140SE7-15	15.61	38.45	4.33	30.94	5.28	0.01	0.33	3.4	0.53	0.82	0.05	0.85	1.24	0.14	170	6.3	31.3	683	391	36.6	99.4	188
S140SE7-16	14.89	40.64	5.28	34.4	1.94	0.006	0.16	1.52	0.5	0.25	0.045	0.4	1.18	0.15	254	7.2	35.7	652	315	38.7	93.4	265
S140SE7-17	14.91	41.58	5.03	35.15	1.99	0.004	0.21	0.13	0.49	0.37	0.056	0.35	1.18	0.14	299	6.89	34.8	646	257	39.6	100	255
S140SE7-18	15.35	40.55	4.89	34.01	3.47	0.004	0.24	0.4	0.49	0.49	0.059	0.85	1.19	0.14	258	6.31	33.1	653	260	35.7	104	265
S140SE7-19	15.5	37.44	4.08	29.19	4.03	0.013	0.39	6.57	0.72	1.47	0.045	0.9	1.28	0.14	153	5.66	25	707	351	35.1	94.7	199
S140SE7-20	13.94	42.07	5.01	34.88	1.83	0.003	0.32	0.47	0.47	0.76	0.06	0.55	1.21	0.14	246	6.79	32.1	701	388	37.9	91.2	281
S140SE7-21	15.82	40.92	4.24	32.45	3.12	0.004	0.46	1.12	0.53	1.26	0.055	1.3	1.26	0.13	168	5.58	26.8	638	336	30.1	72	193
Average	17.7	35.3	3.82	29.28	10.9	0.02	0.27	1.8	0.31	0.39	0.05	0.66	1.21	0.13	295	6.58	30.1	576	360	39.8	114	252
UCC	nd	nd	nd	nd	nd	nd	nd	nd	nd	nd	nd	nd	nd	nd	20	3	11	60	35	10	20	25
CC	nd	nd	nd	nd	nd	nd	nd	nd	nd	nd	nd	nd	nd	nd	14.7	2.2	2.7	9.6	10.3	4	5.7	10.1

Sample	Zn	Ga	Rb	Sr	Zr	Nb	Mo	Cd	In	Sb	Cs	Ba	Hf	Ta	W	Re	Tl	Pb	Bi	Th	U
S140SE7-1	237	68.7	3.98	292	1898	213	3.4	3.13	0.615	4.22	1.17	146	57.6	15.9	4.02	0.042	0.11	58.1	1.41	50.3	25.7
S140SE7-2	190	64.5	2.86	190	1868	185	2.12	2.04	0.654	2.82	1.15	113	51.4	14.6	3.52	0.02	0.058	40.4	1.21	49.7	24.5
S140SE7-3	143	46.4	2.96	282	1487	134	1.73	0.952	0.587	3.51	1.48	141	37.7	8.88	4.44	0.018	0.085	39	1.62	33.4	9.83
S140SE7-4	123	34.9	3.52	256	1284	116	1.8	1.7	0.719	3.7	1.4	122	33.4	8.29	3.44	0.013	0.087	47.4	1.63	28.2	9.71
S140SE7-5	107	23.9	3.2	166	1046	91.6	1.56	1.46	0.457	4.13	1.07	92	26	6.91	3.16	0.014	0.207	61.3	1.22	23.8	8.11
S140SE7-6	100	30.7	2.03	191	1046	90.2	1.22	1.31	0.437	4.23	0.892	89.6	27.1	5.96	2.75	0.007	0.24	53.5	1.15	23.4	8.97
S140SE7-7	109	34.1	3.69	188	1206	109	0.985	1.22	0.494	3.41	1.1	101	30.7	7.6	2.83	0.006	0.103	49.3	1.35	27.9	11.4
S140SE7-8	124	35.9	3.1	232	1324	112	1.04	1.31	0.541	3.91	1.1	115	33.8	7.82	3.66	0.009	0.189	55.2	1.51	29	13.2
S140SE7-9	115	39.5	2.66	194	1392	114	1.03	1.5	0.519	2.23	1.05	99.3	36.9	8.57	3.67	0.006	0.058	34	1.48	30.4	16.6
S140SE7-10	103	35.4	2.46	154	978	77.7	1.45	1.83	0.516	6.79	0.927	79.1	25.4	5.68	2.8	0.01	0.09	75.2	1.14	21.7	12.6
S140SE7-11	192	38.5	5.34	287	1408	129	1.54	2.77	0.6	2.73	1.94	127	36.1	8.79	3.88	0.006	0.093	44	1.73	30	16
S140SE7-12	178	44.9	14.9	325	1467	132	1.68	4.45	0.572	1.71	5.03	141	37.9	8.98	4.05	0.013	0.118	23.6	2.06	31.8	17.7
S140SE7-13	161	37.6	12.5	341	1348	121	1.67	5.14	0.627	2.21	3.52	137	33.2	8.27	3.41	0.006	0.134	33.8	1.69	28.5	14
S140SE7-14	108	19	14.9	503	400	30.7	0.712	18.4	0.211	3.32	1.66	60.3	9.45	2.28	1.7	0.008	0.292	34.9	0.595	8.57	5.46
S140SE7-15	148	36.5	23.7	404	1342	113	1.37	6.96	0.501	2.08	5.27	143	34.9	7.88	3.69	0.006	0.156	26.9	1.52	28.7	14.2
S140SE7-16	170	32.6	6.83	279	1654	151	1.78	3.52	0.628	1.27	1.96	131	41.3	10.4	4.61	0.018	0.071	19.6	1.34	34	17.4
S140SE7-17	166	37	9.09	261	1631	143	0.996	4	0.567	1.68	2.37	131	41.4	10.1	3.76	0.025	0.08	23.8	1.36	32.2	18.5
S140SE7-18	147	35.1	12.1	286	1619	142	1.4	3.55	0.531	1.99	2.25	137	39.8	9.76	3.74	0.013	0.08	26.1	1.39	32	19.5
S140SE7-19	139	33.3	39.5	387	1153	106	1.49	6.92	0.505	2.42	5.33	160	28.8	7.07	3.48	0.036	0.287	21.8	1.39	27	20.1
S140SE7-20	174	35.6	19.8	322	1631	145	1.6	4.14	0.578	1.76	3.18	158	41.4	9.92	4.05	0.009	0.164	23	1.55	34	26.4
S140SE7-21	153	38.7	31.1	323	1405	120	1.48	3.47	0.526	1.5	4.24	165	35	8.39	3.43	0.015	0.236	25.2	1.27	29	23.8
Average	147	38.2	10.5	279	1361	123	1.53	3.8	0.54	2.93	2.29	123	35.2	8.67	3.53	0.01	0.14	38.9	1.41	30.2	15.9
UCC	71	17	112	350	190	25	1.5	0.1	nd	0.2	3.7	550	5.8	2.2	2	0.0004	0.8	20	nd	10.7	2
CC	2.1	2.2	0.1	0.8	7.2	4.9	1	38	nd	14.7	0.6	0.2	6.1	3.9	1.8	35.7	0.2	1.9	nd	2.8	7.9

UCC, the Upper Continental Crust; CC, concentration coefficient of trace elements in the tuff, normalized by average trace element concentrations in UCC [26]; nd, no data.

4.3.1. Scandium

The average content of Sc in tuff samples is 30.1 μg/g, which is close to these of the tuffs from Songzao (29.8 μg/g), Nanchuan (26.3 μg/g) and the mafic rocks (29 μg/g, 1060 samples) [27]. Scandium is immobile during weathering and alteration and thus can be used as a reliable indicator for the source of tonsteins in coal-bearing strata system [28,29].

4.3.2. Vanadium, Cr, Co and Ni

The average contents of V, Cr, Co and Ni in the investigated samples are 576, 360, 39.8, and 114 μg/g, respectively, close to the tuff from Songzao (V, Cr, Co, and Ni being 576, 549, 37.9, and 164 μg/g, respectively) [1,3] and the normal detrital sediments (888 samples) in the south of Sichuan Province surrounding Chongqing (V, Cr, Co, and Ni being 442, 206, 31, and 61 μg/g, respectively) [30]. The contents of V and Cr have the same variations through the seam section, gradually increasing from top to bottom (Figure 6). However, the contents of Co and Ni are higher in the middle relative to the upper and lower portions (Figure 6). The terrigenous source of the inorganic matter in the late Permian coals and normal sediments in southwestern China is the Emeishan Basalt of the Kangdian Upland, which is enriched in V, Cr, Co, and Ni [31,32]. The values of tuff samples in the Zhongliangshan mine are close to those in normal sediments, indicating the normal sediments in southwestern China and tuff in Zhongliangshan have the same magmatic sources (the Emeishan basalt magma enriched in V, Cr, Co, and Ni). Dai *et al.* [1,3,18] suggested that some dark minerals such as basic plagioclase and pyroxene in the basalt rocks could be easily decomposed under weathering conditions and then transported into coal-bearing basin as complex anions.

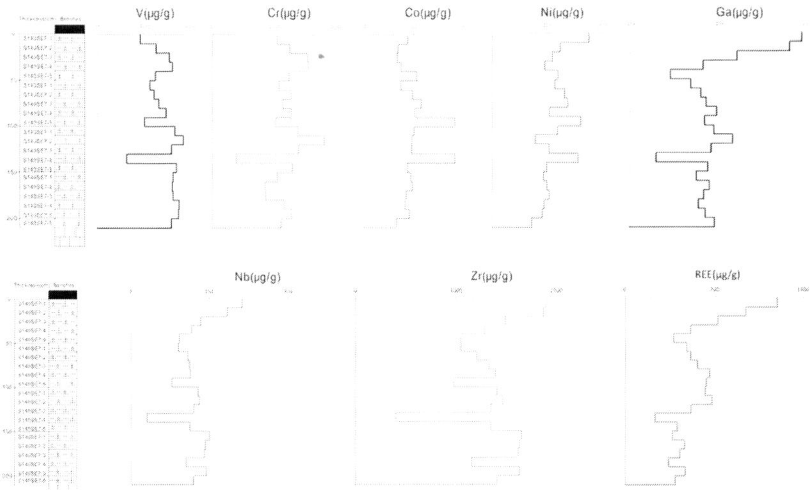

Figure 6. Vertical variations of selected trace elements of the tuff in Zhongliangshan mine, Chongqing.

4.3.3. Niobium, Ta, Zr and Hf

The average contents of Nb, Ta, Zr, and Hf of tuff in Zhongliangshan mine are 123, 8.67, 1361, and 35.2 μg/g, respectively, being close to those of the tuff from Songzao (Nb, Ta, Zr and Hf being 118, 9.46, 1377, and 41.5 μg/g, respectively). The Nb and Zr display a similar trend, both gradually decreasing from top to bottom (Figure 6).

The concentration of $(Nb, Ta)_2O_5$ of tuff in Zhongliangshan mine varies from 47 to 324 μg/g and averages 186 μg/g, lower than the concentration of the late Permian "Nb (Ta)-Zr (Hf)-Ga-REE" polymetallic deposit discovered in eastern Yunnan, southwestern China [11]. $(Zr, Hf)O_2$ varies from 551 to 2632 μg/g and averages 1880 μg/g, which does not meet the minimum industrial grade of the weathering crust type deposit (8000 μg/g) [33].

The common Nb-, Zr-, REE-, and Ga-bearing minerals have rarely been observed in the tuff, and thus it is suggested that these rare metals probably occur as absorbed ions [11,29]. However, Nb may occur as isomorph in the Ti-bearing minerals (Figure 4B,C) and Zr occurs as zircon (Figure 5C,D) in studied samples.

4.3.4. Gallium

The concentration of Ga in Zhongliangshan tuff varies from 19 to 68.7 μg/g and averages 38.2 μg/g, higher than the minimum industrial grade in bauxite (20 μg/g) and coal (30 μg/g) [34], but lower than the concentration of the late Permian "Nb(Ta)-Zr(Hf)-Ga-REE" polymetallic deposit in eastern Yunnan, southwestern China [11]. From top to bottom, the concentration of Ga gradually decreases, consistent with those of the Nb and Zr. Because the geochemical nature of Ga is similar to Al [1,8], it may occur as isomorph in Al-bearing minerals (e.g., kaolinite).

4.4. Rare Earth Elements (REE)

In this study, REE is used to specifically represent the elemental suite La, Ce, Pr, Nd, Sm, Eu, Gd, Tb, Dy, Y, Ho, Er, Tm, Yb, and Lu [35]. The abundances and geochemical parameters of REE in the tuff samples are listed in Table 3 and Table 4, respectively. The concentration of REE varies from 234 to 1189 μg/g, with an average of 548 μg/g. The concentration of REE gradually decreases from top to bottom, similar to that of Nb and Zr. Yttrium is closely associated with lanthanides in nature, because its ionic radius is very similar and its ionic charge is equal to that of Ho [7]. For this reason, yttrium is generally placed between Dy and Ho in normalized REE patterns [36]. Based on Seredin-Dai's classification [7], a three-fold geochemical classification of REE was used in the present study, including light (L-REE: La, Ce, Pr, Nd, and Sm), medium (M-REE: Eu, Gd, Tb, Dy, and Y), and heavy (H-REE: Ho, Er, Tm, Yb, and Lu) REE [7]. Accordingly, three enrichment types are identified, L-type (light REE; $La_N/Lu_N > 1$), M-type (medium REE; $La_N/Sm_N < 1$, $Gd_N/Lu_N > 1$), and H-type (heavy REE: $La_N/Lu_N < 1$), in comparison with the upper continental crust [7]. This classification has been widely adopted and used in recent years [2].

Table 4 and Figure 7 illustrate that the tuff in the Zhongliangshan mine is mainly enriched in heavy REE. Only samples S140SE7-1, S140SE7-3, S140SE7-4, S140SE7-6, and S140SE7-8 are enriched in light REE; and samples S140SE7-2, S140SE7-10, and 140SE7-14 are enriched in medium REE. From top to bottom, the light REE enrichment only occurs in the upper portion of the profile, while the lower portion is enriched in heavy REE and the medium REE enrichment occasionally occurs in the middle portion.

Table 3. Rare earth elements in the tuff samples collected from the Zhongliangshan Mine (μg/g).

Sample	La	Ce	Pr	Nd	Sm	Eu	Gd	Tb	Dy	Y	Ho	Er	Tm	Yb	Lu
S140SE7-1	266	372	55.7	212	34.3	5.38	30	4.92	28.3	142	5.19	15.7	2.38	13.2	1.88
S140SE7-2	180	314	41.8	174	32.2	4.79	25.9	4.22	22.9	111	4.79	12	2.08	10.8	1.67
S140SE7-3	134	313	26.8	93.9	16.9	2.99	15.4	2.94	16.2	76.8	2.99	8.08	1.23	8.02	1.25
S140SE7-4	103	224	17.8	61.1	8.7	1.83	10.4	1.87	10.9	52.7	2.11	5.66	0.929	5.45	0.897
S140SE7-5	74.6	152	12.8	48.1	8.55	1.79	8.32	1.76	9.73	46.7	1.81	5.28	0.843	5.4	0.837
S140SE7-6	91.5	201	18.2	66.4	13.7	2.58	11.2	1.97	9.95	49.5	1.73	5.21	0.796	5.23	0.763
S140SE7-7	95.7	237	16.5	52.9	9.71	2.09	9.78	1.93	9.99	56.6	2.02	5.97	1.03	6.63	0.996
S140SE7-8	108	265	17.6	59.6	10.7	2.05	10.2	2.03	11.1	59.1	2.15	6.5	1.21	7.37	0.986
S140SE7-9	111	338	19.8	67.4	11.9	2.5	11.7	2.24	11.8	62.2	2.27	7.23	1.22	7.69	1.19
S140SE7-10	77.7	293	22.9	107	29	5.19	18.7	2.42	10.9	50	1.87	5.5	0.895	5.65	0.866
S140SE7-11	103	297	22.3	79.7	14.7	2.47	11.4	2.23	12.4	58.3	2.42	7.58	1.28	8.05	1.23
S140SE7-12	106	326	24.4	87.8	15.7	2.91	12.5	2.33	13.6	62.5	2.65	8.38	1.42	9.21	1.43
S140SE7-13	76.5	251	17.7	60.4	10.4	2.46	9.64	2.04	12	52.7	2.26	6.79	1.18	7.91	1.12
S140SE7-14	18.9	101	9.12	45.4	12.7	2.56	7.86	1.23	5.44	22.9	0.907	2.45	0.367	2.45	0.326
S140SE7-15	57.3	184	13	47.7	9	2.22	8.96	1.98	10.7	53.5	2.16	6.45	1.13	7.54	1.04
S140SE7-16	46.2	163	10.7	40.3	9.95	2.44	9.47	2.13	12.2	54.7	2.17	6.58	1.12	7.07	1.06
S140SE7-17	67.2	222	15.4	55.7	9.3	2.59	9.83	2.01	11.9	50.8	2.2	6.38	1.13	6.99	1.08
S140SE7-18	61.8	201	13.7	50.2	9.63	2.64	9.67	1.96	10.5	48.9	2.09	6	1.01	6.68	1.04
S140SE7-19	50.7	158	11.1	37.9	8.35	2.43	7.86	1.45	8.4	41.1	1.56	4.65	0.799	5.21	0.714
S140SE7-20	67.3	241	14.3	49.6	10.3	3.03	10.7	1.73	10.1	46.8	1.86	5.59	0.939	6.37	0.913
S140SE7-21	56.1	197	12.2	43.3	8.68	2.79	8.27	1.46	8.61	40	1.67	4.98	0.817	5.35	0.827
Average	93.0	240	19.7	73.4	14.0	2.84	12.3	2.23	12.3	58.99	2.33	6.81	1.13	7.06	1.05

Table 4. Rare earth elements geochemical parameters of Zhongliangshan tuff.

Sample	REE (μg/g)	LREE (μg/g)	MREE (μg/g)	HREE (μg/g)	L/M	L/H	M/H	$(La/Lu)_N$	$(La/Sm)_N$	$(Gd/Lu)_N$	δCe	δEu
S140SE7-1	1189	940	211	38.4	4.46	24.51	5.49	1.41	1.16	1.26	0.70	0.77
S140SE7-2	942	742	169	31.3	4.40	23.68	5.39	1.08	0.84	1.22	0.83	0.76
S140SE7-3	721	585	114	21.6	5.11	27.10	5.30	1.07	1.19	0.97	1.19	0.85
S140SE7-4	507	415	77.7	15.0	5.34	27.56	5.16	1.15	1.78	0.92	1.19	0.88
S140SE7-5	379	296	68.3	14.2	4.33	20.89	4.82	0.89	1.31	0.78	1.12	0.98
S140SE7-6	480	391	75.2	13.7	5.20	28.47	5.48	1.20	1.00	1.16	1.12	0.96
S140SE7-7	509	412	80.4	16.6	5.12	24.74	4.83	0.96	1.48	0.78	1.36	0.99
S140SE7-8	564	461	84.5	18.2	5.46	25.30	4.64	1.10	1.51	0.82	1.39	0.90
S140SE7-9	658	548	90.4	19.6	6.06	27.96	4.61	0.93	1.40	0.78	1.64	0.97
S140SE7-10	632	530	87.2	14.8	6.07	35.83	5.90	0.90	0.40	1.70	1.58	1.02
S140SE7-11	624	517	86.8	20.6	5.95	25.13	4.22	0.84	1.05	0.73	1.41	0.88
S140SE7-12	677	560	93.8	23.1	5.97	24.25	4.06	0.74	1.01	0.69	1.46	0.95
S140SE7-13	514	416	78.8	19.3	5.28	21.60	4.09	0.68	1.10	0.68	1.56	1.13
S140SE7-14	234	187	40.0	6.5	4.68	28.79	6.15	0.58	0.22	1.90	1.75	1.18
S140SE7-15	407	311	77.4	18.3	4.02	16.98	4.22	0.55	0.96	0.68	1.54	1.14
S140SE7-16	369	270	80.9	18.0	3.34	15.01	4.50	0.44	0.70	0.71	1.67	1.15
S140SE7-17	465	370	77.1	17.8	4.79	20.79	4.34	0.62	1.08	0.72	1.57	1.24
S140SE7-18	427	336	73.7	16.8	4.57	20.00	4.38	0.59	0.96	0.73	1.58	1.26
S140SE7-19	340	266	61.2	12.9	4.34	20.57	4.74	0.71	0.91	0.87	1.52	1.38
S140SE7-20	471	383	72.4	15.7	5.29	24.41	4.62	0.74	0.98	0.93	1.77	1.33
S140SE7-21	392	317	61.1	13.6	5.19	23.25	4.48	0.68	0.97	0.79	1.72	1.51
Average	548	441	88.6	18.4	5.00	24.13	4.83	0.85	1.05	0.94	1.41	1.06

ΣREE, sum of La, Ce, Pr, Nd, Sm, Eu, Gd, Tb, Dy, Y, Ho, Er, Tm, Yb, and Lu; LREE, sum of La, Ce, Pr, Nd, and Sm; MREE, sum of Eu, Gd, Tb, Dy and Y; HREE, sum of Ho, Er, Tm, Yb, and Lu; L/M, ratio of LREE and MREE; L/H, ratio of LREE and HREE; M/H, ratio of MREE and HREE; $(La/Lu)_N$, ratio of $(La)_N$ and $(Lu)_N$; $(La/Sm)_N$, ratio of $(La)_N$ and $(Sm)_N$; $(Gd/Lu)_N$, ratio of $(Gd)_N$ and $(Lu)_N$; $\delta Ce = Ce_N/(La_N \times Pr_N)^{1/2}$; $\delta Eu = Eu_N/(Sm_N \times Gd_N)^{1/2}$; N, REE are normalized by Upper Continental Crust (UCC) [26].

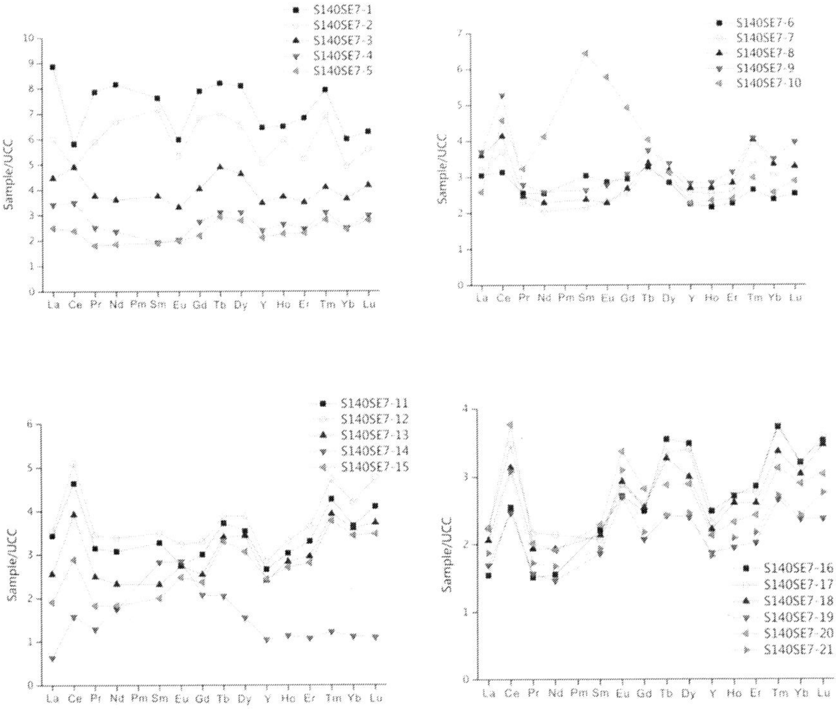

Figure 7. Distribution patterns of REE in the tuff samples from Zhongliangshan mine. REE are normalized by Upper Continental Crust [26].

The Ce-anomaly (expressed as δCe) values vary from 0.70 to 1.77, with an average of 1.41, indicating a well-pronounced Ce positive anomaly. The REE distribution patterns of the tuff display positive Ce anomalies, owing to the in-situ precipitation of Ce^{4+} in the process of weathering, leaching, and eluviation [35]. The Eu-anomaly (δEu) values varying from 0.76 to 1.51, with an average of 1.06, show a slight Eu positive anomaly, indicating the tuff and the Emeishan basalt have the same origin [35]. From top to bottom, δCe and δEu markedly increase. The distribution of REE of the tuff in the Zhongliangshan mine appears as a sawtooth shape, the portions of La-Sm and Gd-Lu occurring gentle and small slope, which indicates that the fractionation of REE is low.

Two reasons may be responsible for the H-REE enrichment of the tuff samples in the Zhongliangshan mine. First, L-REE can be easily leached by groundwater than H-REE; Second, L-REE can be easily adsorbed on the organic matter than the H-REE [37], which may be adsorbed by the coal seam overlying the tuff. The REE enrichment mode in the Zhongliangshan tuff is similar to that of Songzao Coalfield. Some studies have shown that L-REE are more easily to be leached by groundwater and are more apt to be adsorbed by organic matter [38,39,40,41,42].

5. DISCUSSION

5.1. Origin of Tuff

In the late Permian Age, the Dongwu movement, one of the most important tectonic events in southern China, caused the upper Yangtze basin uplifting and the subsequent sea regression, which led to an extensive erosion in the area. The upper part of Maokou limestone of Sichuan Basin had been subjected to a serious erosion, resulting in the formation of a vast weathering residual plain, where peat subsequently accumulated. Meanwhile, the Emeishan basalt volcano began erupting and reached a climax in the early stage of late Permian, leading to a tuff layer overlying the Maokou limestone [17].

Al_2O_3 and TiO_2 are both stable components in the rock and would be little altered during alteration, so the ratio of TiO_2/Al_2O_3 (KAT) would be kept constant and can frequently be used to study the origin of volcanic ash [29,42]. It is suggested that KAT values for silicic volcanic ash are <0.02, and those for mafic and alkali volcanic ashes are >0.08 and between 0.02 and 0.08, respectively [43,44]. The KAT ratios of the tuff in the Zhongliangshan mine are >0.08 (Figure 8), suggesting a mafic volcanic origin. In addition, the tuff samples fall in the area basalt to alkali basalt from the La/Yb-REE diagram (Figure 9), indicating a feature of alkali basalt.

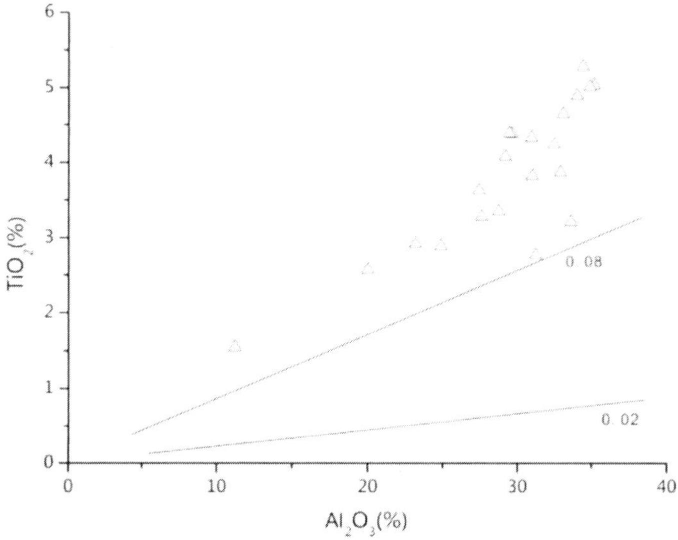

Figure 8. Plot for TiO_2 *vs.* Al_2O_3 of tuff samples in the Zhongliangshan mine, Chongqing.

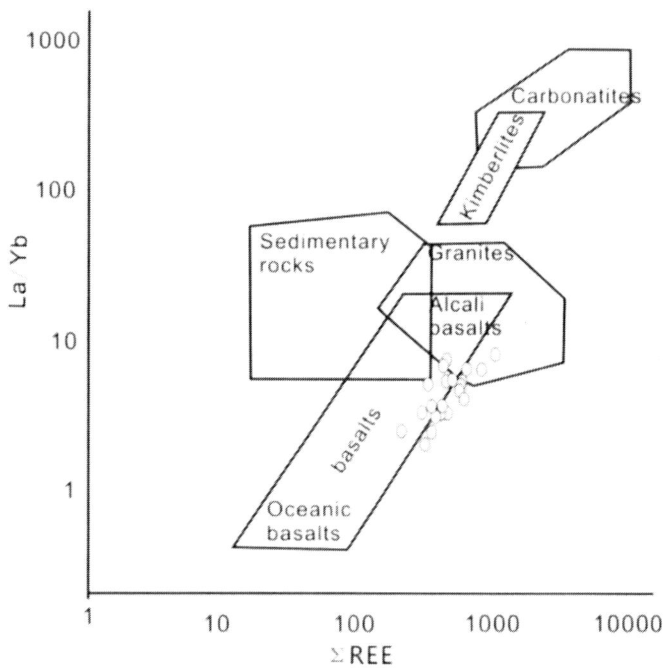

Figure 9. Relation between REE and La/Yb of tuff samples in the Zhongliangshan mine, Chongqing.

5.2. Hydrothermal Solution

Some researchers have shown that there have been activities of low-temperature hydrothermal solutions in the late Permian Age in southwestern China, which resulted in enrichment of trace elements and minerals in some coal [1,45,46,47,48,49,50,51]. Similarly, some minerals of tuff in Zhongliangshan are formed owing to the influence of hydrothermal solution.

In addition to the derivation from volcanic ash, anatase and zircon might have been derived from hydrothermal alteration in the Zhongliangshan tuffs. Anatase of various particle sizes is distributed in the kaolinite matrix (Figure 4B,C). Figure 4A and Figure 5A illustrate that part of anatase could be formed by hydrothermal alteration. Zircon from Figure 5D displays long axis and could be formed by the effect of hydrothermal alteration. Zircon in Figure 5D exclusively contains Zr, Si and O determined by the SEM-EDX. Finkelman [52] has demonstrated that Hf, Th, U, Y and HREE occur in the volcanogenic zircon, but were not identified in authigenic ziron, in accordance with the results of the Zhongliangshan tuff samples.

Florencite, the main carrier of REE in the Zhongliangshan tuff samples, occurs as ellipsoidal in the kaolinite matrix (Figure 5E,F), indicating a syngenetic or early diagenetic hydrothermal origin. Dai *et al.* [14] have also demonstrated that florencite is one of the important carriers of REE in the late Permian coals in southwestern China [3].

Gypsum (Figure 5B) and barite (Figure 4F) occur as crack-fillings, the former occurring as radiating and the latter on the edge of jarosite, indicating an epigenetic hydrothermal origin.

5.3. Preliminary Evaluation of Rare Metals

Coal and coal-bearing strata have recently become alternative sources for recovery of rare metals [2,3,7,8]. The U.S. Department of Energy's National Energy Technology Laboratory has selected 10 projects to receive funding for research in support of the lab's program on recovery of rare earth elements from coal and coal byproducts since 2015 [2,53].

Based on the Chinese industry standards [33], the required $(Nb,Ta)_2O_5$ concentrations for marginal and industrial grade Nb(Ta) ore deposits of weathered crust type are 80–100 and 160–200 μg/g, respectively; equivalent concentrations are 40–60 and 100–120 μg/g for Nb(Ta) ore deposits of river placer type. The concentration of $(Nb,Ta)_2O_5$ varies from 47 to 324 μg/g, with an average of 186 μg/g, higher than the marginal and industrial grade for weathered crust and placer deposit types. Concentration of TiO_2 varies from 1.55% to 5.28% and averages 3.82%, higher than the industrial grade of Chinese industry standard [54]. The average concentration of Ga (38.2 μg/g) is also up to the standards for industrial utilization in bauxite (20 μg/g) and coal mining (30 μg/g) [34]. In addition, the concentrations of REE vary from 234 to 1189 μg/g and averages 548 μg/g, higher than the cut-off grade of Chinese weathering crust ion adsorption type rare earth elements deposits (500 μg/g) [55].

The Nb, Ti, Ga, and REE all exceed their respective industrial grade of China in the tuff of the Zhongliangshan mine. It is considered that the Zhongliangshan tuff is a potential polymetallic ore worth in-depth study.

6. CONCLUSIONS

Compared with the Upper Continental Crust, some trace elements including Li, Be, Sc, V, Cr, Co, Ni, Cu, Zn, Ga, Zr, Nb, Cd, Sb, REE, Hf, Ta, Re, Th, and U are enriched in tuff from Zhongliangshan mine, Chongqing, southwestern China. The minerals mainly include kaolinite, illite, pyrite, anatase, calcite, gypsum, quartz, and traces of minerals such as zircon, florencite, jarosite, and barite. The tuff is of mafic volcanic origin with features of alkali basalt. The H-REE enriched in the tuff due to L-REE being leached easier by groundwater and adsorbed in the organic matter of the coal seam overlying the tuff. Some minerals including florencite, gypsum, barite, and a portion of anatase and zircon are precipitated from hydrothermal solution. It is suggested that Zhongliangshan tuff is a potential polymetallic ore and the opportunity for recovery of these valuable elements needs to be studied in depth.

ACKNOWLEDGMENTS

The authors wish to express their appreciation to Shifeng Dai for revision suggestions and English polishing. We thank Lei Zhao and Lixin Zhao, who helped to identify the minerals under SEM-EDX. We also thank Peipei Wang for the mineral quantitative analysis using Siroquant software. The authors are indebted to three anonymous reviewers for their careful reviews and constructive comments, which greatly improved the manuscript. This research was supported by the National Key Basic Research and Development Program (No.2014CB238902) and National Natural Science Foundation of China (No. 41502162).

AUTHOR CONTRIBUTIONS

Jianhua Zou and Heming Tian collected tuff samples in Zhongliangshan mine. Jianhua Zou conducted the determinations of major-element contents. Tian Li and Heming Tian were responsible for the analysis of trace-element concentrations. Jianhua Zou and Tian Li were responsible for the mineralogy investigation using XRD and SEM-EDX.

CONFLICTS OF INTEREST

The authors declare no conflict of interest.

REFERENCES

1. Dai, S.F.; Ren, D.Y.; Chou, C.-L.; Finkelman, R.B.; Seredin, V.V.; Zhou, Y.P. Geochemistry of trace elements in Chinese coals: A review of abundances, genetic types, impacts on human health, and industrial utilization. Int. J. Coal Geol. 2012, 94, 3–21.
2. Hower, J.C.; Granite, E.J.; Mayfield, D.B.; Lewis, A.S.; Finkelman, R.B. Notes on Contributions to the Science of Rare Earth Element Enrichment in Coal and Coal Combustion Byproducts. Minerals 2016, 6, 32.
3. Dai, S.F.; Chekryzhov, I.Y.; Seredin, V.V.; Nechaev, V.P.; Graham, I.T.; Hower, J.C.; Ward, C.R.; Ren, D.Y.; Wang, X.B. Metalliferous coal deposits in East Asia (Primorye of Russia and South China): A review of geodynamic controls and styles of mineralization. Gondwana Res.2016, 29, 60–82.
4. Dai, S.F.; Wang, P.P.; Ward, C.R.; Tang, Y.G.; Song, X.L.; Jiang, J.H.; Hower, J.C.; Li, T.; Seredin, V.V.; Wagner, N.J.; et al. Elemental and mineralogical anomalies in the coal-hosted Ge ore deposit of Lincang, Yunnan, southwestern China: Key role of N2-CO2-mixed hydrothermal solutions. Int. J. Coal Geol. 2015, 152, 19–46.
5. Johnston, M.N.; Hower, J.C.; Dai, S.; Wang, P.; Xie, P.; Liu, J. Petrology and Geochemistry of the Harlan, Kellioka, and Darby Coals from the Louellen 7.5-Minute Quadrangle, Harlan County, Kentucky. Minerals 2015, 5, 894–918.

6. Seredin, V.V.; Finkelman, R.B. Metalliferous coals: A review of the main genetic and geochemical types. Int. J. Coal Geol. 2008, 76, 253–289.

7. Seredin, V.V.; Dai, S.F. Coal deposits as potential alternative sources for lanthanides and yttrium. Int. J. Coal Geol. 2012, 94, 67–93.

8. Seredin, V.V.; Dai, S.F.; Sun, Y.Z.; Chekryzhov, I.Y. Coal deposits as promising sources of rare metals for alternative power and energy-efficient technologies. Appl. Geochem. 2013, 31, 1–11.

9. Dai, S.F.; Seredin, V.V.; Ward, C.R.; Jiang, J.H.; Hower, J.C.; Song, X.L.; Jiang, Y.F.; Wang, X.B.; Gornostaeva, T.; Li, X.; et al. Composition and modes of occurrence of minerals and elements in coal combustion products derived from high-Ge coals. Int. J. Coal Geol. 2014, 121, 79–97.

10. Seredin, V.V. From coal science to metal production and environmental protection: A new story of success. Int. J. Coal Geol. 2012, 90–91, 1–3.

11. Dai, S.F.; Zhou, Y.P.; Zhang, M.Q.; Wang, X.B.; Wang, J.M.; Song, X.L.; Jiang, Y.F.; Luo, Y.B.; Song, Z.T.; Yang, Z.; et al. A new type of Nb (Ta)–Zr(Hf)–REE–Ga polymetallic deposit in the late Permian coal-bearing strata, eastern Yunnan, southwestern China: Possible economic significance and genetic implications. Int. J. Coal Geol. 2010, 83, 55–63.

12. Dai, S.F.; Seredin, V.V.; Ward, C.R.; Hower, J.C.; Xing, Y.W.; Zhang, W.G.; Song, W.J.; Wang, P.P. Enrichment of U-Se-Mo-Re-V in coals preserved within marine carbonate successions: Geochemical and mineralogical data from the Late Permian Guiding Coalfield, Guizhou, China. Miner. Deposita 2015, 50, 159–186.

13. Hower, J.C.; Eble, C.F.; O'Keefe, J.M.K.; Dai, S.F.; Wang, P.P.; Xie, P.P.; Liu, J.J.; Ward, C.R.; French, D. Petrology, Palynology, and Geochemistry of Gray Hawk Coal (Early Pennsylvanian, Langsettian) in Eastern Kentucky, USA. Minerals 2015, 5, 592–622.

14. Dai, S.F.; Zhao, L.; Hower, J.C.; Johnston, M.N.; Song, W.J.; Wang, P.P.; Zhang, S.F. Petrology, mineralogy, and chemistry of size-fractioned fly ash from the Jungar power plant, Inner Mongolia, China, with emphasis on the distribution of rare earth elements. Energy Fuels 2014, 28, 1502–1514.

15. Zhuang, X.G.; Su, S.C.; Xiao, M.G.; Li, J.; Alastuey, A.; Querol, X. Mineralogy and geochemistry of the Late Permian coals in the Huayingshan coal-bearing area, Sichuan Province, China. Int. J. Coal Geol. 2012, 94, 271–282.

16. Dai, S.F.; Yang, J.Y.; Ward, C.R.; Hower, J.C.; Liu, H.D.; Garrison, T.M.; French, D.; O'Keefe, J.M.K. Geochemical and mineralogical evidence for a coal-hosted uranium deposit in the Yili Basin, Xinjiang, northwestern China. Ore Geol. Rev. 2015, 70, 1–30.

17. China Coal Geology Bureau. Sedimentary Environments and Coal Accumulation of Late Permian Coal Formation in Western Guizhou, Southern Sichuan and Eastern Yunnan, China; Chongqing University Press: Chongqing, China, 1996. (In Chinese).

18. Dai, S.F.; Liu, J.J.; Ward, C.R.; Hower, J.C.; French, D.; Jia, S.H.; Hood, M.M.; Garrison, T.M. Mineralogical and geochemical compositions of Late Permian coals and host rocks from the Guxu Coalfield, Sichuan Province, China, with emphasis on enrichment of rare metals. Int. J. Coal Geol. 2016.

19. Standardization Administration of the People's Republic of China. Sampling of Coal Seams; Chinese Standard GB/T 482–2008. Standardization Administration of the People's Republic of China: Beijing, China, 2008. (In Chinese).

20. ASTM International. Test Method for Ash in the Analysis Sample of Coal and Coke from Coal; ASTM D3174–11; ASTM International: West Conshohocken, PA, USA, 2011.

21. Taylor, J.C. Computer programs for standardless quantitative analysis of minerals using the full powder diffraction profile. Powder Diffr. 1991, 6, 2–9.

22. Rietveld, H.M. A profile refinement method for nuclear and magnetic structures. Appl. Crystallogr. 1969, 2, 65–71.

23. Ward, C.R.; Spears, D.A.; Booth, C.A.; Staton, I.; Gurba, L.W. Mineral matter and trace elements in coals of the Gunnedah Basin, New South Wales, Australia. Int. J. Coal Geol. 1999,40, 281–308.

24. Ward, C.R.; Matulis, C.E.; Taylor, J.C.; Dale, L.S. Quantification of mineral matter in the Argonne Premium coals using interactive Rietveld-based X-ray diffraction. Int. J. Coal Geol.2001, 46, 67–82.

25. Ruan, C.D.; Ward, C.R. Quantitative X-ray powder diffraction analysis of clay minerals in Australian coals using Rietveldmethods. Appl. Clay Sci. 2002, 21, 227–240.

26. Taylor, S.R.; McLennan, S.M. The Continental Crust: Its Composition and Evolution; Blackwell: Oxford, UK, 1985; p. 312.

27. Chi, Q.H.; Yan, M.C. Handbook of Elemental Abundance for Applied Geochemistry; Geological Publishing House: Beijing, China, 2007; pp. 1–148. (In Chinese)

28. Zhou, Y.P.; Ren, Y.L. Element gochemistry of volcanic ash derived tonsteins in late Permian coal-bearing formation of eastern Yunnan and western Guizhou, China. Acta Sedimentol Sin.1994, 12, 123–132. (In Chinese)

29. Zhou, Y.P.; Bohor, B.F.; Ren, Y.L. Trace element geochemistry of altered volcanic ash layers (tonsteins) in late Permian coal-bearing formations of eastern Yunnan and western Guizhou provinces, China. Int. J. Coal Geol. 2000, 44, 305–324.

30. Sichuan Bureau of Coal Geology; Sichuan Institute of Coal Geology. Sedimentary Environment and Coal Accumulating Regulations of Late Permian Coal-Bearing Formation in Southern Sichuan; Guizhou Sceinece and Technology Press: Guiyang, China, 1994. (In Chinese)

31. Dai, S.F.; Chou, C.-L.; Yue, M.; Luo, K.L.; Ren, D.Y. Mineralogy and geochemistry of a Late Permian coal in the Dafang Coalfield, Guizhou, China: Influence from siliceous and iron-rich calcic hydrothermal fluids. Int. J. Coal Geol. 2005, 61, 241–258.

32. Dai, S.F.; Ren, D.Y.; Zhou, Y.P.; Chou, C.-L.; Wang, X.B.; Zhao, L.; Zhu, X.W. Mineralogy and geochemistry of a superhigh-organic-sulfur coal, Yanshan Coalfield, Yunnan, China: Evidence for a volcanic ash component and influence by submarine exhalation. Chem. Geol.2008, 255, 182–194.

33. DZ/T 0203–2002. Geology Mineral Industry Standard of P.R. China: Specifications for Rare Metal Mineral Exploration; Geological Press: Beijing, China, 2002. (In Chinese)

34. Mineral Resources Industry Requirements Manual Editorial Board. Mineral Resources Industry Requirements Manual; Geological Press: Beijing, China, 2010. (In Chinese)

35. Dai, S.F.; Graham, I.T.; Chou, C.-L.; Ward, C.R. A review of anomalous rare earth elements and yttrium in coal. Int. J. Coal Geol. 2016, 159, 82–95.

36. Bao, Z.W.; Zhao, Z.H. Geochemistry of mineralization with exchangeable REY in the weathering crusts of granitic rocks in South China. Ore Geol. Rev. 2008, 33, 519–535.

37. Dai, S.F.; Li, D.; Chou, C.-L.; Zhao, L.; Zhang, Y.; Ren, D.Y.; Ma, Y.W.; Sun, Y.Y. Mineralogy and geochemistry of boehmite-rich coals: New insights from the Haerwusu Surface Mine, Jungar Coalfield, Inner Mongolia, China. Int. J. Coal Geol. 2008, 74, 185–202.

38. Eskenazy, G.M. Rare earth elements in a sampled coal from the Pirin Deposit, Bulgaria. Int. J. Coal Geol. 1987, 7, 301–314.

39. Crowley, S.S.; Stanton, R.W.; Ryer, T.A. The effects of volcanic ash on the maceral and chemical composition of the C coal bed, Emery Coal Field, Utah. Org. Geochem. 1989, 14, 315–331.

40. Hower, J.C.; Ruppert, L.F.; Eble, C.F. Lanthanide, yttrium, and zirconium anomalies in the fire Clay coal bed, Eastern Kentucky. Int. J. Coal Geol. 1999, 39, 141–153.

41. Dai, S.F.; Ren, D.Y.; Chou, C.-L.; Li, S.S.; Jiang, Y.F. Mineralogy and geochemistry of the No. 6 coal (Pennsylvanian) in the Jungar Coalfield, Ordos Basin, China. Int. J. Coal Geol. 2006, 66, 253–270.

42. Dai, S.F.; Li, T.J.; Jiang, Y.F.; Ward, C.R.; Hower, J.C.; Sun, J.H.; Liu, J.J.; Song, H.J.; Wei, J.P.; Li, Q.Q.; et al. Mineralogical and geochemical compositions of the Pennsylvanian coal in the Hailiushu Mine, Daqingshan Coalfield, Inner Mongolia, China: Implications of sediment-source region and acid hydrothermal solutions. Int. J. Coal Geol. 2015, 137, 92–110.

43. Addison, R.; Harrison, R.K.; Land, D.H.; Young, B.R.; Davis, A.E.; Smith, T.K. Volcanogenic tonsteins from tertiary coal measures, East Kalimantan, Indonesia. Int. J. Coal Geol. 1983, 3, 1–30.

44. Burger, K.; Zhou, Y.P.; Ren, D.Y. Petrography and geochemistry of tonsteins from the 4th Member of the Upper Triassic Xujiahe Formation in southern Sichuan Province, China. Int. J. Coal Geol. 2002, 49, 1–17.

45. Dai, S.F.; Zhang, W.G.; Seredin, V.V.; Ward, C.R.; Hower, J.C.; Song, W.J.; Wang, X.B.; Li, X.; Zhao, L.; Kang, H.; et al. Factors controlling geochemical and mineralogical compositions of coals preserved within marine carbonate successions: A case study from the Heshan Coalfield, southern China. Int. J. Coal Geol. 2013, 109–110, 77–100.

46. Dai, S.F.; Zhang, W.G.; Ward, C.R.; Seredin, V.V.; Hower, J.C.; Li, X.; Song, W.J.; Wang, X.B.; Kang, H.; Zheng, L.C.; et al. Mineralogical and geochemical anomalies of Late Permian coals from the Fusui Coalfield, Guangxi Province, southern China: Influences of terrigenous materials and hydrothermal fluids. Int. J. Coal Geol. 2013, 105, 60–84.

47. Ren, D.Y.; Zhao, F.H.; Dai, S.F.; Zhang, J.Y.; Luo, K.L. Geochemistry of Trace Elements in Coal; Science Press: Beijing, China, 2006. (In Chinese)

48. Zhou, Y.P.; Ren, Y.L. Distribution of arsenic in coals of Yunnan Province, China, and its controlling factors. Int. J. Coal Geol. 1992, 20, 85–98.

49. Dai, S.F.; Chou, C.-L. Occurrence and origin of minerals in a chamosite-bearing coal of Late Permian age, Zhaotong, Yunnan, China. Am. Mineral. 2007, 92, 1253–1261.

50. Wang, X.B.; Dai, S.F.; Chou, C.-L.; Zhang, M.Q.; Wang, J.M.; Song, X.L.; Wang, W.; Jiang, Y.F.; Zhou, Y.P.; Ren, D.Y. Mineralogy and geochemistry of Late Permian coals from the Taoshuping Mine, Yunnan Province, China: Evidences for the sources of minerals. Int. J. Coal Geol. 2012, 96–97, 49–59.

51. Dai, S.F.; Tian, L.W.; Chou, C.-L.; Zhou, Y.P.; Zhang, M.Q.; Zhao, L.; Wang, J.M.; Yang, Z.; Cao, H.Z.; Ren, D.Y. Mineralogical and compositional characteristics of Late Permian coals from an area of high lung cancer rate in Xuan Wei, Yunnan, China: Occurrence and origin of quartz and chamosite. Int. J. Coal Geol. 2008, 76, 318–327.

52. Finkelman, R.B. Modes of Occurrence of Trace Elements in Coal; US Geological Survey Open-File Report; United States Geological Survey: Reston, VA, USA, 1981; No. 81–99; p. 322.

53. The U.S. Department of Energy. Available online: http://www.energy.gov/fe/articles/doe-selects-projects-enhance-its's-research-recovery-rare-earth-elements-coal-and-coal.2015(accessed on 2 December 2015).

54. DZ/T 0208–2002. Geology Mineral Industry Standard of P.R. China: Specifications for Placer (Metallic Mineral) Exploration; Geological Press: Beijing, China, 2002. (In Chinese)

55. DZ/T 0204–2002. Geology Mineral Industry Standard of P.R. China: Specifications for Rare Earth Mineral Exploration; Geological Press: Beijing, China, 2002. (In Chinese)

10

CHAPTER

Morphology Engineering of Porous Media for Enhanced Solar Fuel and Power Production

Silvan Suter[1] and Sophia Haussener[1]

[1]*Institute of Mechanical Engineering, E´ cole Polytechnique Fe´de´rale de Lausanne, 1015 Lausanne, Switzerland*

ABSTRACT

The favorable and adjustable transport properties of porous media make them suitable components in reactors used for solar energy conversion and storage processes. The directed engineering of the porous media's morphology can significantly improve the performance of these reactors. We used a multiscale approach to characterize the changes in performance of exemplary solar fuel processing and solar power production reactors incorporating porous media as multifunctional components. The method applied uses imaging-based direct numerical simulations and digital image processing in combination with volume averaging theory to characterize the transport in porous media. Two samples with varying morphology (fibrous vs. foam) and varying size range (mm vs. μm scale), each with porosity between 0.46 and 0.84, were characterized. The obtained effective transport properties were used in continuum-scale models to quantify the performance of reactors incorporating multifunctional porous media for solar fuel processing by photoelectrochemical water splitting or power production by solar thermal processes.

INTRODUCTION

The direct conversion of solar energy into a storable, high-energy density fuel via solar thermochemistry or photoelectrochemistry and the solar production of power via solar thermal processes are promising renewable fuel processing and power production routes. The essential requirements for the processes' impact on our fuel and power economy are their sustainable, efficient, stable, and economic implementation via solar reactors and their assembly into practical large-scale systems.1–4 The engineering of solar reactors needs to address various issues including optimal design and operational conditions for enhanced coupled multiphysics transport and the integration and optimization of multiscale components, e.g., porous media. The latter is of special interest as porous media exhibit favorable and tunable transport properties. Particularly interesting is the observed fourfold increase in efficiency of a solar reactor for the thermochemical splitting of water and CO_2 into synthesis gas when changing the porous absorber and reactant morphology from a sintered backed bed to a highly porous foam morphology.5,6 Similar influences on performance are expected for solar receivers used for thermal power production, which rely on porous absorber and heat exchangers of various morphologies,7,8 or for photoelectrochemical fuel production devices relying on microstructured to nanostructured photoelectrodes9–13 or separators.14

The systematic understanding of how the transport properties of porous media change for distinct base morphologies (foam or fibrous structure) but varying morphological characteristics, i.e., varying porosity, is of practical interest due to the straightforward synthesis of such porous media. Systematic direct numerical studies on how morphology and its tailoring can influence the transport properties have been conducted for micron-sized sintered packed beds,15 reticulate porous ceramics and packed beds,16,17 and foams.18

We used a multiscale approach to quantify the changes in the performance of solar reactors incorporating porous media with varying morphologies (foams and fibers) and morphological properties (characteristic size and porosity). This multiscale approach consisted of (1) the transport quantification in porous structures by applying tomography-based approaches that incorporate the exact morphology into pore-scale numerical simulations19–22 in combination with image processing techniques,18 and (2) the subsequent use of the transport characteristics in continuum-scale models of reactors processing fuel via photoelectrochemical processes or power via solar thermal processes.

METHODOLOGY

Multiscale Approach

Continuum-Scale Models of Reactors Incorporating Multifunctional Porous Media

Photoelectrochemistry is a promising direct solar-to-fuel processing route. It uses the photon energy to generate electron–hole pairs in a semiconductor absorber. The electron–hole pairs are separated via an electric field and used at liquid–solid interfaces to drive catalyzed electrochemical reactions such as water electrolysis described by

$$2H_2O + 4h^+ \rightarrow 4H^+ + O_2, \tag{1}$$

$$4H^+ + 4e^- \rightarrow 2H_2, \tag{2}$$

in an acidic environment. Ions are transported through an ion-conducting phase to conserve the charge. This is schematically shown in Fig. 1a for a solar electrochemical reactor incorporating a multifunctional porous slab.

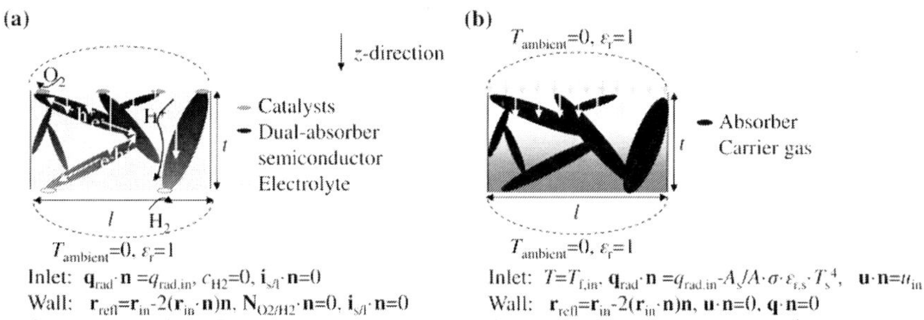

Figure 1. Schematics of the 1-D slabs of multifunctional porous media to be used for solar (a) electrochemical fuel and (b) thermal power processing, and the corresponding boundary conditions for the continuum-scale models

Solar thermal processes for renewable power production use the solar energy as sensible heat to heat a (pressurized) heat transfer fluid, which is used in subsequent power cycles.8This is schematically shown in Fig. 1b for a solar receiver incorporating a multifunctional porous slab.

Radiative Heat Transfer

The radiative transport in porous media, composed of a transparent and an opaque phase, was given by the volume-averaged radiative transfer equation23

$$\mathbf{s} \cdot \nabla I = -\beta_e I + \varkappa_e I_b + \frac{\sigma_{scat,e}}{4\pi} \int_0^{4\pi} \Phi_{scat,e} I \mathrm{d}\Omega \tag{3}$$

β_e, $\sigma_{scat,e}$, and $\Phi_{scat,e}$ were the effective extinction and scattering coefficients, and scattering phase function, to be determined by direct numerical simulations. Collimated solar irradiation with 1.5 AM spectral distribution was irradiating the boundary at the top of the slab. The lateral walls were assumed perfectly specular; the inlet and outlet ambient boundaries were black surrounding at 0 K. The reflected R and the transmitted T fraction of the incident radiation were calculated by integrating the fluxes over the inlet and outlet boundary. The absorbed radiation, $A = 1 - R - T$, contributed to the divergence of the radiative flux. We used a path-length-based Monte Carlo method for solving the volume-averaged radiative transfer equations.24

Photoelectrochemical Fuel Processing

The solar-to-fuel efficiency of a photoelectrochemical fuel processing reactor incorporating multifunctional porous medium.25

$$\eta = \frac{iE_0}{Q_{sol,in}} \frac{i - nFD_{H_2,e}c_{sat}/t}{i} \tag{4}$$

was calculated using equivalent circuit models.26,27 i represented the superficial current density of the cell, which was calculated as the intersection between the load curve of the cell,

$$\Phi = E_0 + \eta_a + |\eta_c| + i\frac{t}{\kappa_e} + i\frac{t}{\sigma_e} \tag{5}$$

and the power curve of the dual absorber material,

$$i = i_{ph} - i_0\left(\exp\left(\frac{q\Phi}{kT}\right) - 1\right) \tag{6}$$

The parameters in Eq. 6 were calculated for the absorbed fraction of the irradiation (Eq. 3) using the Shockley-Queisser limit28 for a dual-absorber tandem porous medium. This approach neglected possible bulk and surface recombination mechanism in the absorber. The effective ionic and electronic conductivities, κ_e and σ_e, were calculated via the direct numerical simulation technique, and the reaction overpotentials at the anode and cathode, η_a and η_c, were approximated by Tafel expressions assuming that reaction occurred at the top and bottom solid phase boundary. The second fraction on the right side of Eq. 4described the fraction of the current density that was not lost due to product crossover via diffusion through the membrane. The effective hydrogen diffusivity was calculated via direct numerical simulations and the worst-case scenario for crossover was assumed, i.e., the hydrogen evolution side completely saturated with hydrogen and the oxygen evolution side completely hydrogen starved.14

Solar Thermal Power Production

We used a quasi-one-dimensional (1-D) model to solve the steady-state mass, momentum, and energy conservation equations at local thermal nonequilibrium in the solar thermal power production reactor incorporating a multifunctional porous media. The volume-averaged energy conservation equations were given by

$$0 = \nabla \cdot \left(k_{\mathrm{e,s}} \nabla T_{\mathrm{s}}\right) - h_{\mathrm{sf}} A_0 (T_{\mathrm{s}} - T_{\mathrm{f}}) + q_{\mathrm{r}} \tag{7}$$

$$\rho c_p \nabla \cdot (\mathbf{u} T_{\mathrm{f}}) = \nabla \cdot \left(k_{\mathrm{e,f}} \nabla T_{\mathrm{f}}\right) + h_{\mathrm{sf}} A_0 T_{\mathrm{f}} (T_{\mathrm{s}} - T_{\mathrm{f}}) \tag{8}$$

where the temperature variables represented intrinsic averages. It was assumed that only the solid fraction of the inlet and outlet boundary (A_s) reradiated, describing the boundary condition for the inlet and outlet boundaries

$$Q_{\mathrm{rerad}} = (1 - r) A_{\mathrm{s}} \sigma T_{\mathrm{s}}^4 \tag{9}$$

where r was the directional-hemispherical reflectivity of the material. The lateral walls were symmetric boundaries. The carrier gas was modeled as an ideal gas with temperature-dependent properties. The solar-to-power efficiency of the slab of multifunctional porous medium was described by

$$\eta = \frac{\dot{m}(h_{\text{outlet}} - h_{\text{intlet}})}{Q_{\text{sol,in}} + W_{\text{pump}}} \eta_{\text{Carnot},T_{\text{f}}|_{z=t}} \tag{10}$$

The pumping work was given by the volume-averaged momentum conservation29

$$W_{\text{pump}} = \Delta p u_{\text{f}} A = \left(\frac{\mu_{\text{f}}}{K} u_{\text{f}} + \rho_{\text{f}} F u_{\text{f}}^2\right) t u_{\text{f}} A \tag{11}$$

with u_{f} the superficial velocity average. The effective transport properties permeability $K$and the Forchheimer coefficient F were determined via the direct numerical simulations. An inlet mass flow and an outlet pressure were set at the corresponding boundaries.

Pore-Scale Transport Characterization

Radiative Heat Transfer

β_{e}, $\sigma_{\text{scat,e}}$, and $\Phi_{\text{scat,e}}$ were calculated by solving the pore-scale radiative heat transfer equations. Radiative distribution functions were used to determine the effective radiative properties30 using a collision-based Monte Carlo technique. An opaque solid phase and a transparent fluid phase were assumed. The laws of geometrical optics were applicable as $\pi d/\lambda \gg 1$.

Conductive Heat Transfer

$k_{\text{e,s}}$ and $k_{\text{e,f}}$ were calculated by solving the steady-state conduction equations within the void and solid phases20,22,31,32 in a representative cubical sample of porous medium applying predefined inlet and outlet temperatures, insulating lateral wall boundaries, and providing continuity in temperature and heat flux at the solid–fluid phase boundary. For the orthotropic fibrous sample, the temperature boundary conditions were applied along the z-direction. $k_{\text{e,s}}$ and $k_{\text{e,f}}$ were determined as the asymptotic cases where the heat conductivity of the fluid phase was significantly smaller than the heat conductivity of solid phase ($k_{\text{f}} \ll k_{\text{s}}$) and vice versa ($k_{\text{s}} \ll k_{\text{f}}$).

Convective Heat Transfer

The Nu correlations were calculated by solving the heat, mass, and momentum (incompressible and laminar) conservation equations within the fluid phase of the porous media for an isothermal solid phase. The fluid phase of cubical porous samples within a square duct with inlet and outlet regions was used as computational domain.21,22,31,32Uniform inlet velocity and temperature, constant outlet pressure, symmetry at the lateral duct walls, and no-slip and constant temperature at the solid–fluid interface were assumed. Inlet velocities in the range of 0.0008–0.5 m/s (corresponding to $d_{\text{void,foam}}$-based Re between 0.1 and 100) and Pr numbers of 0.5, 1, and 10 were simulated. For the orthotropic fibrous sample, the inlet and outlet boundaries were applied along the z-direction.

Mass Transfer

K and F were determined by using the previously calculated pressure and velocity vector fields.22 For the orthotropic fibrous sample, the pressure boundary conditions were applied along the z-direction. $D_{\text{H2,e}}$ was calculated by33,34

$$D_{\mathrm{H_2,e}} = D_{\mathrm{H_2}} \frac{\varepsilon}{\tau^2_{\mathrm{mean}}} \tag{12}$$

The mean tortuosity, $\tau_{\text{mean}} = l_{\text{path,mean}}/t$, was numerically determined using the previously calculated velocity vector field and determining the lengths of 3600 streamlines with uniformly distributed starting points at the medium inlet for Re = 0.1.22

Charge transfer

κ_e and σ_e were calculated by solving the charge transport equations, i.e., Ohm's law, in the ionic and electronic phases. The methodology was analogous to the methodology used for the determination of the effective thermal conductivity in porous media.20,22 Again, the asymptotic cases were considered, i.e., κ_e at $\kappa_s \ll \kappa_f$ and σ_e at $\sigma_f \ll \sigma_s$, respectively. For the orthotropic fibrous sample, the potential boundary conditions were applied along the z-direction.

Morphology

Exemplary, we used two distinct morphologies with continuous, percolating solid phases: (1) foam-like structures and (2) fibrous-like structures. The foam-like samples have been manufactured by a template method using a ceria-based slurry and 10 ppi polyurethane sponges.6 The fibrous samples were ceria felts manufactured by Zircar Zirconia, Inc. (Florida, NY, USA). These structures were imaged via x-ray computed tomography. The initial samples can be subjected to fabrication variations, e.g., increasing strut and fiber sizes by additional applications of slurry or coating. This variation in the fabrication process and the corresponding changes in the sample's morphology were virtually reproduced by digital image processing, i.e., dilation operations with spherical structuring elements.

The three-dimensional (3-D) rendering of the two base morphologies and varying porosities, i.e., increasing numbers of applied coating steps, i.e., dilation operations with increasing size of the structural element, are depicted in Fig. 2. The *z*-direction described one of the two symmetry axis of the orthotropic fibrous sample, perpendicular to the fiber alignment.

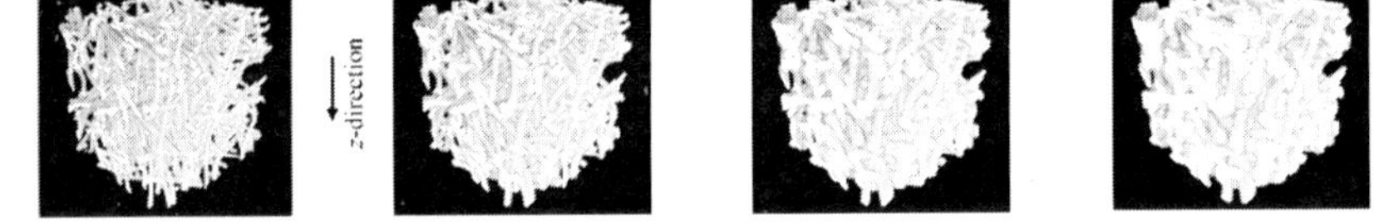

Figure 2. Foam-like morphology with sample edge length = 20 mm (a–d),18 fibrous morphology with sample edge length = 0.49 mm (e–h), for porosities 0.84 (a, f), 0.72 (b, f), 0.59 (c, g), and 0.46 (d, h)

The characteristic sizes of the two original samples differed by two orders of magnitude. To allow for a direct comparison and a meaningful size range for the photoelectrochemical application, the foam-like sample was shrunk by a factor of 100 to represent a third type of media exhibiting the same morphology as the original foam but in the same size range as the fibrous material. We thus investigated two foam-like samples with two distinctly different size ranges (foam original and foam shrunk) and one fiber sample.

The mean pore diameters, represented as the largest spherical pore within the pore space, were calculated via opening operations and were $d_{\text{void,foam,o}}$ = 2.6, 2.3, 2.0, and 1.7 mm for the original foam-like samples for decreasing porosity. The shrunk foam-like sample's pore diameters were $d_{\text{void,foam,s}}$ = 26, 23, 20, and 17 μm for decreasing porosity. The mean fiber and pore diameter were calculated via digital image processing, i.e., connectivity calculations, and they were $d_{\text{solid,fiber}}$ = 24, 36, 48, and 59 μm and $d_{\text{void,fiber}}$ = 95, 83, 71, and 59 μm for decreasing porosity, respectively. The specific surface areas were calculated via two-point correlations and fitted to a second-order polynomial functions

$$A_0 = -117740\varepsilon^2 + 147250\varepsilon - 1116 \tag{13}$$

$$A_0 = (-2277\varepsilon^2 + 2533\varepsilon + 262)\frac{d_{\text{void}}}{d_{\text{void,o}}} \tag{14}$$

for fibrous and foam-like (original and shrunken) samples, respectively. The anisotropy of the samples was characterized via mean intercept lengths35 and showed a degree of anisotropy of 0.23 and 0.55 for the foam and the fibrous samples, respectively.

RESULTS

Transport Characterization

Normalized β_e of the foam (original and shrunken) and fibrous samples with different porosities is depicted in Fig. 3. The increase in β_e with decreasing porosity was linear and is in accordance with the decreasing characteristic pore size for decreasing porosity. The increase in $\beta_e/\beta_{e,\varepsilon=0.84}$ with decreasing porosity was more pronounced for the foam sample.

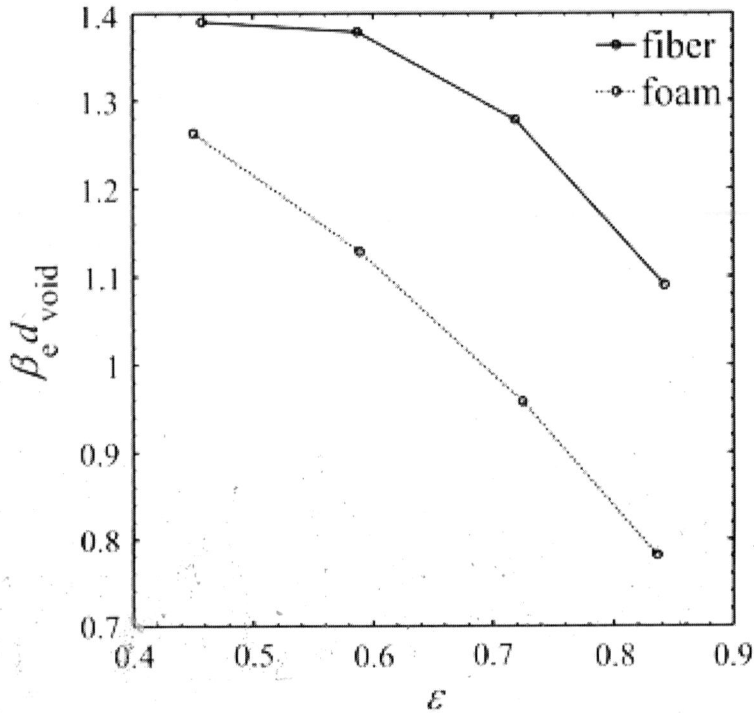

Figure 3. Normalized effective extinction coefficient as a function of the sample porosity for original and shrunken foam-like samples (dotted line), and fibrous samples (solid line)

$\Phi_{scat,e}$ for diffusely reflecting phase boundaries showed little dependence on the morphology and porosity and were fitted to a second-order polynomial

$$\Phi_{scat,e} = 0.58\mu_{scat}^2 - 1.40\mu_{scat} + 0.81 \quad (15)$$

$$\Phi_{scat,e} = 0.59\mu_{scat}^2 - 1.41\mu_{scat} + 0.80 \quad (16)$$

for fibrous and foam-like (original and shrunken) samples, respectively.

The Nusselt numbers were given by

$$\mathrm{Nu} = 5.54 + (0.71\varepsilon^2 - 0.63\varepsilon + 0.30)\mathrm{Re}^{\sqrt{1.70-1.39\varepsilon}}\mathrm{Pr}^{0.6} \quad (17)$$

$$\mathrm{Nu} = 6.55 + (1.35\varepsilon^2 - 1.29\varepsilon + 0.62)\mathrm{Re}^{\sqrt{0.93-0.55\varepsilon}}\mathrm{Pr}^{0.6} \quad (18)$$

for fibrous and foam-like (original and shrunken) samples, respectively. The foam-like samples showed an almost constant Nu at low Re, independent of porosity, while for Re > 10, the Nu was highly porosity dependent.18 The Nu numbers of the fibrous sample showed almost no porosity dependence for the whole Re range investigated.

The normalized effective conductivities are depicted in Fig. 4. The effective thermal and ionic conductivities of the fluid phase of the fiber and foam-like samples were similar, i.e.,$\kappa_{e,foam}/\kappa_{e,fib} = k_{e,f,foam}/k_{e,f,fib}$ = 0.95–1.05. σ_e/σ_s and $k_{e,s}/k_s$ of the fibers were significantly lower than for the foam-like samples, i.e., $\sigma_{e,foam}/\sigma_{e,fib} = k_{e,s,foam}/k_{e,s,fib}$ = 2–25, with the largest differences for high porosities. The decreased effective electronic and thermal conductivities of the solid phase of the fibrous samples were explained by the more perpendicular arrangement of the fibers to the electrical potential and heat flux, respectively.

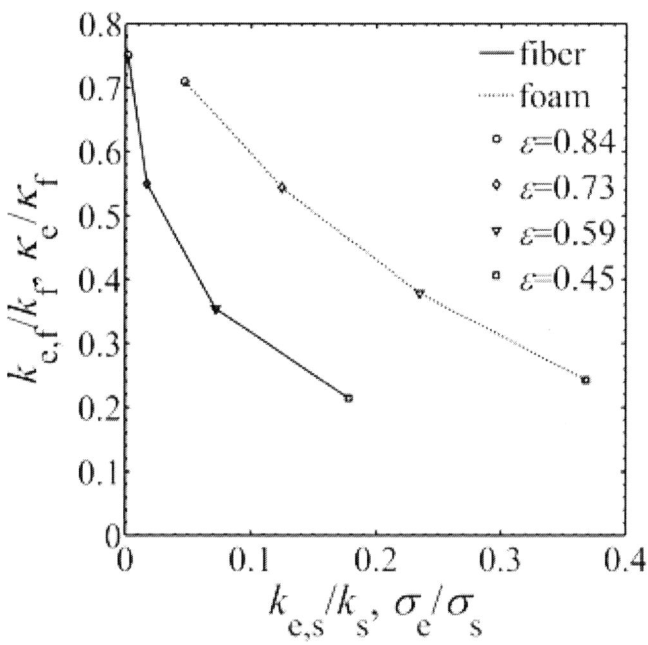

Figure 4. Normalized effective thermal and ionic conductivity of the fluid phase as a function of the normalized thermal or electrical conductivity of the solid phase for fibrous (solid line) and original and shrunken foam-like (dotted) samples for varying porosities

The normalized K and F of the fibrous and foam-like samples for various porosities are depicted in Fig. 5. K of the original foam samples was three orders of magnitude larger than K of the fibrous samples. F differs by almost two orders of magnitude. The samples' permeabilities can be accurately (RMS = 2%) described by exponential functions in porosity ($K = a\exp(b\varepsilon)$), i.e., lumping all the other morphological characteristics into two constants.

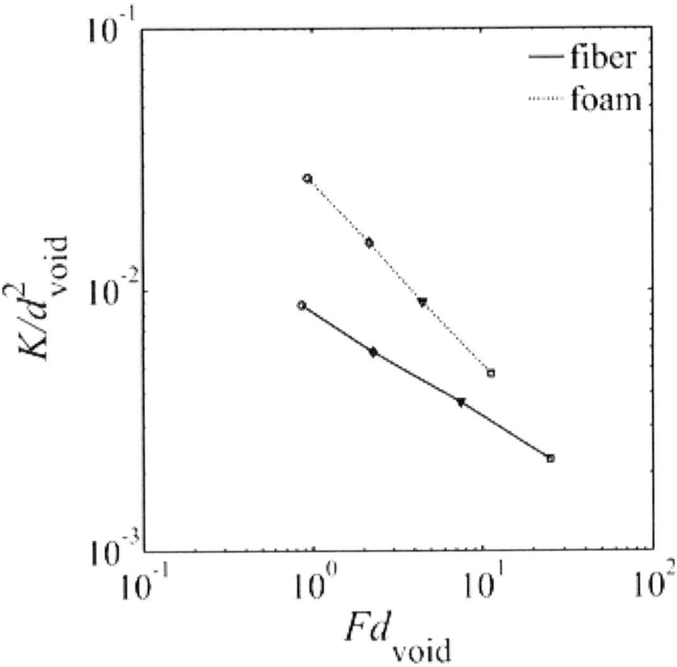

Figure 5. Permeability and Forchheimer coefficient for fibrous (solid line) and original and shrunken foam-like (dotted, dashed) samples for varying porosities

τ_{mean} of the fiber and foam-like samples is given in Table I. $\tau_{mean,fiber} > \tau_{mean,foam}$ especially at lower porosities as dead ends and recirculation zones were more frequent. Consequently, the effective (hydrogen) diffusivity of the foam-like samples was larger than the effective diffusivity of the fibrous sample at similar porosities.

Table I. Mean tortuosity for the fiber and original and shrunken foam-like samples for various porosities

ε	**0.84**	**0.72**	**0.59**	**0.46**
$\tau_{mean,fiber}$	1.269	1.394	1.550	1.746
$\tau_{mean,foam}$	1.150	1.234	1.342	1.482

Application to Solar Fuel Processing and Solar Thermal Power Generation

The pure porosity-dependent absorption behavior of the fibrous and foam-like samples (original and shrunken) with $r = 0.1$ is depicted in Fig. 6. The foam-like samples with largest porosity reached maximal absorptance ($T < 0.005$) within $t/d_{void,foam} \geq 7$, while the fibrous samples required $t/d_{void,fiber} \geq 5$. The maximal absorption achievable was only dependent on the sample porosity and not the sample morphology ($(A_{fiber} - A_{foam})/A_{fiber} \leq 0.3\%$). The fibrous and the shrunken foam samples achieved the maximal achievable$A$ within a sample thickness of only a few hundred microns; the original foam sample required thicknesses larger than 20 cm.

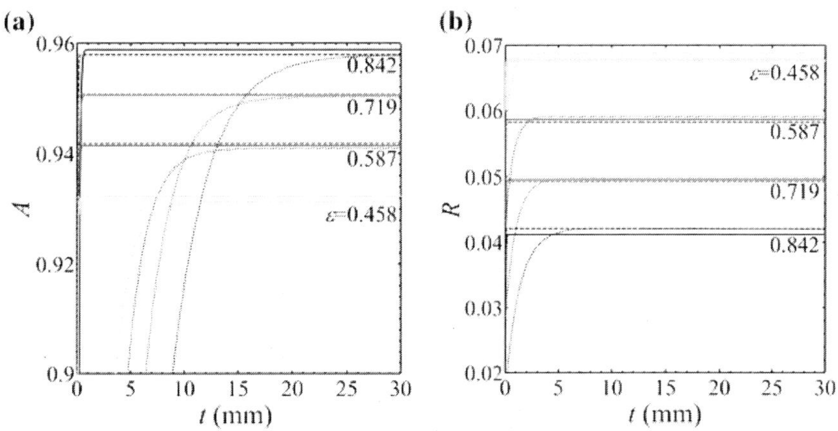

Figure 6.(a) Absorptance and (b) reflectance as a function of layer thickness for fibrous (solid lines), original (dotted lines), and shrunk (dashed lines) foam-like samples with $r = 0.1$ and varying porosity

Photoelectrochemical Fuel Processing

We used Pt and RuO_2 catalysts for the hydrogen and oxygen evolution reactions with $i_{0,a} = 10^{-8}$ A/cm^2, $i_{0,c} = 10^{-3}$ A/cm^2, $A_a = 35$ mV/dec, and $A_c = 30$ mV/dec.36 The area of the porous slab modeled was 25 cm^2. The solar irradiation was 1 kW/m^2, collimated and 1.5 AM. The energy band gaps of the dual absorber were 1.6 eV and 1.0 eV, corresponding to a current matching absorber combination.25 The ambient temperature was assumed ($T = 300$ K) and the hydrogen saturation concentration in water was 0.78 mol/m^3. We used $\kappa_l = 40$ S/m, which corresponds to the ionic conductivity of 1 M sulfuric acid. $\sigma_s$varied significantly depending on the materials and their doping; we assumed $\sigma_s = 1$ S/m.

The efficiency of the multifunctional porous slab used for photoelectrochemical fuel production using the two different base morphologies (fiber and shrunken foam) and four different porosities is given in Fig. 7. The initial gain in efficiency resulted from the increase in absorbed solar radiation. The decrease in efficiency for larger thicknesses was a result of the dominance of the ohmic losses, especially the losses due to conduction in the solid absorber phase. The decrease in efficiency was observed at smaller sample thicknesses for the fibrous samples than the shrunken foam samples as $\sigma_{e,fiber} < \sigma_{e,foam,s}$. For samples with larger porosities, the thickness at which the ohmic losses started to dominate did not yet allow for complete radiation absorption, and consequently, the observed maximal efficiency was significantly lower than the maximal achievable efficiency for maximal absorption. The absorption-limited efficiency was only reached for the fibers with $\varepsilon = 0.46$ and the shrunken foam samples with $\varepsilon = 0.46$–0.719.

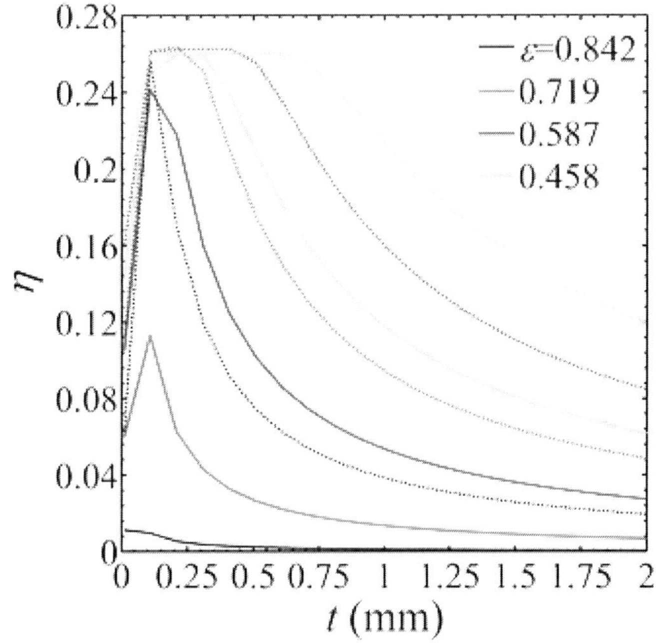

Figure 7. Solar-to-fuel efficiency of a multifunctional porous slab for photoelectrochemical solar fuel processing. The efficiency is shown as a function of layer thickness for fibrous (solid lines) and shrunk foam-like (dashed lines) samples with varying porosity

Solar Thermal Power Production

A sample of area 0.503 m^2 was irradiated by a collimated solar flux of 1000 kW/m^2. Air was the heat transfer fluid assumed, modeled as an ideal gas with temperature-dependent properties37 and inlet mass flow of m˙m˙ = 0.4 kg/s, initial temperature of 298 K, and an absolute pressure of 10 bars. These boundary conditions were an estimate of a 1 MW pressurized air receiver for solar-driven gas turbines.8 Sintered silicon carbide was used as the solid absorber with r = 0.1.37 The efficiency of the multifunctional 1-D slab used for solar thermal power production for the three different morphologies and four different porosities is given in Fig. 8. The fibrous samples reached the maximal absorptance for slab thicknesses of several hundred microns, and the highest efficiency was achieved for samples with highest porosities as their reflectance were the smallest. The shrunken foam-like samples reached their maximum efficiencies within the first 100 μm due to the larger extinction coefficients; however, a higher reflectance for low porosity led to slightly lower maximum values. Increasing the sample thickness led to a decrease in efficiency because the pressure losses through the slab became significant. The influence of the pump work onto the efficiency of the fibrous samples was less significant but still apparent. In contrast, the original foam-like samples showed no decrease in efficiency with increasing thickness. The original foam samples at comparable porosity required larger sample thicknesses to reach the maximal efficiency due to significantly lower extinction coefficients. The maximum efficiency of a fibrous sample and the shrunk foam-like samples was always larger compared to the original foam-like sample with the same porosity mainly due to their larger specific surface area and enhanced convective heat transfer.

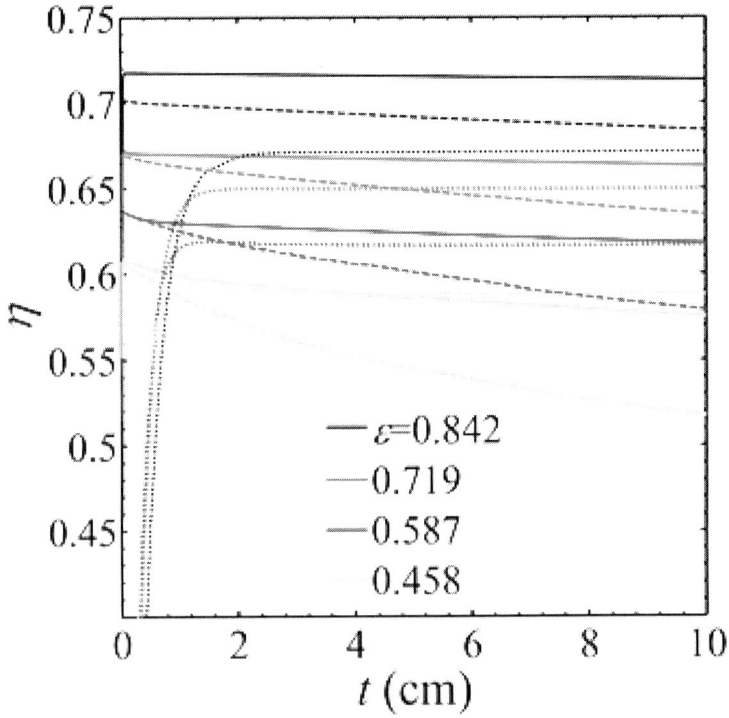

Figure 8. Overall efficiency of a multifunctional porous slab for solar thermal power production. The efficiency is shown as a function of layer thickness for fibrous (solid lines), original (dotted lines), and shrunk (dashed lines) foam-like samples with varying porosity

CONCLUSIONS

A multiscale experimental–numerical methodology has been used to quantify the gain in performance due to engineering of the morphology of porous media used as absorber, heat exchanger, charge conductor, and reaction site in solar reactors. Two base morphologies have been investigated, i.e., fibrous and foam-like samples. Their exact morphologies were experimentally obtained via x-ray computed tomography and subsequently manipulated by digital image processing to vary characteristic morphological sizes and porosity, both at a constant base morphology. The obtained morphologies (fibrous, original, and shrunken foam samples) were used in direct numerical simulations for the determination of their effective heat, mass, and charge transport properties, which were used in continuum-scale models of reactors incorporating 1-D multifunctional porous slabs performing photoelectrochemical hydrogen generation and solar thermal power production.

We showed that heat, mass, and charge transport in porous media cannot accurately be described based on only two morphological characteristics, e.g., porosity and a characteristic length. For the particular cases of fibrous, original foam, and shrunk foam-base morphologies, we determined the porosity-dependent effective extinction coefficients, effective scattering phase functions, permeabilities and Forchheimer coefficients, effective diffusivities and tortuosity distributions, Nu correlations, and effective ionic, electronic, and thermal conductivities.

The developed continuum-scale models of reactors incorporating 1-D slabs of porous media used in a photoelectrochemical hydrogen generation process and a solar thermal power production process were used to quantify the influence of morphology on process performance. We observed that the morphology leading to optimized efficiencies was dependent on the application of the porous media. The efficiency of photoelectrochemical fuel processing could be increased by a factor of 23 when using a shrunken foam morphology instead of a fibrous morphology at high porosities. The efficiency of solar thermal power production increased by a factor of 1.2 and the required slab thickness was reduced by a factor of 12 when using a highly porous fibrous sample with fiber diameters in the μm range instead of a low porosity foam sample with pore diameters of a few mm. The study showed—in the limit of the validity of the continuum model's assumptions—that the morphology, scale, and porosity of the multifunctional porous components incorporated in solar reactors significantly influence their performances.

ACKNOWLEDGEMENTS

We acknowledge Jan Marti from ETH Zurich performing initial studies on transport in fibrous materials at the Professorship of Renewable Energy Carriers and Prof. Steinfeld heading this Professorship.

REFERENCES

1. S. Chu and A. Majumdar, *Nature* 488, 294 (2012).
2. N. Lewis and D. Nocera, *Proc. Natl. Acad. Sci.* 103, 15729 (2006).
3. J. Newman, P.G. Hoertz, C.A. Bonino, and J.A. Trainham, *J. Electrochem. Soc.* 159, A1722 (2012).
4. J.A. Turner, *Science* 285, 687 (1999).
5. P. Furler, J.R. Scheffe, and A. Steinfeld, *Energy Environ. Sci.* 5, 6098 (2012).
6. P. Furler, J. Scheffe, M. Gorbar, L. Moes, U. Vogt, and A. Steinfeld, *Energy Fuels* 26, 7051 (2012).
7. M. Röger, L. Amsbeck, B. Gobereit, and R. Buck, *J. Sol. Energy Eng.* 133, 031009 (2011).
8. I. Hischier, P. Leumann, and A. Steinfeld, *J. Sol. Energy Eng.* 134, 021003 (2012).
9. M. Graetzel, *Curr. Opin. Colloid Interf. Sci.* 4, 314 (1999).
10. A. Kay, I. Cesar, and M. Grätzel, *J. Am. Chem. Soc.* 128, 15714 (2006).
11. R. van de Krol, Y. Liang, and J. Schoonman, *J. Mater. Chem.* 18, 2311 (2008).
12. K. Sivula, F. Le Formal, and M. Grätzel, *Chem. Mater.* 21, 2862 (2009).
13. C. Liu, J. Tang, H.M. Chen, B. Liu, and P. Yang, *Nano Lett.* 13, 2989 (2013).
14. S. Haussener, C. Xiang, J.M. Spurgeon, S. Ardo, N.S. Lewis, and A.Z. Weber, *Energy Environ. Sci.* 5, 9922 (2012).
15. S. Haussener and A. Steinfeld, *Materials* 5, 192 (2012).
16. A. Akolkar and J. Petrasch, *Int. J. Heat Mass Trans.* 54, 4775 (2011). A. Akolkar and J. Petrasch, *Transport Porous Media*, 1 (2012).

17. S. Suter, A. Steinfeld, and S. Haussener, *Int. J. Heat Mass Trans.* (submitted).

18. J. Petrasch, P. Wyss, and A. Steinfeld, *J. Quant. Spectrosc. Radiat. Trans.* 105, 180 (2007).

19. J. Petrasch, B. Schrader, P. Wyss, and A. Steinfeld, *J. Heat Trans.* 130, 032602 (2008).

20. J. Petrasch, F. Meier, H. Friess, and A. Steinfeld, *Int. J. Heat Fluid Flow* 29, 315 (2008).

21. S. Haussener, P. Coray, W. Lipinski, P. Wyss, and A. Steinfeld, *J. Heat Trans.* 132, 023305 (2010).

22. W. Lipinski, J. Petrasch, and S. Haussener, *J. Quant. Spectrosc. Radiat. Trans.* 111, 253 (2010).

23. J. Petrasch, S. Haussener, and W. Lipinski, *J. Quant. Spectrosc. Radiat. Trans.* 112, 1450 (2011).

24. S. Haussener, S. Hu, C. Xiang, A.Z. Weber, and N. Lewis, *Energy Environ. Sci.* (2013).

25. R.E. Rocheleau and E.L. Miller, *Int. J. Hydrogen Energy* 22, 771 (1997).

26. S. Hu, C. Xiang, S. Haussener, A.D. Berger, and N.S. Lewis, *Energy Environ. Sci.* 6, 2984 (2013).

27. W. Shockley and H.J. Queisser, *J. Appl. Phys.* 32, 510 (1961).

28. S. Whitaker, *The Method of Volume Averaging, Theory and Applications of Transport in Porous Media* (Dordrecht, the Netherlands: Kluwer Academic Publisher, 1999).

29. M. Tancrez and J. Taine, *Int. J. Heat Mass Transf.* 47, 373 (2004)

30. H. Friess, S. Haussener, A. Steinfeld, and J. Petrasch, *Int. J. Numer. Methods Eng.* 93, 1040 (2013).

31. Ansys Inc., *Ansys 14* (Canonsburg, PA: ANSYS, Inc., 2013).

32. W. Ullman and R. Aller, *Limnol Oceanogr.* 27, 552–556 (1982).

33. J. Van Brakel and P. Heertjes, *Int. J. Heat Mass Trans.* 17, 1093 (1974)

34. Arganda-Carreras, F.P. Cordelieres, R.P. Dougherty, J.S. Jackson, B. Schmid, J.R. Hutchinson, M. Doube, M.M. Klosowski, and S.J. Shefelbine, *Bone* 47, 1076 (2010).

35. M.G. Walter, E.L. Warren, J.R. McKone, S.W. Boettcher, Q. Mi, E.A. Santori, and N.S. Lewis, *Chem. Rev.* 110, 6446 (2010).

36. AIChE Design Institute for Physical Properties, *DIPPR Project 801—Full Version*(New York, NY: Design Institute for Physical Property Research/AIChE, 2012).

11

CHAPTER

Whisker Formation on SAC305 Soldered Assemblies

S. Meschter,[1], P. Snugovsky,[2] Z. Bagheri,[2] E. Kosiba,[2] M. Romansky,[2] J. Kennedy,[2] L. Snugovsky,[3] and D. Perovic[3]

[1]*BAE Systems, Endicott, NY 13760, USA.*

[2]*Celestica, Toronto, ON M3C-1V7, Canada*

[3]*University of Toronto, Toronto, ON M5S-2J7, Canada*

ABSTRACT

This article describes the results of a whisker formation study on SAC305 assemblies, evaluating the effects of lead-frame materials and cleanliness in different environments: low-stress simulated power cycling (50–85°C thermal cycling), thermal shock (−55°C to 85°C), and high temperature/high humidity (85°C/85% RH). Cleaned and contaminated small outline transistors, large leaded quad flat packs (QFP), plastic leaded chip carrier packages, and solder balls with and without rare earth elements (REE) were soldered to custom designed test boards with Sn3Ag0.5Cu (SAC305) solder. After assembly, all the boards were cleaned, and half of them were recontaminated (1.56 $\mu g/cm^2$ Cl^-). Whisker length, diameter, and density were measured. Detailed metallurgical analysis on components before assembly and on solder joints before and after testing was performed. It was found that whiskers grow from solder joint fillets, where the thickness is less than 25 μm, unless REE was present. The influence of lead-frame and solder ball material, microstructure, cleanliness, and environment on whisker characteristics is discussed. This article provides detailed metallurgical observations and select whisker length data obtained during this multiyear testing program.

INTRODUCTION

After the European Union RoHS directive deadline was enacted in July 2006, the electronics industry began a transition to manufacturing with Sn-Ag-Cu (SAC) near-eutectic alloys to avoid the use of Pb-based solders. This change has had little impact on most commercial consumer products because of the limited life requirements and relatively benign use environments. However, lead-free implementation for the aerospace and defense industry requirements, which includes long service lifetimes, rugged operating environments, and high consequences of failure, requires additional evaluation to assure safe systems operation.

The new Pb-free and high Sn (about 96%) alloys may be susceptible to spontaneous growths of filament-like tin structures called tin whiskers. Pb is well known as an effective suppressant of whisker formation. Without Pb, the military, aerospace, and other high-reliability industries have experienced increased risks of system failure due to tin whiskers that cause electrical short circuits, impact high-frequency circuits, and create loose debris.1,2 The majority of recent tin whisker research has concentrated on coupons or components with an emphasis on the Sn plating composition, thickness, grain size, grain orientations, and Ni underlayer.2–12 These studies produced a great deal of knowledge; however, they do not consider the very real situation in which components are assembled on circuit boards using Pb-free solder pastes. Recent observations have demonstrated the ability of Pb-free solder joints to grow whiskers.13,14 One challenge accompanying real electronic systems is the interplay among the component design and materials, manufacturing process, and service environments that contribute directly or conspire with one another to exacerbate whisker growth.

This article is focused on highlights from the Whisker Testing and Modeling project15funded by the U.S. Strategic Environmental Research Program. The objective of this project is to perform systematic tin whisker testing and Monte Carlo risk modeling on lead-free electronic assemblies in order to improve the reliability of military electronic systems. The observations described in this article are based on an extremely large set of data; 129724 whiskers were counted and 10985 whiskers were measured.

EXPERIMENTAL

Approach

The experimental portion of this study was designed to study the role of selected manufacturing process variables in the growth of tin whiskers under certain isothermal and thermal cycling environments. The main assembly variables include package type and lead/ball materials as well as component and assembly cleanliness. Electronic components were attached to printed circuit boards to form a functional printed circuit assembly. Component leads were electrically and mechanically fixed to the board Cu pads with molten SAC305 solder. During reflow, the solder starts melting at 217°C. The solder then wicks up the leads and dissolves the electroplated Sn part. The rest of the electroplated Sn melts at 232°C and continues mixing with the molten phase. During cooling, this mixture solidifies. The temperature range of the solder joints across the board varies with the minimum acceptable temperature of 232°C and the maximum of 255°C. Therefore, after Pb-free reflow there is no original electroplated Sn coverage of the component leads. The modified component tin or tin alloy coverage after assembly might be thicker or thinner than electroplated Sn before assembly, and it has an elemental composition closer to the original SAC solder or pure Sn depending on the component type (length of the leads), location, and reflow parameter and solidified microstructure.

Test Vehicle

Three custom boards were designed to evaluate the key piece parts of interest. The small outline transistor (SOT) board has three, five, and six leaded SOT parts with C194 Cu alloy and alloy-42 alloy leads. The quad flat pack (QFP) board has C7025 Cu alloy high pin count QFPs with two lead pitches and a C151 Cu alloy plastic leaded chip carrier (PLCC). The ball grid array (BGA) board contains solder alloy balls that evaluate whisker growth on Ce rare earth element (REE) additions to SAC. The test vehicles for this study are small boards 6 cm × 6 cm in size processed together 12-up on a panel (Fig. 1). The boards were separated after surface-mount soldering, inspected, and then subjected to the environments. The parts were distributed across three different board designs, designated as the SOT board, the QFP board, and the BGA board. The board was found to be well suited for inspection by scanning electron microscopy (SEM) because the board fit easily into the SEM's sample chamber. The placement area was limited to 5 cm × 5 cm to ensure all leads of interest were contained in a region that could be viewed by the

SEM. The boards are 0.24 cm thick and are a double-layer design manufactured in accordance with IPC-6012 Class 3, Type 2 with high-temperature woven E glass epoxy laminate (IPC-4101/126), Cu foil (IPC-4562/3 Type E Grade 3 [HTE]), solder mask over bare Cu (IPC-SM-840 Class H liquid photoimageable), with immersion tin over Cu finish (IPC-6012, Type IT) (IPC Association Connecting Electronics Industries, Bannockburn, IL).

Figure 1. Test vehicle assembly soldering panel

The SOT board has 64 alloy-42 lead material SOT3 Fairchild 2N7002 parts, 40 C194 lead material SOT5 Fairchild NC7S08M5X parts, and 17 alloy-42 lead material SOT6 Diodes Inc. 2N7002DW-7-F parts (Diodes Inc., Plano, TX). The lead composition for the alloy-42 is Fe-42Ni and the Cu alloy C194 is Cu2.1-2.6Fe-0.015-0.15P-0.05-0.2Zn. The QFP board has the following Cu alloy lead material (Practical Components Inc., Los Alamitos, CA) and Amkor Technology (Chandler, AZ) parts (Practical Components Inc.,): six C7025 lead material QFP44 A-QFP44-10 mm-0.8-3.2-DC-Sn (0.8-mm pitch) parts, four C151 lead material PLCC20 A-PLCC20T-DC-Sn (1.25-mm pitch) parts, and four C7025 lead material QFP64 A-LQFP64-0.7 mm-0.4 mm-2.0-DC-Sn (0.4-mm pitch) parts. The lead composition for the C7025 is Cu2.2-4.2Ni0.25-1.2Si0.05-0.3 Mg and for the C151 is Cu0.1Zr. Half of the SOT3, SOT5, QFP44, and QFP64 parts contained some leads with a 5-V bias and the remainder was grounded. The SOT6 and PLCC parts were not connected electrically and are contained in the zero volt bias data presented in this study.

The BGA board contains provisions for two types of BGA components (100 ball 0.8-mm pitch Amkor CABGA and a 97 ball 0.4-mm pitch Amkor CVBGA. In the present evaluation, BGA devices were not soldered to the boards because of the difficulty of inspecting for whiskers under the package; only balls were populated. Two different ball alloys were studied on the land pattern area SAC105 and SAC105 with Ce REE added. These were compared with pads printed with SAC305 paste.

Assembly

Before assembly components were divided into two groups. The first group was cleaned using the method developed in screening experiments.16 The samples were cleaned twice to get the level of contamination 10 times below typical acceptable industry level and did not exceed 0.062 $\mu g/cm^2$ (0.4 $\mu g/in^2$) Cl. The second group was contaminated with NaCl using the method developed earlier.16 The piece part level of chlorine contamination 0.465 $\mu g/cm^2$ (3 $\mu g/in^2$) was selected to be within the industry levels encountered (no standards exist).

Both component groups cleaned and intentionally contaminated were assembled using SAC305 solder paste. Assembled boards with clean and contaminated components were divided into two groups each. The first group was cleaned after assembly and the second group was purposely contaminated after assembly. The assembly level of equivalent chlorine contamination was selected to be within the J-STD-001 (IPC Association Connecting Electronics Industries, Bannockburn, IL) industry limits typically used to provide satisfactory electrical function—approximately 1.56 $\mu g/cm^2$ (10 $\mu g/in^2$) Cl. The contamination levels are designated as: 0-0 = no part contamination-no board contamination; 0-1 = no part contamination-board contamination; 1-0 = part contamination-no board contamination; and 1-1 = part contamination-board contamination.

For the assembly of the BGA boards, 0.4-mm (0.0157 in) diameter spheres were ordered from a major solder manufacturer. The target concentration was 0.5% Ce. The ball manufacturer had difficulty attaining this level and it was found to be <0.01% Ce. Then, after soldering to the board with SAC305 paste, the Ce concentration was further reduced.

Experimental Environments and Testing

Four environments were selected to assess whisker growth. There were two isothermal high humidity environments; a long-term low-temperature high humidity (LTHH) and a short-term high-temperature high humidity (HTHH). In addition, there were two thermal cycling conditions used: a simulated power cycling thermal cycle (PCTC) and a thermal shock cycle (TC). In high-humidity conditions, 85°C has been shown to promote whisker growth in lead-free solder alloys.12–14 The two isothermal high-humidity environments employed were 25°C/85% RH (3 years) and 85°C/85% RH (4,000 h). The PCTC cycle was +50°C to 85°C with 15-min ramps and dwells in air simulating the conditions in Ref. 17that grew whiskers when rework flux contamination was present. The TC cycle was –55°C to 85°C with three cycles per hour air-to-air thermal shock as defined in JESD201*Environmental Acceptance Requirements for Tin Whisker Susceptibility of Tin and Tin Alloy Surface Finishes* (JEDEC Solid State Technology Association, Arlington, VA) and has representative temperature limits but temperature ramp rates that are much greater than moderate aerospace service conditions. Note that in the PCTC testing, the actual measured chamber temperature was 48–88°C and humidity was not controlled, but it was measured to be between 25%RH at 88°C to 10% RH at 48°C. For the TC cycle, the humidity was not controlled or measured.

Whisker Inspection

The assemblies were removed for inspection at progressively increasing exposure times. For example, the PCTC samples were first examined after 250 cycles, returned to cycling for an additional 500 cycles and inspected at 750 cycles (250 + 500), then returned for an additional 1000 cycles and inspected at 1750 cycles (250 + 500 + 1000). The industry whisker inspection and measurement standard JESD22A121A *Test Method for Measuring Whisker Growth on Tin and Tin Alloy Surface Finishes* (JEDEC Solid State Technology Association) developed for part-level whisker measurement was adapted for the assembled parts. The parts were randomly selected for inspection and approximately 50–100 leads were examined for each part condition. The same parts and leads were inspected by the same person for all environments to monitor the whisker growth progression on the leads and minimize operator related inspection variation. Inspections were performed at increasingly longer intervals with the boards being returned to the chamber. The whisker length criterion in JESD22A121A was used to perform the whisker length measurements in the SEM. Whiskers longer than 10 μm were measured for all experiments except for the HTHH 4000-h inspection, where only whisker lengths greater than approximately 30–40 μm were measured to concentrate limited inspection resources on the longest whiskers. A unique aspect of measuring whiskers on assemblies is that the whisker growth surfaces are nonplanar and the sample must be continually rotated and tilted during the inspection. In the current work, the reported whisker length measurements are conservative. During the length measurement procedure, the SEM inspection azimuth axis was adjusted to be aligned to within approximately 30° from perpendicular to the whisker. The SEM axis misalignment from the whisker normal results in a potential whisker length underreporting of up to 15%. All whisker lengths and diameters are reported in microns.

The detailed inspection was performed using an SEM. The occurrence of a whisker was counted if it was observable at 100 times magnification in the SEM. There was no differentiation between odd shaped eruptions and whiskers in the occurrence count. For the whisker-length measurements, an SEM magnification higher or lower than that used for inspection may be required, such that the whisker being measured approximately fills the field of view at the selected magnification. The numbers of whiskers were recorded along with the amount of surface area that was inspected as well as the region of the lead where the measurements were made. These were the same regions that were used to obtain the whisker length. For selected whiskers, the whisker diameter or thickness was measured by drawing a line perpendicular to the whisker axis.12

Metallurgical Analysis

Sequential whisker cross sectioning was performed on representative whiskers with SEM inspection at various slice depths. A metallurgical evaluation was performed to determine the tin grain morphology, intermetallic (IMC) layer thickness and morphology, and IMC particle distribution at the base of the whisker and in the whisker itself. Optical and SEM images and EDX elemental spectrum were used to record findings similar to the results published by authors in 2008 and 2012.16,17

RESULTS AND DISCUSSION

Characterization of Components Before and After Assembly

SOT and QFP board parts were analyzed before assembly in the as-received condition, after cleaning and after intentional contamination.

The assembled boards were cross sectioned to study the solder microstructure, component coverage, IMC characteristics, and contamination distribution.

As-Received Parts

There were no significant anomalies in the plating of the part leads relating to solderability. Some defects important for whiskering that are not commonly reported (Fig. 2) such as exposed base metal, thin plating from lead forming, voids, roughness, and deep surface grooves and some ionic contamination such as Cl, S, Si, Na, and Br were detected. The QFP44, QFP64, and PLCC20 leads exhibited some instances of plating cracks. The SOT3 part was also noted to have some silicon contamination present before cleaning. The distribution of the contamination on the cleaned and re-contaminated part was distributed into the plating grooves and valleys similar to the part before cleaning.16Cross sectioning of the PLCC demonstrated that the Sn plating of the leads is extremely uneven. The XRF measurements reported Sn thicknesses of 8–10 μm; however, the actual thicknesses observed during cross sectioning varied from 1 to 38 μm. Because thin Sn plating is more prone to whisker growth, an 8-μm minimum tin plating thickness is recommended for whisker mitigation.1

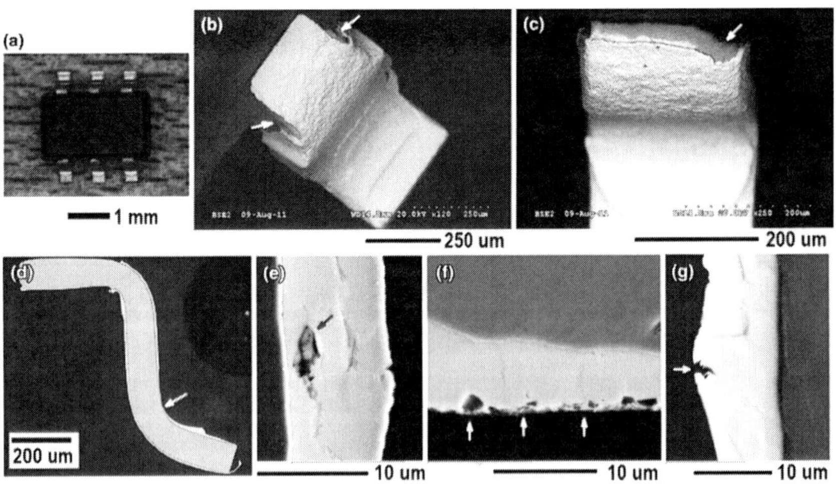

Figure 2. As-received part lead conditions contributing to whisker growth; (a) SOT6 overall part optical view, (b) SOT5 lead SEM image showing exposed Cu on lead near package body, (c) SOT6 lead SEM image showing exposed alloy-42 base metal near package, (d) SOT6 overall lead optical cross-section view showing thin tin plating caused by lead forming. SEM images of tin plating cross sections show (e) a void within the tin plating, (f) voids and a rough surface, and (g) rough surface with deep grooves

Assembly Characterization

After soldering, the solder coverage and IMCs were characterized using cross sectioning. A typical cross-section evaluation is shown in Fig. 3. The SOT3 and SOT6 and QFP64 leads were fully covered in solder, but the solder did not always extend to the top of the leads on the SOT5. The QFP44 and PLCC leads were partially covered with solder. The reflowed Sn lead finish beyond the solder no longer exhibited the fine-grain structure of the initial plating. The reflowed Sn and solder grain sizes were typically more than 25 μm. The thinnest IMC was observed on the alloy-42 leads (0.3 μm). With the alloy-42 leaded parts, the board $(Cu, Ni)_6Sn_5$ IMC was thicker (2.6 μm) than the lead side intermetallic. The SOT3 part was found to have Cl trapped at the thinnest part of the solder joint between the lead and Cu pad.

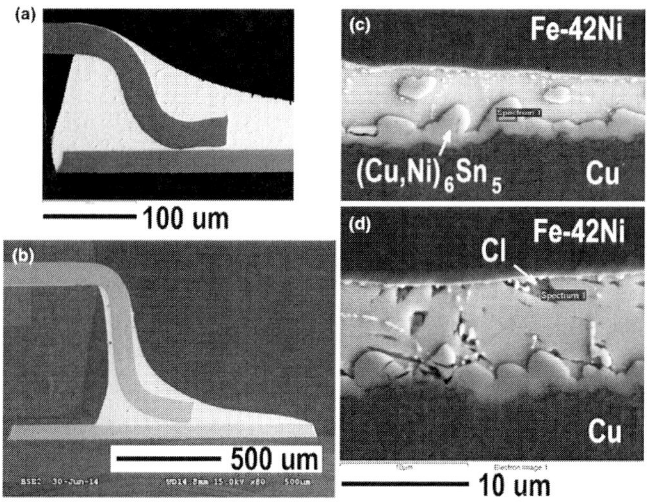

Element	Weight%	Atomic%
Ni K	13.37	19.81
Cu K	26.24	35.92
Sn L	60.40	44.27
Totals	100.00	

Element	Weight%	Atomic%
Cl K	1.21	3.78
Fe K	2.07	4.08
Ni K	0.95	1.78
Cu K	1.59	2.77
Sn L	94.18	87.59
Totals	100.00	

Figure 3. Alloy 42 leaded SOT solder joint cross-section SEM images after assembly; (a) SOT3 overall, (b) SOT6 overall, (c) $(Cu, Ni)_6Sn_5$ IMC on board, and (d) Cl contamination trapped in solder joint

Whisker Characterization After Environmental Testing

Highlights of the whisker length, growth location, and diameter are presented next. Currently the testing and data analysis are almost finished with the exception of the long term high humidity testing. The vast majority of measurement work was done on SOT boards because similar packages with different thermal coefficient of expansion (CTE) alloys were present. Table I summarizes the subset of whiskers counted and measured on the SOT boards.

Table 1. Summary of whisker measurements made on SOT boards

	Counted	#Meas.	Counted	#Meas.
Comp.	PCTC, 1750 cycles		HTHH, 1000 h	
SOT3	216	216	2438	867
SOT6	179	179	1106	486
SOT5	No	No	12020	3388
Total	395	395	15564	4741
	TC, 2100 cycles		HTHH, 4000 h	
SOT3	9130	691	24639	197
SOT6	9866	1606	37641	185
SOT5	5883	63	13106	107
Total	24879	2360	75386	489

The thermal cycling environments promoted whisker growth on the alloy-42 lead terminations, whereas the high-humidity environments caused greater whisker growth on Cu-leaded terminations. The results that follow are for leads that were either grounded or not connected. Preliminary inspection of powered devices indicated that bias did not have a significant impact on whisker growth, but the sample size was small.

PCTC

Whisker inspections were performed after 250, 750 (250 + 500), and 1750 (250 + 500 + 1000) cycles. No whiskers were observed at the 250 cycle inspection. At the 750 and 1750 cycle inspections, whisker nucleation and growth were observed (Fig. 4). The maximum whisker length was 19.4 μm on the SOT3 part with the 1-1 contamination level. Contamination level had a slight impact on whisker length and density. The whiskers grew in the region where the lead exits the solder fillet and the solder is thin (Fig. 5). Although the whiskers in this group were the shortest of all the experiments, it is believed that nucleation and growth would continue with more cycles as was observed in Ref. 17.

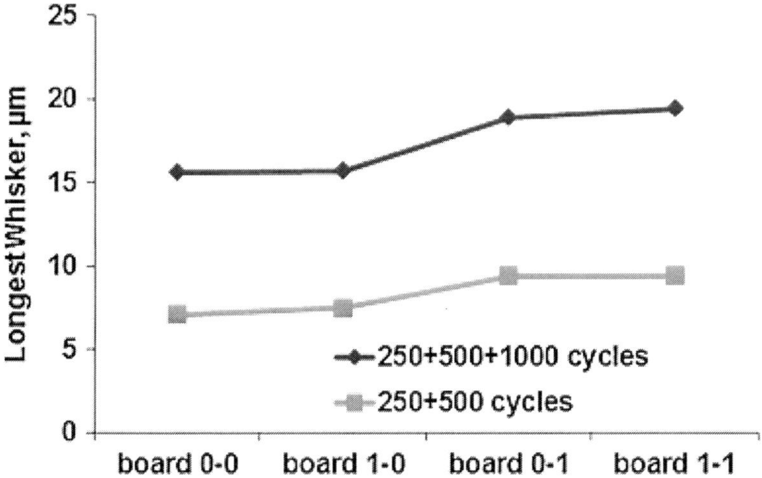

Figure 4. SOT3 maximum whisker length results after a total of 1750 PCTC cycles

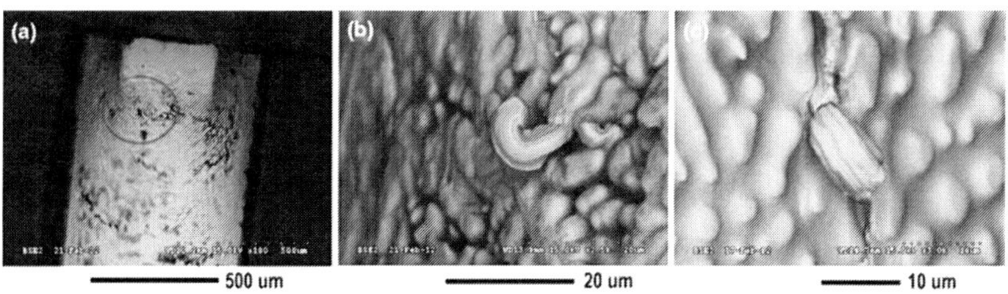

Figure 5. SEM images showing whisker growth on alloy-42 lead solder joint after a total of 1750 PCTC cycles; (a) and (b) overall views highlighting whisker growth where the lead exits the top of the solder joint with (b) and (c) showing higher magnification whisker growth images

TC

Whisker inspections were performed at 500 and 2110 (500 + 1610) cycles. The greatest whisker nucleation and growth occurred from the alloy-42 leads. During thermal cycling, the whisker growth is driven by the CTE differences between the low CTE alloy-42 and the higher CTE solder. The longest whiskers were on the SOT6 termination; 32 μm after 500 cycles and 115 μm after 2,110 (500 + 1610) (Fig. 6). The whiskers on the alloy-42 lead terminations predominantly grew from regions where the solder was thin near the top of the main solder fillet; however, there were also massive tin eruptions indicating a significant amount nonwhisker stress relaxation (Fig. 7).

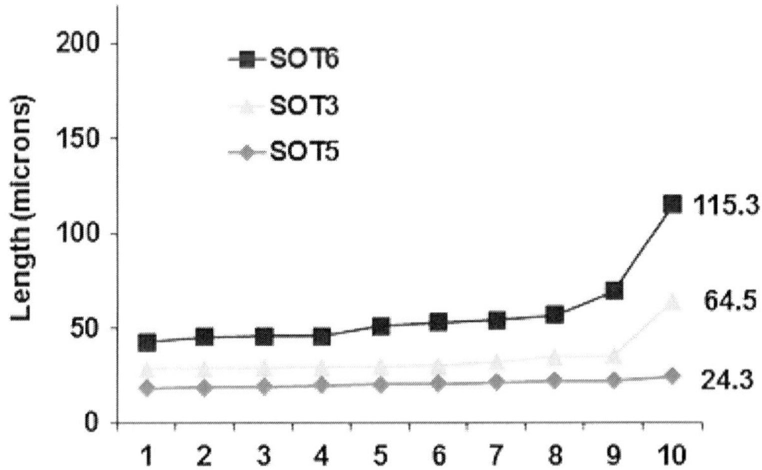

Figure 6. Top 10 longest whiskers after 2110 TC cycles

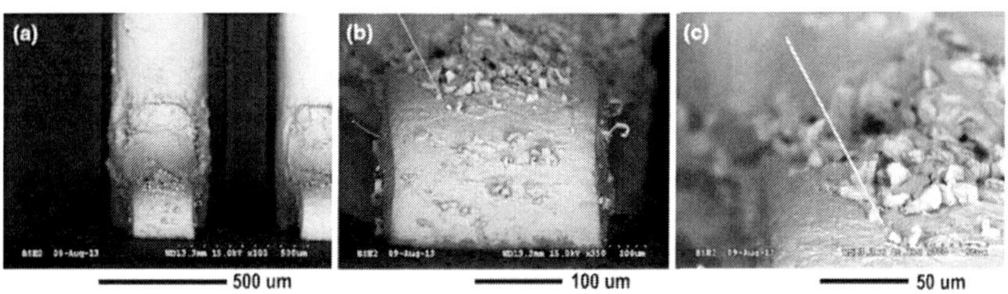

Figure 7. SEM images showing whiskers growing from alloy-42 lead after 2110 TC cycles; (a) overall of lead with (b) and (c) showing whisker region at increasing magnifications

HTHH

Whisker inspections were performed at 1000 and 4000 (1000 + 3000) cycles. The HTHH environment yielded the longest whisker growth (Fig. 8) when compared with the PCTC and TC environments (note the 3-year LTHH is not yet complete). The HTHH assembly 1000-h exposure resulted in whisker lengths up to 186 μm, more than three times longer than the 45-μm JESD201 piece part test limit for 4000 h (55°C/85% RH).

After 4000 h, the longest whisker was 214 μm on the SOT5 solder joints at the Cu board pad edge. While not part of the main measurement set, the QFP64 terminations also exhibited long tin whisker growth (Fig. 9). The intermediate contamination levels (0-1 and 1-0) exhibited the longest mean whisker length, but the 1-1 contamination grew the longest whiskers. The Cu-Cu (lead-pad) couple exhibited the greatest whisker nucleation and growth (Table I). The alloy-42/Cu couple exhibited significantly less nucleation and growth after 1000 h than the Cu/Cu couple. However, as the exposure time increased, the whisker length difference was significantly reduced. During HTHH, the whisker growth is driven by oxidation and corrosion (discussed later). Although it is possible that additional contamination accumulated near the board, the cleaned assembly also exhibited significant whisker growth along the board pad edge. The majority of whiskers were between 1 μm and 5 μm in diameter regardless of alloy or contamination level. After 4000 h, the whisker density varied from 50 to 500 whiskers per termination across the lead material and contamination levels.

There was a marked difference in whisker growth location between the thermal cycling and HTHH. In PCTC, whisker growth was at the top of the solder fillet (Fig. 5) and in TC the whiskers grew from the thin solder on the lead, whereas during HTHH, the majority of whiskers grew from the thin solder regions on the Cu board pad edge (Fig. 10).

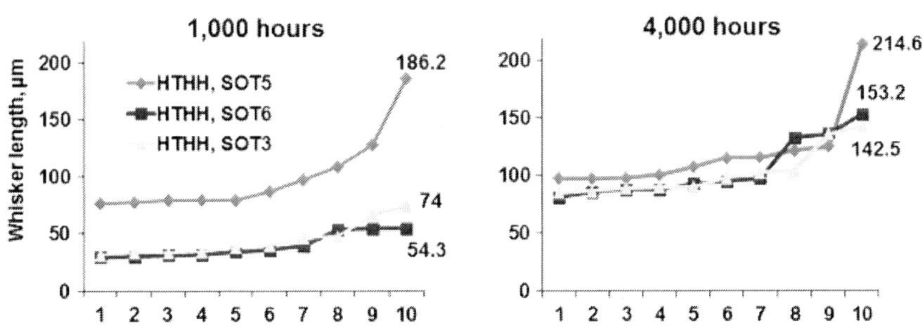

Figure 8. Comparison of top 10 longest whiskers after 1000 and 4000 HTHH hours

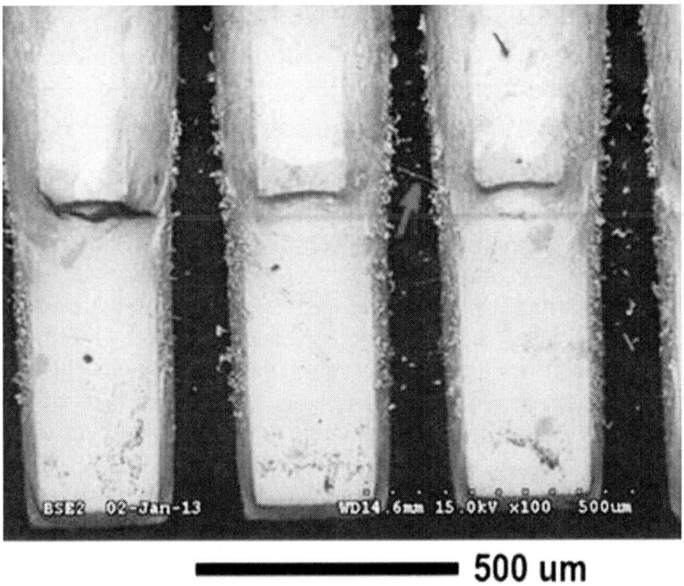

Figure 9. SEM image showing whisker growth from QFP64 board pads after 4000 HTHH hours

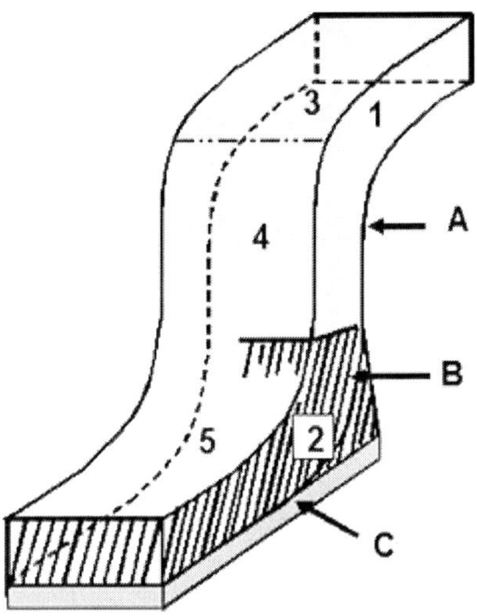

Figure 10. Primary whisker growth regions after TC and HTHH environments. The thin solder in region A exhibited the majority of the whisker growth after TC of alloy-42 lead terminations. The thick solder in region B displayed little or no whisker growth. The thin solder over the copper board pad in region C produced the most whisker growth after HTHH

Solder Microstructure and Whiskers

Recrystallization and Grain Change

The dynamic recrystallization theory proposed by Vianco and Rejent18,19 suggests that there is an optimal stress for whisker growth, above which other nonwhisker growth stress relaxation mechanisms occur. In the current study, varying degrees of recrystallization and stress relaxation were observed. Recrystallized grains adjacent to whiskers have been observed after PCTC and TC (Fig. 11). In some cases, the recrystallization was below the surface (Fig. 12). Possible evidence of recrystallization and/or grain growth was found on the increasing or decreasing whisker diameters observed on the TC samples (Fig. 13). Note that whisker growth angle changes were also observed, which could be due to tin mass transport changes at the whisker base20 or grain rotation during cycling. In contrast to observations by Susan et al.,20 sometimes substantial whisker growth occurred after the whisker base diameter increased. The reflowed SAC alloy generally has relatively large tin grains after reflow so recrystallization processes transforming large grains to small grains as part of whisker nucleation and growth are expected.

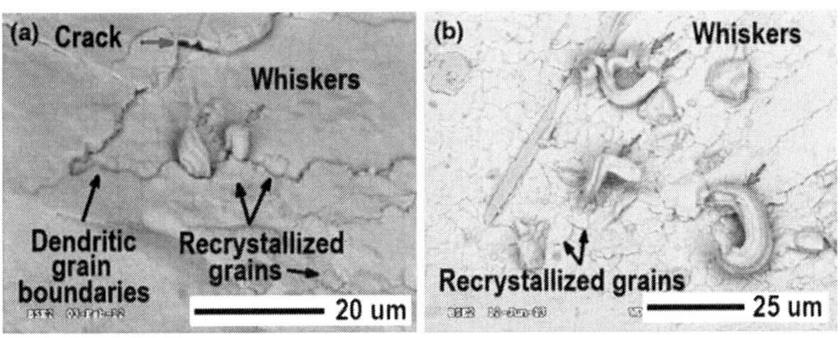

Figure 11. SEM image showing whisker growth near recrystallized grains and dendritic grain boundaries on the SAC solder fillets after (a) 1750 PCTC cycles and (b) 2110 TC cycles

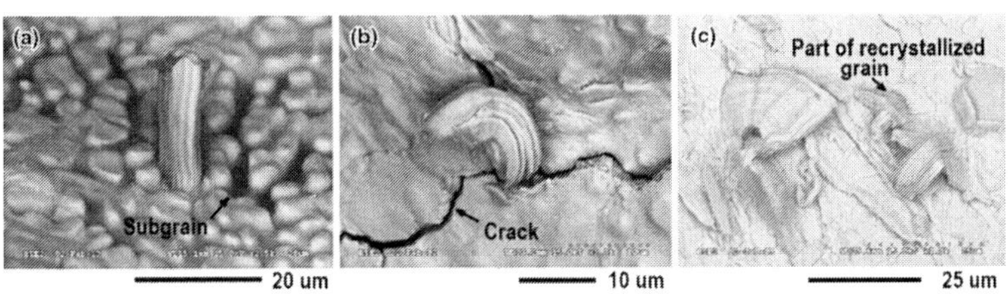

Figure 12. Subsurface grain recrystallization: (a) from subgrain region after 1750 PCTC cycles and (b) from primary tin dendrite triple junction after 1750 PCTC cycles. (c) A whisker on top of an underlying recrystallized grain after 2110 TC cycles

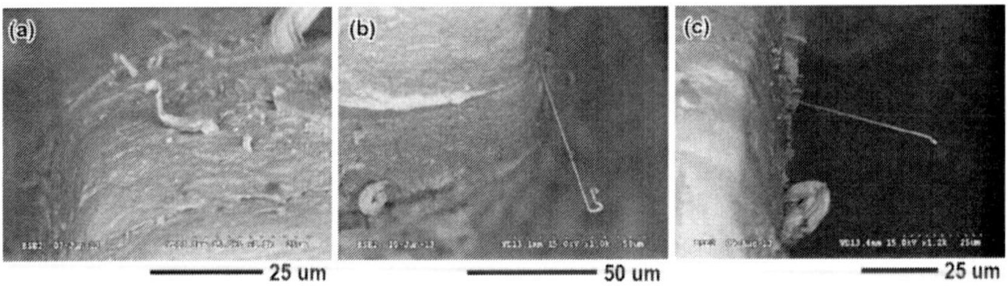

Figure 13. SEM image showing whisker diameter change during TC exposure on alloy-42 SOT6 with 1-0 contamination; (a) increasing, (b) and (c) decreasing whisker diameters

IMC

Primary IMCs such as Ag_3Sn and Cu_6Sn_5 particles promote whisker nucleation by increasing local stress in temperature cycling through differences in CTE, shear bands, and misoriented deformation zones around hard particles (Fig. 14). The growth of copper-tin IMC between the substrate and tin can also drive whisker growth.3 During PCTC, the IMC on lead was observed to grow much faster than expected. The SOT3 IMC before cycling was 0.3 μm and after 1,750 PCTC cycles was 1.6 μm—more than a five times increase in thickness. In contrast, the IMC on boards before cycling was 2.6 μm and after cycling was 3.0 μm. In addition, large CuNiSn IMCs were observed in the solder fillets next to whiskers. It is postulated that enhanced Cu diffusion from the board pad to component during cycling promotes IMC formation that increases whisker growth. As will be discussed next, IMCs also contribute to oxidation and corrosion.

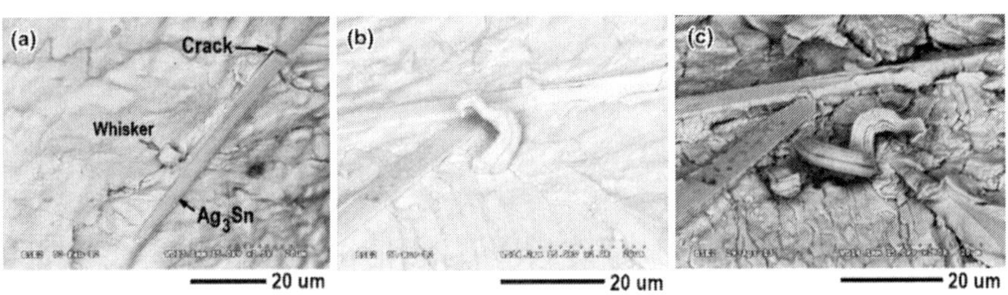

Figure 14. SEM image showing whisker on alloy-42 lead adjacent to Ag_3Sn IMC plate: (a) after 1750 PCTC cycles and (b) after 500 TC cycles. (c) Continued growth of whisker in (b) and new whisker growth after 2110 TC cycles

Oxidation/Corrosion

IMC particles contribute to uneven oxidation along the primary Sn grain boundaries, which creates local stress and promotes whisker formation (Fig. 15). The oxidation/corrosion often begins in the Sn interdentric spaces and can penetrate deeply into the solder. The solder fillet region was observed to be partially oxidized during the HTHH exposure with some oxide/corrosion regions traversing the entire solder fillet thickness to the substrate (Fig. 16). When all the tin around the whisker base becomes oxidized/corroded, whisker will stop growing. In addition, as oxidation/corrosion proceeds, whisker ductility is reduced and it is easier for the whiskers to become detached from the substrate. Corrosion was also observed adjacent to a whisker after the very low-humidity PCTC environment. The compressive stress caused by the 29–34% volume increase of the Sn oxides compared with the βSn promotes whisker growth. These findings are consistent with other investigators assessing the role of humidity on tin whisker growth on electronic component leads21–23 and on lead-free assemblies.5

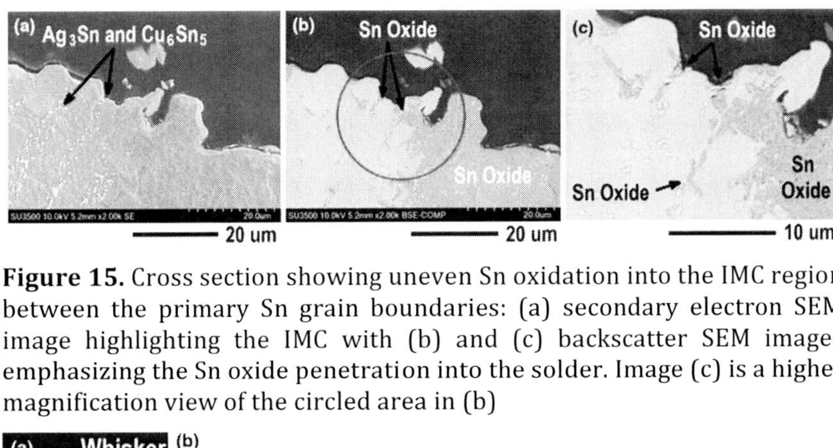

Figure 15. Cross section showing uneven Sn oxidation into the IMC region between the primary Sn grain boundaries: (a) secondary electron SEM image highlighting the IMC with (b) and (c) backscatter SEM images emphasizing the Sn oxide penetration into the solder. Image (c) is a higher magnification view of the circled area in (b)

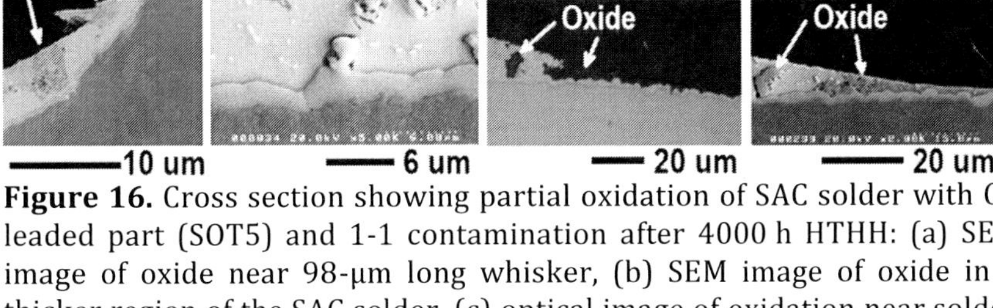

Figure 16. Cross section showing partial oxidation of SAC solder with Cu leaded part (SOT5) and 1-1 contamination after 4000 h HTHH: (a) SEM image of oxide near 98-μm long whisker, (b) SEM image of oxide in a thicker region of the SAC solder, (c) optical image of oxidation near solder wetting line, and (d) SEM image of region shown in (c)

It is important when studying whisker growth of solder joints that all the material interaction be considered. In alloy-42 solder joints, Cu_6Sn_5 interfacial layer at the board pads and solder particles contain Ni, which reduces the galvanic potential between the IMCs and the Sn, slowing the oxidation rate and the accompanying whisker nucleation. Ni from alloy-42 retarded whisker nucleation from the board Cu pads (Fig. 17). This is due to the Ni in the alloy-42 promoting the formation of the $(CuNi)_6Sn_5$ ternary IMC on the board Cu pad.

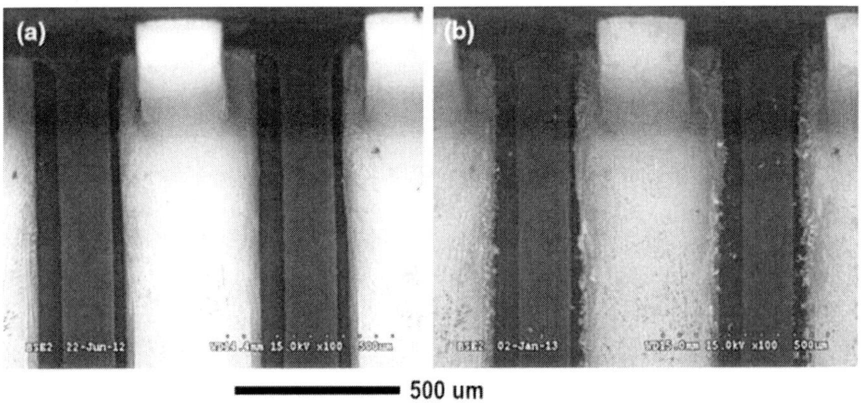

Figure 17. SEM image showing board Cu pad region whisker nucleation inhibition due to alloy-42 lead material: (a) 1000 h and (b) 4000 h of HTHH

Surface roughness of the solder also contributes to corrosion-induced whiskering. The roughness is partly a result of the shrinkage of the liquid between the primary tin dendrites. The roughness tends to trap contamination, which is difficult to clean (Fig. 12a).

Broken Whiskers

There are two possible explanations for the large amount of broken whiskers found during the examination at 4000 HTHH hours. Some of the thin regions of solder were completely oxidized and had no whiskers on the surface (Fig. 16). This finding suggests that oxidation/corrosion propagating under a whisker can make the whisker attachment to the base more brittle. In another cross-section of a whisker through its base, voids and a crack were observed (Fig. 18).

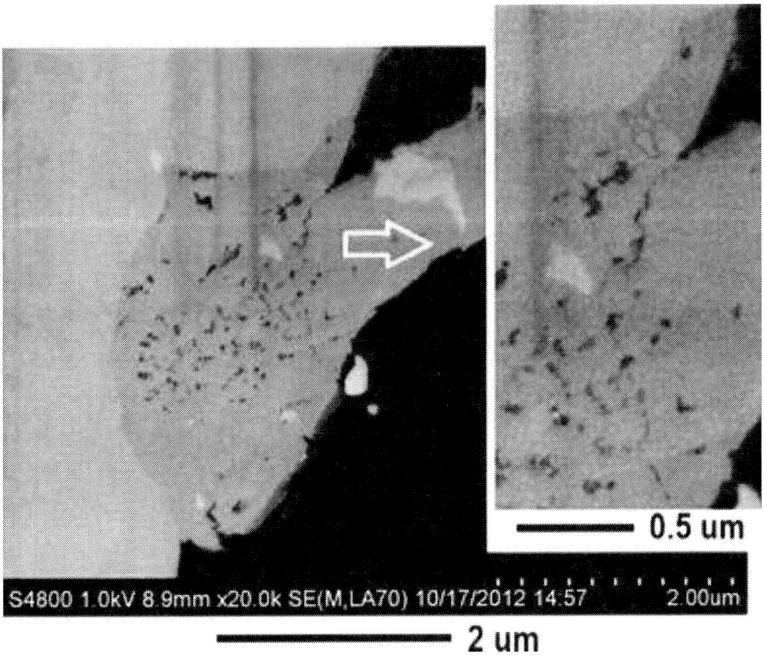

Figure 18. SEM image showing voids and cracks observed in the whisker base after 4000 HTHH hours

Nonwhisker Growth Stress Relaxation

In some cases, nonwhisker growth stress relaxation occurred. During PCTC in the thicker solder region, faceted crystal growth was observed (Fig. 19). Under the higher stress conditions of TC, the alloy-42 solder joints eventually exhibited massive cracking (Fig. 20) and massive eruptions consisting of many recrystallized grains (Fig. 7).

Figure 19. SEM image showing faceted grains on alloy-42 SOT3 with 1-1 contamination after 1750 PCTC cycles: (a) overall view and (b) high magnification view of circled region in (a)

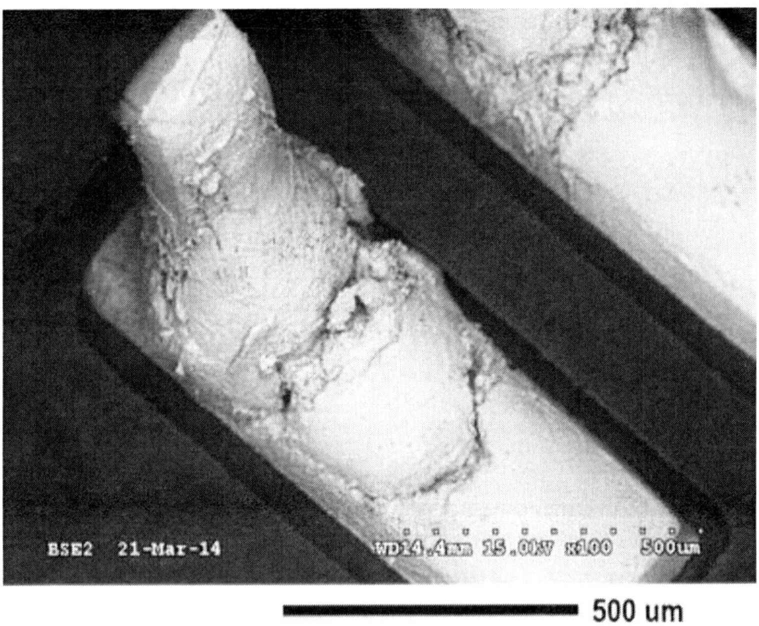

Figure 20. SEM image showing an alloy-42 termination solder joint with massive cracking after 2110 TC cycles

Test Considerations

An important test observation was that the intermediate SEM inspections stopped some whiskers from further growth after resuming environmental exposure. This condition was most pronounced on the HTHH samples where the majority of whiskers inspected at 1000 h stopped growing further when inspected at 4000 h, but new whiskers formed (Fig. 21). The interruption of whisker growth by inspection during thermal cycling (TC or PCTC) was less pronounced and existing whiskers continued to grow (Fig. 14). Thus, periodic inspection will cause an underreporting of whisker length. By progressively increasing time intervals or having longer uninterrupted tests, longer whisker growth would be obtained.

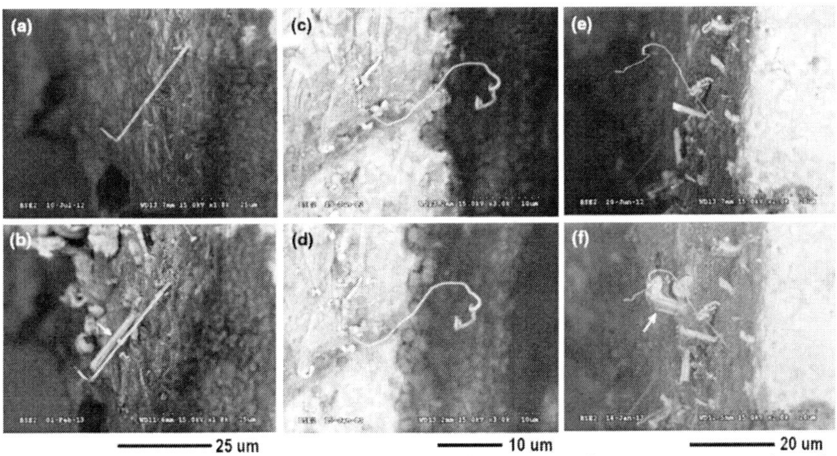

Figure 21. SEM images showing whisker growth interruption by the inspection process during HTHH testing. Red arrows indicate whiskers that stopped growing and yellow arrows indicate new whisker growth: (a) SOT5 1-0, U26, lead 1, 1000 h; (b) SOT5 1-0, U26, lead 1, 4000 h; (c) SOT5 1-1, U37, lead 5, 1000 h; (d) SOT5 1-1, U37, lead 5, 4000 h; (e) SOT5 1-1, U38, lead 2, 1000 h; and (f) SOT5 1-1, U38, lead 2, 4000 h

REE Oxidation

The SAC solder with REE was included in the test matrix because REE additions have increased the SAC solder whisker growth propensity.24,25 Ce was chosen over La and Y because results have been published indicating that Ce has better oxidation resistance,25which would be promising for incorporation in lead-free solder. The preliminary experiment in the current work26 had a different result; Ce was at least as effective as Y and La in whisker formation. Therefore, to discourage industry from even considering Ce additions to electronic solder, the BGA board used Ce at a concentration much lower than the 2% being considered for improving drop shock performance and whisker growth was still observed (Fig. 22). The manner in which RESn compound oxidation contributes to whisker growth is (I) liberation of Sn and (II) development of compressive stress by the RE oxide. Because REEs are less noble than Sn, RE in $RESn_3$ is more prone to oxidation, which results in rapid oxidation of $CeSn_3$ (or YSn_3 and $LaSn_3$). In addition, the surface active nature of REEs results in a condition where RESn IMC tends to segregate to the surface giving it improved access to oxygen. RESn IMCs can also be exposed to air if cracks form during solder joint thermal cycling.

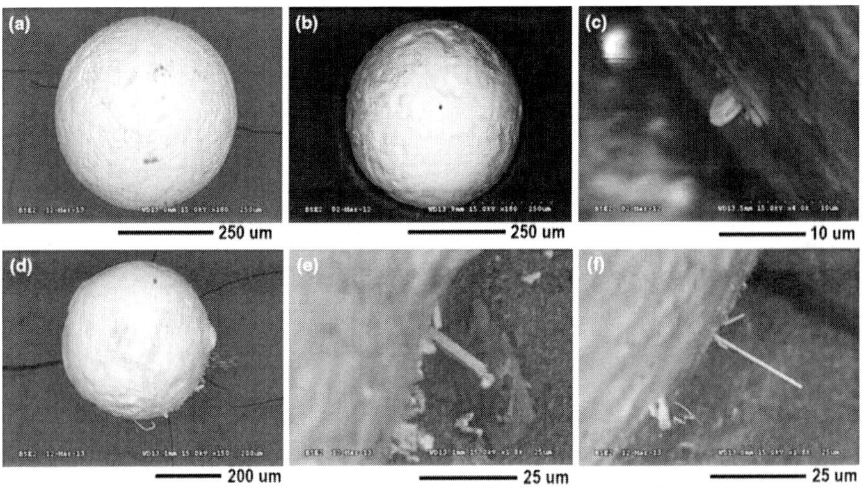

Figure 22 .SEM images comparing SAC105 (a) with SAC105 + 0.01Ce (b–f) solder balls attached to BGA board pads with SAC305 solder paste after 4000 HTHH hours (a, d–f) and after 1,750 PCTC cycles (b and c) at a 0-0 contamination level; (a) highlights no whisker growth on a SAC105 ball. Image (e) shows whisker highlighted by arrow in (d)

CONCLUSION

Based on a systematic study of electronic assemblies with leaded components covered with SAC305 solder, the following conclusions can be made. Whiskers grow from solder fillets around component leads and Cu pads, particularly where the solder is less than 25 μm thick. Very long whiskers can grow from SAC305 fillets. Whisker risks must be considered for high-reliability equipment. Thermal cycling promotes whisker growth when tin is on low CTE base materials. The stresses in the mild +50°C to +85°C PCTC cycle were sufficient to initiate recrystallization of the lead-free solder and nucleation of whiskers. Even though the −55°C to 85°C cycling does not create optimal stress conditions for whisker growth from alloy-42 terminations in SAC305 assembly, the alloy-42 leads still exhibited whisker lengths *longer* than the JESD201 piece part test limits. High humidity promotes whisker growth when thin SAC is on Cu. The 85°C/85 RH 1,000-h exposure on an assembly resulted in whisker lengths *longer* than the JESD201 piece part test limits for 4000 h. All the solder joint materials are important for ascertaining whisker growth. Although Ni tended to retard nucleation on Cu pads in HTHH, it tended to promote whisker growth by forming thick CuNiSn IMC on alloy-42 leads during PCTC. Assembly contamination is an important consideration for whisker growth in harsh service environments. REE additions to lead-free solder significantly increase whisker growth. Last, interruption of high-humidity conditions during interim inspections can cause an underreporting of whisker length.

ACKNOWLEDGEMENTS

The work was performed under the Strategic Environmental Research and Development Programs, a partnership between the U.S. Department of Defense, Environmental Protection Agency, and the Department of Energy, Project WP1753. The authors would like to thank the following individuals from Celestica: David Yang and Andre Delhaise for whisker SEM inspections and analysis, Russell Brush and Alon Walk for ATC testing and data analysis, Andrea Rawana for DSC analysis, and the following people from BAE Systems: Stephen McKeown, Clive Morris, Craig Sabin, and John Dion for their help with the tin whisker evaluation plan, and Don Hewett for his help with the test vehicle design.

REFERENCES

1. GEIA-STD-0005-2, *Standard for Mitigating the Effects of Tin Whiskers in Aerospace and High Performance Electronic Systems* (Warrendale, PA: SAE Int., 2012).
2. G.T. Galyon, *IEEE Trans. Electron. Packag. Manuf.* 28, 94 (2005).
3. K.N. Tu, *Phys. Rev. B* 49, 2030 (1994).

4. M. Dittes, P. Oberndorff, P. Crema, and V. Schroeder, *IEEE Proceedings of the Electronic Package Technology Conference* (Piscataway, NJ: IEEE, 2003), pp. 822–826.

5. J.W. Osenbach, R.L. Shook, B.T. Vaccaro, B.D. Potteiger, A.N. Amin, K.N. Hooghan, P. Suratkar, and P. Ruengsinsub, *IEEE Trans. Electron. Packag. Manuf.* 28, 36 (2005).

6. F. Pei, N. Jadhav, and E. Chason, *Appl. Phys. Lett.* 100, 221902 (2012).

7. W.J. Boettinger, C.E. Johnson, L.A. Bendersky, K.-W. Moon, M.E. Williams, and G.R. Stafford, *Acta Mater.* 53, 5033 (2005).

8. E.R. Crandall, G.T. Flowers, P. Lall, and M.J. Bozack, *Proceedings of the 57th IEEE Holm Conference Electronic Contacts* (Piscataway, NJ: IEEE, 2011), pp. 1–5.

9. P. Sarobol, Y. Wang, W.H. Chen, A.E. Pedigo, J.P. Koppes, J.E. Blendell, and C.A. Handwerker, *JOM* 65, 1350 (2013).

10. T. Woodrow, *Proceedings of the Surface Mount Technologies Association International Conference* (Edina, MN: SMTA, 2006), pp. 674–723.

11. T. Fang, M. Osterman, and M. Pecht, *Microelectron. Reliab.* 46, 846 (2006).

12. L. Panashchenko and M. Osterman, *Proceedings of the 59th IEEE Electronic Components Technology Conference* (Piscataway, NJ: IEEE, 2009), pp. 1037–1043.

13. A. Baated, K.S. Kim, K. Suganuma, S. Huang, B. Jurcik, S. Nozawa, and M. Ueshima,*J. Mater. Sci.* 21, 1066 (2010).

14. K. Sweatman, J. Masuda, T. Nozu, M. Koshi, and T. Nishimura, *Proceedings of the 2010 Technology Conference IPC APEX EXPO* (Bannockburn, IL: IPC, 2010), pp. S11_02-1-S11_02-5.

15. S. Meschter, S. McKeown, P. Snugovsky, and J. Kennedy, *SMTA J.* 24, 23 (2011).

16. P. Snugovsky, S. Meschter, Z. Bagheri, E. Kosiba, M. Romansky, and J. Kennedy, *J. Electron. Mater.* 41, 204 (2012).

17. P. Snugovsky, Z. Bagheri, and M. Romansky, *SMTA International Conference Soldering and Reliability* (Edina, MN: SMTA, 2008).

18. P.T. Vianco and J.A. Rejent, *J. Electron. Mater.* 38, 1815 (2009).

19. P.T. Vianco and J.A. Rejent, *J. Electron. Mater.* 38, 1826 (2009).

20. D. Susan, J. Michael, R.P. Grant, B. McKenzie, and W.G. Yelton, *Metall. Mater. Trans. A* 44, 1485 (2013).

21. J.W. Osenbach, J.M. DeLucca, B.D. Potteiger, A. Amin, R.L. Shook, and F.A. Baiocchi,*IEEE Trans. Electron. Packag. Manuf.* 30, 23 (2007).

22. P. Oberndorff, M. Dittes, P. Crema, P. Su, and E. Yu, *IEEE Trans. Electron. Packag. Manuf.* 29, 239 (2006).

23. P. Su, J. Howell, and S. Chopin, *IEEE Trans. Electron. Packag. Manuf.* 29, 246 (2006).

24. T.-H. Chuang and S.-F. Yen, *J. Electron. Mater.* 35, 1261 (2006).

25. M.A. Dudek and N. Chawla, *J. Electron. Mater.* 38, 210 (2009).

26. D. Perovic, L. Snugovsky, J. Rutter, P. Snugovsky, Z. Bagheri, and S. Meschter, *SMTA Interantional Conference Soldering and Reliability* (Edina, MN: SMTA, 2011).

12

CHAPTER

Apatite Biominerals

Christèle Combes [1], Sophie Cazalbou [2] and Christian Rey [1]

[1]CIRIMAT, Université de Toulouse, CNRS, INPT, UPS, ENSIACET, 4 allée Emile Monso, CS 44362, 31030 Toulouse cedex 4, France

[2]CIRIMAT, Université de Toulouse, CNRS, INPT, UPS, Université Paul Sabatier, Faculté de Pharmacie, 35 Chemin des Maraichers, 31062 Toulouse cedex 9, France

ABSTRACT

Calcium phosphate apatites offer outstanding biological adaptability that can be attributed to their specific physico-chemical and structural properties. The aim of this review is to summarize and discuss the specific characteristics of calcium phosphate apatite biominerals in vertebrate hard tissues (bone, dentine and enamel). Firstly, the structural, elemental and chemical compositions of apatite biominerals will be summarized, followed by the presentation of the actual conception of the fine structure of synthetic and biological apatites, which is essentially based on the existence of a hydrated layer at the surface of the nanocrystals. The conditions of the formation of these biominerals and the hypothesis of the existence of apatite precursors will be discussed. Then, we will examine the evolution of apatite biominerals, especially during bone and enamel aging and also focus on the adaptability of apatite biominerals to the biological function of their related hard tissues. Finally, the diagenetic evolution of apatite fossils will be analyzed.

Keywords: biominerals; calcium phosphate apatites; bone; enamel; dentine; crystallization; chemical composition; evolution; surface reactivity

1. INTRODUCTION

The existence of calcium phosphate biominerals has been reported in living creatures from unicellular organisms to vertebrates [1]. Most of these calcium phosphate biominerals exist as amorphous phase in primitive organisms. However, in evolved organisms and especially in vertebrates, they exist mainly as apatite structures, although a variety of other crystallized calcium phosphate phases (whitlockite, brushite, and octacalcium phosphate) may form in uncontrolled pathologic calcifications [2,3]. Compared to other crystalline biominerals such as calcium carbonates, calcium phosphate apatites exhibit undeniably larger biological adaptability. Such adaptability is notable in specific physico-chemical and structural properties, rendering them useful for a large variety of biological uses such as in the protection of internal organs (shells, scales, and flat bones), internal skeleton (bones), sensors (bones of internal ears, rostrum of whales, otolithes in some species [4] or tympanic bullae [5]) and in the organs of attack and defense (antlers and tusks). In addition, other functional roles have been attributed to apatite biominerals like homeostasis or the inactivation of toxic elements. Several reviews have been published on this subject with different approaches focusing on crystal structures, and calcium phosphate precursors [6], the role of OH-channels [7], the role of unstable amorphous precursors [8], the role of polyphosphates [9], general view including the organic matrices [10], apatite biominerals and biomimetic processing and materials [11], and mineralogical oriented reviews [12]. This review aims to present and discuss the specific characteristics of biomineral apatites and illustrate how these characteristics and the resulting mineral properties have been employed in the adaptation for specific biological functions, in the case of vertebrate hard tissues (bone, dentine and enamel). To start, the structure and composition of biomineral apatites will be reviewed, followed by a description of their biological formation, evolution and maturation particularities and finally their influence on the properties of tissues and their biological behavior will be summarized.

2. STRUCTURE AND CHEMICAL FORMULAS

The first structural identifications of calcium phosphate biominerals in vertebrates using X-ray diffraction were obtained by de Jong [13], who established that they corresponded to an apatite structure. Excellent reviews on the apatite structure have been published and interested readers may report to these for more in-depth information [14,15]. Since then, stoichiometric hydroxyapatite (HA):

$$Ca_{10} \cdot (PO_4)_6 \cdot (OH)_2 \quad (1)$$

has been generally used as a model for bone mineral and tooth enamel. However, unlike pure stoichiometric HA which crystallizes in the monoclinic $P2_1/b$ space group, biological apatites are generally indexed in the hexagonal $P6_3/m$ space group [15]. The main difference between the HA model and biological apatites is the presence of significant amounts of carbonate ions in all mineralized biological tissues, including pathological apatite biominerals. Early studies on synthetic carbonated apatites established that carbonate ions could in fact be part of the apatite structure [16,17]. Detailed studies indicated that carbonate ions could be located in the two anionic sites of the apatite structure: in the PO_4^{3-} sites (type B carbonated apatite) and the OH^- sites (type A carbonated apatite). Another important characteristic of biological apatites is their non-stoichiometry, often referred to as calcium deficiency, although it appears more complex. Even if it has been the subject of much controversy, the use of non-stoichiometric carbonated apatite as a model for the biological calcification of vertebrate hard tissues is accepted nowadays, with some alterations taking into account the multiplicity of carbonate sites in apatites related to coupled substitutions and interactions [15,17,18,19,20].

Another specificity of biological apatites, which was more recently established, is the presence of hydrogen phosphate (HPO_4^{2-}) ions in PO_4^{3-} sites [21,22,23]. These two types of bivalent ions substituting for PO_4^{3-} (type B CO_3^{2-} and HPO_4^{2-}) have been shown to correspond to the formation of calcium deficient apatites, the chemical formulas of which have been the subject of several works, essentially based on the composition of model minerals proposed by mineralogists or synthetic analogues. A general chemical formula proposed by Winand for HPO_4^{2-}-containing apatites was [24]:

$$Ca_{10-x} \cdot (PO_4)_{6-x} \cdot (HPO_4)_x \cdot (OH)_{2-x} \text{ with } 0 \leqslant x \leqslant 2 \quad (2)$$

and a similar one by Labarthe *et al.* for carbonate-containing apatites [16]:

$$Ca_{10-x} \cdot (PO_4)_{6-x} \cdot (CO_3)_x \cdot (OH)_{2-x} \text{ with } 0 \leqslant x \leqslant 2 \quad (3)$$

These formulas exhibit similar behavior for the bivalent ion substitution of trivalent phosphates and the necessary maintenance of the structural neutral electrical charge, *i.e.*, the loss of a negative charge due to these substitutions is compensated by the creation of a cationic vacancy and an anionic vacancy in monovalent sites. These chemical formulas are consistent with the limit composition observed ($x = 2$) and the decrease of the OH^- content when the amount of carbonate and/or HPO_4^{2-} in the apatite increases [25]. Other chemical formulas have been proposed to take into account the existence of other charge compensation mechanisms related to the condition of formation of these non-stoichiometric apatites, such as an excess of calcium (u), usual in carbonate apatites obtained in alkaline media [16] or an intracrystalline hydrolysis of phosphate groups (y). One of the most general one proposed by Rey *et al.* [26] is however of little relevance for biological apatites:

$$Ca_{10-x+u} \cdot (PO_4)_{6-x-y} \cdot (HPO_4^{2-} \text{ or } CO_3^{2-})_{x+y} \cdot (OH)_{2-x+2u+y} \text{ with } 0 \leq x \leq 2 \text{ and } 0 \leq 2u + y \leq x \quad (4)$$

Biological apatites are best approximated by the simple combination of the two previous chemical formulas (2) and (3), taking into account the possible existence of type A carbonates:

$$Ca_{10-x} \cdot (PO_4)_{6-x} \cdot (HPO_4 \text{ or } CO_3)_x \cdot (OH \text{ or } 1/2\,CO_3)_{2-x} \text{ with } 0 \leq x \leq 2 \quad (5)$$

Many ionic substitutions are possible in apatites, involving for example, trivalent cations (e.g., rare earth elements, actinides) or monovalent cations (especially Na^+) or other bivalent cations for Ca^{2+} or tetravalent or bivalent ions replacing PO_4^{3-} in addition to trivalent ones, and bivalent or monovalent ions replacing OH^-. Several charge compensation mechanisms have been proposed. The composition of biological apatites will be developed in Section 4.

3. NON-APATITIC ENVIRONMENTS AND THE HYDRATED LAYER

More recently, these models, which are based on well-crystallized apatites, were re-examined due to the discovery, using mostly spectroscopic techniques (Fourier transform infrared (FTIR), Raman and solid-state nuclear magnetic resonance (NMR) spectroscopies), of the existence of specific spectral lines in the spectra of biological nanocrystalline apatites which do not appear for well-crystallized apatites and which have been designated as "non-apatitic environments" of the mineral ions [22,23,27,28]. These "non-apatitic" phosphate and carbonate environments have been shown to appear more clearly in the ν_4 PO_4 and ν_2 CO_3 domains of FTIR spectra. Synthetic models of nanocrystalline apatites mimicking the main characteristics of biological apatites have been prepared and studied [29,30,31]. These "non-apatitic" phosphate and carbonate environments have been shown, using ion exchange experiments, to share the same surface domain corresponding to a structured hydrated layer on apatite nanocrystals [32]. The structure of the hydrated layer seems very sensitive to its ion content and state of hydration: a loss of the original fine structural details revealed by spectroscopic techniques, in wet state, is observed on drying or when specific ions like magnesium are present, leading to line broadening and amorphization [33,34]. Several rapid and reversible ion exchange reactions have been reported [35,36,37,38] and it has been shown that the adsorption of several ionized organic molecules corresponded to ion exchanges with mineral ions of the hydrated layer [39,40,41]. Consequently, the global chemical formulas of apatites reported in the previous paragraph, which do not take into account the existence of hydrated surface domains, should be re-examined. A definitive formula still seems out of reach due to uncertainties regarding the composition of the hydrated layer and the apatite core.

Based on complementary investigations, using spectroscopic techniques and analytical chemistry, several features of biological apatites and their synthetic analogues have been identified:

1) Apatite nanocrystals contain non-apatitic anionic and cationic chemical environments,

2) These environments strongly interact with hydrated domains,

3) Immature samples (freshly precipitated nanocrystalline apatites) show FTIR fine-band substructure changes upon drying without leading to long-range order modifications,

4) In the early stages of formation, this fine substructure shows striking similarities to the FTIR spectroscopic signature of octacalcium phosphate (OCP), which is constituted by alternating "apatitic" and "hydrated" layers [33,42].

These features allowed for the proposition of a model in which apatite nanocrystals are covered with a rather labile but structured hydrated surface layer, containing relatively mobile ions (mainly bivalent anions and cations: Ca^{2+}, HPO_4^{2-}, CO_3^{2-}) in "non-apatitic" environments [43]. However, the exact structure and composition of this hydrated layer is still under investigation. A schematic representation of this hydrated surface layer model of apatite nanocrystals in aqueous medium is given in Figure 1.

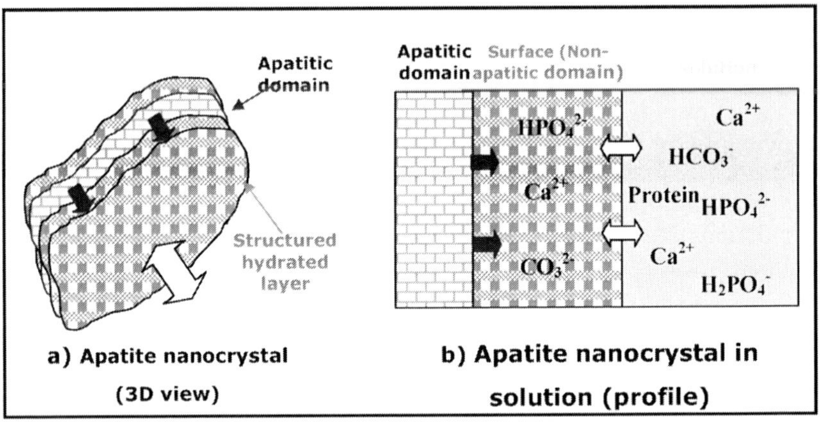

Figure 1. Schematic representation of the surface hydrated layer model for poorly crystalline apatite nanocrystals (Reprinted from *Nanocrystalline apatite based biomaterials: synthesis, processing and characterization*; Copyright (2009), Eichert D., Drouet C., Sfihi H., Rey C. and Combes C. [43] with permission from Nova Science Publishers Inc.).

As for any other kind of nanomaterial, nanocrystalline apatites, whether biological or synthetic, exhibit a high surface to volume ratio leading to consider experimental results (spectra, physico-chemical or biological properties) as a combination of bulk and surface contributions. The latter is of particular importance, since numerous functions of the bone mineral involve processes or phenomena at the interface between bone apatite nanocrystals and their surroundings. However the role of the apatite core seems determining in the manifestation of some properties. Studies of synthetic analogues of biological nanocrystalline apatites reveal that they exhibit strong reactivity related to this structure, especially a specific ageing process, frequently called maturation in reference to bone. The driving force of this process is the relative instability of the hydrated layer compared to that of apatite domains. Thus, the apatite domains develop slowly at the expense of the hydrated layer. Two main routes of progression have been recognized, depending on the composition of the solution (Figure 2) [30].

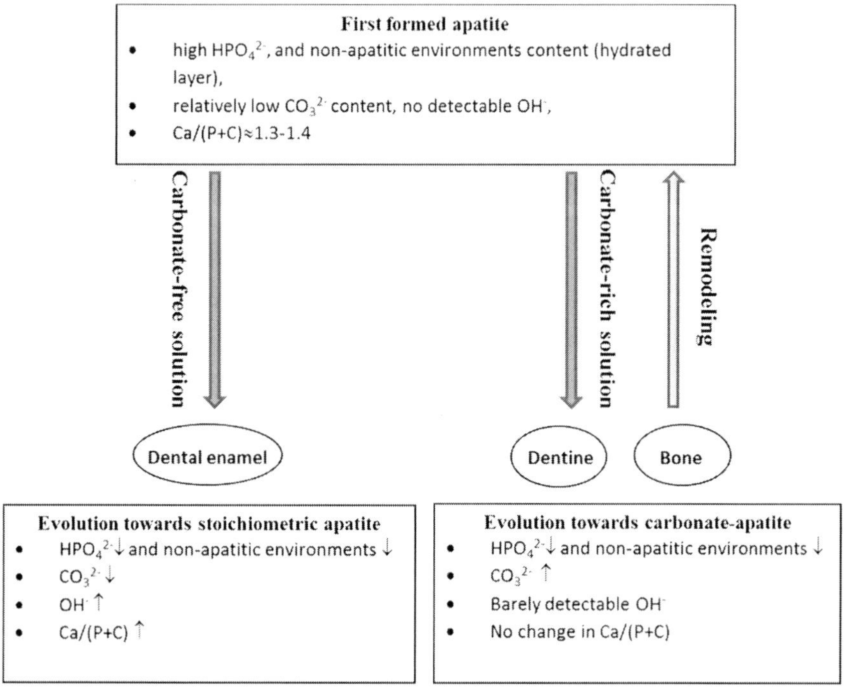

Figure 2. Maturation pathways of nanocrystalline apatites depending on the composition of the medium (the arrows, after the ions, represent an increase (or decrease) of the species considered during maturation).

When the solution contains carbonate ions, the initial precipitates at time zero contain only a few carbonate ions, mainly in the hydrated layer, and a large amount of HPO_4^{2-} ions. On ageing, the amount of carbonate ions increases in the hydrated layer, as the amount of labile HPO_4^{2-}decreases, leaving the Ca/(P + C) atomic ratio almost unchanged. This change in the hydrated layer is accompanied by an increase of carbonate ions in the growing apatite domains, in both type A and type B sites, at an apparent constant ratio during ageing at physiological pH.

When the solution does not contain carbonate ions a different maturation process is observed. Although the hydrated layer converts progressively into apatite on ageing, the Ca/P ratio increases and higher amounts of OH^- ions are incorporated in the growing apatite domains, whereas the total amount of HPO_4^{2-} decreases. This evolution corresponds to a formation of hydroxyapatites closer to stoichiometry.

In both cases, the growing apatite domains show a different composition to that of the hydrated layer and their growth is associated with a release of protons and/or carbonate [44].

4. COMPOSITION OF THE MAIN MINERALIZED TISSUES

Biomineral apatites always exist in association with organic matrices, and even if the elemental composition of whole hard tissues can be determined it is often difficult to know the precise composition of each component, especially regarding trace elements. The mean elemental composition of the three main human hard tissues (bone, dentine, enamel) is given in Table 1. Conventionally, three types of elements are distinguished by biologists: main, minor and trace. These elements can be part of very different molecular or crystalline structures [45,46,47,48,49].

Table 1. Mean values of elemental composition of dried human main hard tissues (determined from Iyengar and Tandon compilation [46], completed by other works [45,47,48,49,50,51,52,53].

Elements or Ions	Bone	Dentine	Dental Enamel	References
	Major elements (wt %)			
C (total)	16.7	11.8	1.4	bone [41]; dentine and enamel [49]
CO_3^{2-}	5.6	4.6	3.2	bone [53]; dentine and enamel [45]
N	4.9	4.0	0.32	dentine and enamel [49]
Ca	25.4	26.9	36.6	dentine and enamel [45]
P	11.6	13.2	17.7	dentine and enamel [45]
	Minor elements (wt %)			
Cl	0.13	0.065	0.37	dentine and enamel [46–49,51,52]
K	0.0047	0.024	0.070	dentine and enamel [46,47,51]
Mg	0.27	0.74	0.29	dentine and enamel [46–48,51,52]
Na	0.53	0.76	0.77	dentine and enamel [47–49,51,52]
S	0.08	0.070	0.021	dentine and enamel [47]
	Main trace elements (ppm)			
Al	29	210	55	dentine and enamel [46,49,51]
B	22	-	11	enamel [45]
F	400	215	50	all values from Ishiguro *et al.* [50]
Fe	76	44	34	dentine and enamel [46,47,51,52]
Pb	4.4	15	17	all values computed from [46]
Sr	70	145	173	dentine and enamel [46–49,52]
Zn	205	148	170	dentine and enamel [46–49,51,52]

4.1. Main Elements

Nitrogen is considered to belong mainly to proteins and is often considered to represent the organic matrix, although non-protein organic molecules may also constitute a fraction of the organic matrix. Calcium is essentially located in the apatite structure and is considered to represent the mineral content.

Different mineral-organic associations result in tissues with very different characteristics: tooth enamel with a high content of mineral is the hardest tissue of vertebrates, with good resistance to compression and wear, but it is also rather brittle. Dentine, found underneath the thin enamel layer in teeth, appears much less mineralized than enamel and due to its collagen matrix offers both high compressive and tensile strengths. Bone seems generally less mineralized than dentine, although it is quite similar in composition, with excellent compressive and tensile strengths and a high adaptability to mechanical stress. The mineralization ratio in bone can vary considerably. In some intramuscular fish bones like those found in herrings, for example, mineralization is very low and progresses very slowly within the bone organ, in relation to the growth of the animal [54]. This is also the case for turkey tendons, which transform into bone very slowly and offer, like herring intramuscular bone, a possibility to follow a mineralization process in slow motion. Rather strong variations of mineralization ratios are also found in humans depending on the age and type of bone (trabecular or cortical bone). In infants, for example, bones are much less mineralized than in adults. The mineralization ratio of the tissue should not be confused with the bone mass. In elderly people, the global mass of the skeleton decreases and partly the mineralization ratio of the bone tissue [55]. More recently, a strong adaptation of bone tissue to mechanical stresses has been highlighted during space travels, which can have a strong influence on the bone mass [56]. Hypermineralized bones have been described such as in the rostrum of some whales (up to 85%–90% mineral), where the collagen matrix is progressively lysed and replaced by apatite [57].

Carbon can be found in the organic matrices and also in apatite minerals in the carbonate anions. Oxygen and hydrogen can be found in the organic components and in minerals as components of carbonates, phosphates, hydrogen-phosphate and hydroxide ions, as well as water molecules associated to the organic matrices and minerals. One of the main characteristics of all apatite biominerals is the presence of significant amounts of carbonate species and several authors have suggested banning the term hydroxyapatite in reference to carbonate-apatites, in the case of dentine and bone. The carbonate ions in biological apatites have been located in the apatite structure in both trivalent anionic sites and monovalent sites [17,58,59]. A third location of carbonate ions in the hydrated surface layer of biological nanocrystals has also been identified [58]. These carbonate species present different spectroscopic characteristics in Raman, FTIR or solid-state NMR spectroscopies [27,58,60,61].

Phosphorus essentially exists in bones as orthophosphate ions associated with the apatite crystals. Two main species have been identified: PO_4^{3-} and HPO_4^{2-}. These species exist on trivalent anionic sites and, as in the case of carbonate, a surface location has also been identified. Hydrogen-phosphate ions have been detected using FTIR and solid-state NMR spectroscopies [22,29,62,63,64]. The bivalent anion content, carbonate and hydrogen phosphate, of apatite is associated with a calcium and hydroxide ion deficiency. The amount of bivalent species varies with the tissue.

Average compositions have been derived for biological apatites neglecting the existence of the hydrated layer and its mineral content. These average compositions of mineral in human bone, dentine and enamel are based on the simplified model of calcium deficient apatites [21] presented above, which do not account for the sodium content, the presence of type A carbonate replacing OH^- ions and all minor and trace elements except Mg:

$$\text{Human bone apatite}: Ca_{8.1}\cdot Mg_{0.2}\cdot (PO_4)_{4.3}\cdot (HPO_4)_{0.5}\cdot (CO_3)_{1.2}\cdot (OH)_{0.3} \quad (6)$$

$$\text{Human dentine apatite}: Ca_{8.0}\cdot Mg_{0.4}\cdot (PO_4)_{4.4}\cdot (HPO_4)_{0.7}\cdot (CO_3)_{0.9}\cdot (OH)_{0.4} \quad (7)$$

$$\text{Human enamel apatite}: Ca_{8.8}\cdot Mg_{0.1}\cdot (PO_4)_{4.9}\cdot (HPO_4)_{0.6}\cdot (CO_3)_{0.5}\cdot (OH)_{0.9} \quad (8)$$

This presentation highlights the differences in bone and dentine apatite vacancies content compared to enamel: in humans but also in animals, the amount of vacancies in bone apatites is close to the highest amount of vacancies possible in the apatite structure. In enamel, on the contrary, the amount of vacancies is much lower.

One of the most controversial differences between biological apatite compositions is the hydroxide ion content. In tooth enamel the OH^- ions can be directly observed by spectroscopic techniques (FTIR, Raman scattering and solid-state NMR) and the denomination of hydroxyapatite is justified for this biomineral. However, OH^- lines in bone or dentine, when they are detected, show very weak intensity, and the reported estimation of the hydroxide content varies considerably, depending on the study and on the bone samples used. Taylor *et al.* [65] report 50%

OH^- content of stoichiometric hydroxyapatite ($Ca_{10}\cdot(PO_4)_6\,(OH)_2$) in ox bone, Cho and Ackerman found 20% in chicken bone using solid-state NMR [66]. It should be noted that these values appear in large excess considering the average bone composition of chemical formula (6) and others failed to detect any OH^- by Raman or FTIR spectroscopies [67,68]. Recent estimations on different kinds of fresh and freeze-dried bones suggest a null or extremely low OH^- content (a few percent) [41]. These discrepancies have been attributed to a possible internal hydrolysis process between water molecules trapped in the apatite lattice and PO_4^{3-} groups:

$$H_2O + PO_4^{3-} \rightarrow HPO_4^{2-} + OH^- \tag{9}$$

Such reactions have been observed in synthetic apatite nanocrystals heated at moderate temperatures (<200 °C) [69]. At higher temperatures an increase of OH^- content in calcined bone is always observed. It can be assigned to the decomposition of carbonate species and the hydration of the resulting oxide ions:

$$\begin{aligned} CO_3^{2-} &\rightarrow CO_2 + O^{2-} \\ O^{2-} + H_2O &\rightarrow 2\,OH^- \end{aligned} \tag{10}$$

Concerning the OH^- content, Pasteris *et al.*, reported a decrease of OH^- Raman line intensities in spectra of ground stoichiometric hydroxyapatite with increased crystals strains and decreased crystal sizes [67]. The apatite crystal size or the level of strain does not however necessarily alter the observation of OH^- lines, considering that nanocrystalline apatites with clear measurable OH^-lines can be prepared [69].

On ageing, several alterations of the bone composition have been reported as will be discussed. In the very early stages, in embryonic bones, the amount of HPO_4^{2-} is rather high and the carbonate content is very low, leading to calcium deficient apatites with low Ca/P ratios [32,70]. On ageing however, the Ca/P ratio and the carbonate content increases. In fact, these events are linked: as the counter ion of CO_3^{2-} ions is calcium, the increase in carbonate leads inevitably to an increase of the related calcium ions and of the Ca/P ratio. However, this does not mean that there is an evolution of bone apatite composition towards stoichiometry as is reported sometimes in publications. The Ca/(P + C) atomic ratio, where C stands for carbonate species, remains remarkably constant at about 1.3–1.4, regardless of the age or the animal species [21,70]. This consistency of the Ca/(P + C) ratio simply reveals that the amount of bivalent species (HPO_4^{2-} or CO_3^{2-}) is close to constant in bone and that carbonate ions replace hydrogen-phosphate ones upon ageing. On the contrary, in teeth enamel, the carbonate content remains at a low level and during its formation an increase of the Ca/P and Ca/(P + C) ratios is observed, which is related to the presence of hydroxide ions and type A carbonate [30].

4.2. Minor Elements

Although sulfur may enter in the apatite structure this element, present in most biological tissues, is essentially associated with polysaccharides moieties. Strong variations of potassium levels have been found in bone, which could possibly be related to the release of internal cell fluid and can depend on preparation methods. However, the potassium content of enamel appears higher than that of bone and it is possible that this element may be trapped in the enamel apatite mineral. Variations in the chlorine and sodium content of bone and probably in dentine can be attributed in part to the associated biological tissue fluids and may also be related to the tissue preparation methods employed. However, the sodium/chlorine ratio in these tissues appears much higher than in biological extracellular fluid, suggesting that sodium is being incorporated into the mineral crystals. Sodium incorporation could be related to that of carbonate as hypothesized by different researchers [71]. Solid-state NMR spectroscopy analysis confirms that sodium belongs essentially to the mineral phase [72]. A higher content of chlorine in enamel compared to dentine and bone is observed. The incorporation of chlorine as chloride ions in enamel is supported by the occurrence of Cl–OH hydrogen bonding which can be clearly discerned in FTIR and Raman spectra (shift of the OH^- line towards lower wavenumbers: 3500 cm^{-1} instead of 3570 cm^{-1} in hydroxyapatite) [73]. The reason for the incorporation of chlorine as a substitute for hydroxide ions in enamel apatite, compared to bone or dentine apatite is most probably related to the relatively low level of bivalent ions in enamel apatite and the higher amount of occupied monovalent ionic sites in accordance with the proposed chemical formulas (6)–(8). The magnesium content of bone decreases with increasing age. Although this element can enter into the calcium phosphate apatites as a substitute for calcium, it has also been reported to attach to the surface of the crystals [74].

4.3. Trace Elements

Numerous trace elements are found in hard tissues that are generally divided as essential elements, (*i.e.*, elements necessary for the living organism) such as As, F, Mn, Cu, Zn, and Sr; toxic elements (Al and Pb); and elements without known biological effects. These categories are, however, not clearly separated and several elements may be essential at low concentrations and become toxic at higher ones. In some cases, Al, As, F, Pb and Sr for example, the trace element concentration is related to diet intake, and some of these elements, Al, As, F and Pb for example, are considered to show additive accumulation. The identification of locations of trace elements in hard tissues is a complex task and methods involving the selective destruction of the organic part (hydrazine or hypochlorite treatments) or the mineral component (acid or Ethylenediaminetetraacetic acid (EDTA) dissolutions) do not prevent parallel or post treatment reactions, which may distort the validity of the collected data. Only in a few reported cases has the clear location of trace elements been partly determined, often using spectroscopic techniques. Silicon has been located in the organic matrix in the vicinity of the mineralization front [75]. On the contrary, fluorine has been located in the apatite mineral as a fluoride ion [76,77]. It is important to note that these data do not mean that traces of silicon or fluoride do not exist in other compounds, in other forms. Except for a few elements, like fluoride ions, trace elements are not believed to exhibit a significant physico-chemical effect on the mineral, although most of them can show a biological effect depending on concentration.

One of the most emblematic trace elements is fluorine, existing in the form of fluoride ions, and mostly found in the apatite biomineral (99% of the F body burden). The fluoride content varies strongly depending on dietary uptake, though in normal bone it constitutes 0.05% to 0.1%. A high fluoride content (above 0.3%) is indicative of fluorosis disease, which results in an increase in bone mineral density and in cancellous bone volume and enhanced bone fragility [78]. During the formation of teeth, fluorosis disturbs the formation of enamel. In teeth enamel, the fluoride content has been shown to vary strongly, with a higher content on the surface than in the inner enamel. Globally, the content of F in enamel is lower than that in dentine or bone (Table 1). In addition, in these tissues, the distribution of fluoride and other bone seeking elements is not homogeneous and the zones which are the best vascularized (cancellous bone, dentinal pulp) show a higher fluoride content [79]. This inhomogeneous distribution of trace elements is also found for other bone seeking elements like lead, aluminum and strontium [80]. Some elements like Sr, Si, and Zn have shown interesting biological properties.

5. CRYSTAL PHYSICAL CHARACTERISTICS

The mineralization process is controlled and occurs within delimited tissue boundaries with some degree of order. Bone exhibits elongated plate-like nanocrystals (about 50 nm long, 25 nm width and 10 nm thick) [10,81]. The length corresponds to the c axis of the hexagonal unit-cell, which is parallel to the axis of the collagen fibers. The crystals form agglomerates of parallel platelet crystals. This common orientation begins at the very early stages and has been observed, for example, in mineralizing turkey tendons [82]. The orientation of bone apatite crystals along the collagen fibers has long been considered as an indication of strong interactions between the two main constituents of the bone composite, however recent data obtained by rotational-echo double resonance (REDOR) solid-state NMR failed to find proof of any significant and stable chemical bonds between collagen and bone apatite [83]. One of the most important organic constituents bound to bone crystals are citrate ions, which constitute an important fraction of the organic matter and have been shown to adsorb strongly onto apatite crystals [84]. The citrate ions are the only organic molecules that exist in quantities high enough in bone to have a chemical influence on the crystals. A recent report suggests that these ions could be responsible for the inter-crystalline bonding of the bone crystals and their arrangements in bundles of parallel crystals [85]. Such arrangements have however also been described in synthetic preparations without citrate ions and are considered to result in interactions between the hydrated layers of the nanocrystals [64]. These data do not necessarily exclude the existence of inter-crystalline citrate bonding, but this role has to be confirmed and precisely defined [86]. Interestingly, another common organic acid, lactic acid, has been shown to be stored at significant amounts as lactate in bone tissues and shells of turtles under hypoxic conditions [87], although this buffering effect of bone involves probably an alteration of the mineral, its effect is not yet clarified.

The orientation of crystals in the collagen array does not seem related to strong chemical interactions. A templating effect has been suggested, based on the fact that apatite crystal formation occurs preferably when the fibers are organized in their regular "quarter stagger array". Disruption of the collagen organization due to diseases (osteogenesis imperfecta for example) [88] is often associated with mineralization impairment. The growth of apatite crystals, towards the free available space left vacant by collagen fibers arrangement, appears as a natural process to consider. Several reports mention that the plate-like crystals of bone appear bent or even folded either in native form in calcifying vesicles, for example [89], in bones [90] or when extracted from the organic matrix of calcifying cartilage [91]. These observations could possibly be related to preparation artifacts altering the very thin bone crystals. Recently McNally *et al.* [91], analyzing dark field images, concluded that these mineral structures are most probably polycrystalline associations rather than bent single crystals.

Unlike dentine or bone, enamel is composed of needle-like crystals elongated along the hexagonal c axis of the unit-cell. The crystals appear as irregular hexagons, with different dimensions in different species: 50–60 nm wide and 25–30 nm thick for rats mature enamel [92] and 20–180 nm wide, 10–90 nm thick for humans [93]. The crystal length is difficult to determine with accuracy. The crystals are organized in rods of densely grouped parallel crystals with the needle axis perpendicular to the enamel surface. They are believed to be as long as the enamel thickness. Only very few non-collagenous proteins (enamelins and taftelins) remain in mature enamel (about 1%) [94]. Enamel is an acellular tissue that is subject to abrasion and dissolution on contact with acidic foods or acids generated by oral bacteria.

6. FORMATION

Apatite calcium phosphates are mainly involved in mineral ions metabolism (Ca, P, but also Mg [81] and possibly Zn [95]) in living organisms through dissolution/precipitation/crystallization processes. The fundamental physico-chemical principles involved in crystal formation from solution are: solution supersaturation, nucleation and crystal growth. In the case of biomineral formation, these steps are not only determined by the local ion concentration in the medium but also by the nature of the interfaces present in the biological environment (mineral-organic matrix and -biological environment). In addition, the mineralization process occurs in partly regulated extracellular space that cannot reach thermodynamic equilibrium due to continuous remodeling and circadian variations of main mineral ions concentrations. In the case of bone mineral formation, the extracellular collagen protein matrix constitutes a supramolecular framework to support, delimit and control apatite formation. The formation of bone mineral crystals has been extensively studied *in vivo* (in embryos, in bone defects [81], in slow mineralizing systems such as turkey tendons [82] or fish bones [54,96]), or *in vitro*, in osteoblast cell cultures [97]). In all cases, bone tissue formation begins with the extracellular deposition of a structured collagen matrix that exhibits a specific arrangement of the fibrils resulting in the creation of hole zones between aligned fibrils and interfibrillar space for parallel fibrils [98]. The mineral formation begins in the hole zone of the collagen fibrils and is then progressively extended to the inter- and intrafibrillar space. Vesicular mineralization, reported by several authors, seems to play a very limited role in the bone formation of vertebrates, as discussed in the review of Christoffersen and Landis [99].

The composition of bone extracellular fluid has only been the object of a few studies. Most researchers consider that it is close to blood plasma and some of them use simulated body fluid [100] as a model although it is certainly exaggeratedly oversaturated as this model does not consider mineral ions associations with organic molecules of blood [41]. The evaluation of blood plasma free calcium and phosphate concentrations reported by Eidelman *et al.* seem more realistic [101] and suggest a concentration of these main ions close to that of the solubility product of octacalcium phosphate. This does not establish however the mineral ions concentrations at the time and place of mineral crystals formation.

According to the classical theory of crystallization, nucleation represents an activation barrier to the spontaneous precipitation of a solid phase from a supersaturated solution, which is essentially related to the creation of interfacial energy. Crystallization at surfaces (heterogeneous nucleation and crystal growth) may be induced at supersaturations lower than those required for spontaneous precipitation ranging in the domain of metastability. In the presence of a foreign body or surface, the overall free energy change associated with the formation of a critical nucleus under heterogeneous conditions can be less than the corresponding free energy change associated with homogeneous nucleation. Another parameter to consider is the presence of adsorbed molecules that may change the interfacial energy in solution and stabilize smaller nuclei compared to solutions without adsorbate (Figure 3) [102,103]. The stabilization of smaller nuclei favors their formation. Furthermore, the rapid growth of these nuclei is prevented due to the adsorbed molecules at their surface. As a result, fast multiplication of nuclei is observed until most of the molecules are adsorbed, at this stage, the nuclei continue to grow without inhibition and a burst of crystal growth can be eventually observed, with crystal growth rates several orders of magnitude larger than in the case of adsorbate-free nucleation (Figure 3A) [102,103]. Such a process of nuclei multiplication can theoretically occur with any adsorbing molecules, lowering the interfacial energy (Figure 3B). The advantage, regarding an organic matrix mineralization, is to facilitate the control of crystal size through the multiplication of nuclei and multisite growth and the preservation of mineral ion diffusion capabilities in the matrix toward the inner growth sites by delaying a crystal growth obstruction of the diffusion paths. These very classical effects can be related to the non-classical crystallization concept involving transient phase(s) proposed by several authors (amorphous intermediates, dense liquid calcium phosphate clusters, and polymer-induced liquid-precursor phase) (Figure 4) [104,105,106,107,108] and to the action of some ionic additives such as magnesium.

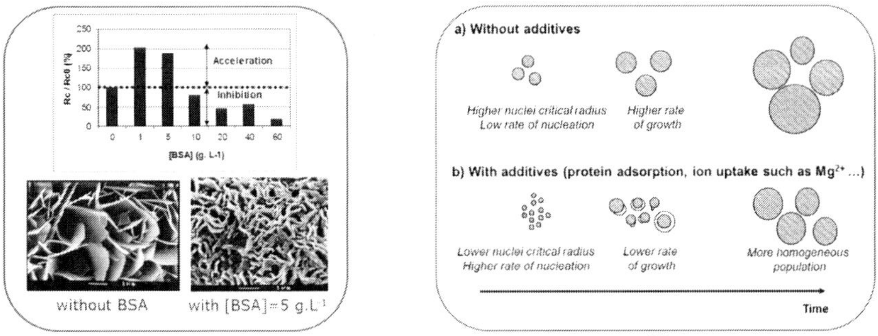

(A) (B)

Figure 3. Example of the effect of adsorbed molecules on crystal growth [102,103]. (**A**) Effect of bovine serum albumin (BSA) on octacalcium phosphate crystal growth: kinetics results (relative rate of crystal growth of octacalcium phosphate (OCP) in %) and SEM micrographs of OCP crystal growth on collagen in the presence of various concentrations of bovine serum albumin (BSA) using the constant composition crystal growth technique (adapted from [102,103]). At low albumin concentrations, when the solution reaches depletion due to nucleation and adsorption, a burst rate of crystal growth occurs exceeding that observed in the absence of albumin. At high albumin concentrations, however, there is never depletion of free BSA molecules in solution and only the well-known crystal growth inhibitory effect of the molecule on OCP crystals is evidenced. SEM micrographs showed smaller but more numerous OCP crystals in the presence of BSA. (**B**) Schematization of the effect of nucleation and growth without (**a**) or with (**b**) the presence of adsorbing molecules/ions on crystal growth depending on their solution concentration. Adsorbing species may have positive effect on crystal nucleation by stabilizing nuclei with a critical radius smaller than the critical radius in the absence of adsorbing species, due to a decrease of interfacial energy. As a result a multiplication of nuclei is observed in the presence of such adsorbing species. Nuclei grow together slowly and new nuclei form until depletion is reached, a "burst" of growth is then observed when a multitude of nuclei have formed and no crystal growth inhibitory species remains in solution: this leads to small but numerous crystals and provides a relatively homogeneous population of crystals. With larger amounts of adsorbing species, only an inhibiting crystal growth effect dominates.

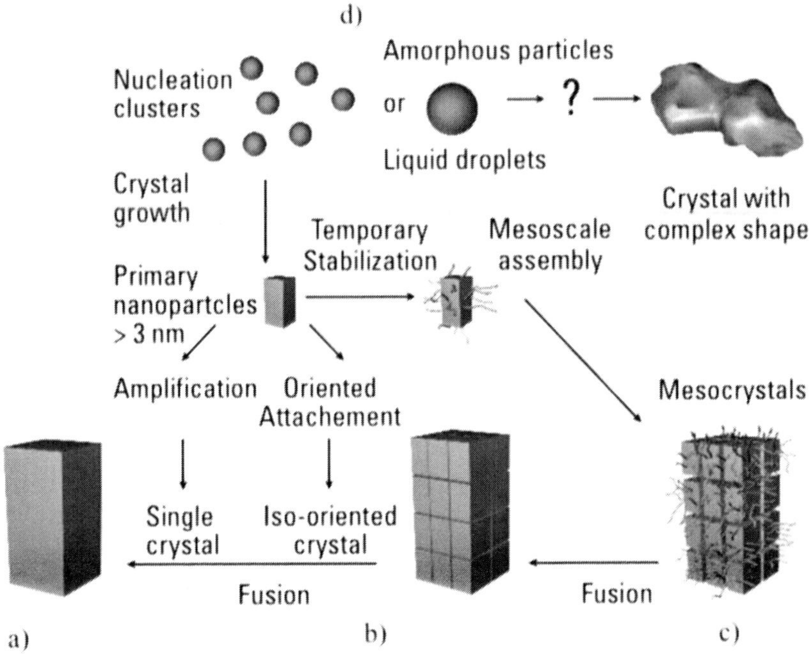

Figure 4. Schematic representation of classical and non-classical theory of crystallization: classical nucleation and crystal growth pathway (**a**); iso-oriented crystal pathway (**b**); mesoscale assembly pathway in the presence of polymer or additive (**c**); and transformation of amorphous precursor particles pathway (**d**) (from [108]. Copyright 2008 John Wiley & Sons. Reproduced with permission).

Other than the classical crystallization theory, as illustrated by pathway (a) in Figure 4, several authors [104,105,106,107,108] have also proposed that biomimetic mineralization could take place following a non-classical theory of crystallization, proceeding via an intermediate self-assembly, pathways (b) and (c), prior to crystal fusion or transformation to a metastable or amorphous precursor phase/particle, (d) pathway. These non-classical crystallization routes could especially apply to systems far from thermodynamic equilibrium for which crystal morphology and size cannot be predicted considering classical thermodynamics.

From a kinetic and thermodynamic point of view, if we consider the non-classical crystallization process following the (d) pathway compared to the classical pathway (a) in Figure 4, several authors propose a precursor multistep pathway ((B) pathway in Figure 5) involving a sequence of formed phases rather than a single step pathway (pathway A in Figure 5) for various biomineralizations such as bone mineral formation [104,105,106,107]. This concept is based on Oswald's step rule, which empirically predicts that when a solution is supersaturated with respect to more than one phase, the first phase formed is the least stable/most soluble/least dense phase due to the lower energy barrier compared to pathway A. The metastable phase formed could then be successively transformed into a more thermodynamically stable phase down to the ultimate step, leading to a more stable phase (HA in the case of calcium phosphate-based biomineralizations). However, as already discussed in the previous sections of this review, we should bear in mind that biological apatites are far from the stoichiometric HA in terms of composition, defects, solubility and thus thermodynamic properties [109]. Consequently, the position of synthetic and biological non-stoichiometric apatites in a free enthalpy variation diagram (Figure 5) would not be that of HA, but would probably be between that of HA and OCP and probably closer to that of OCP [110]. In addition there are uncertainties of the supersaturation ratio at the time and place of mineral formation.

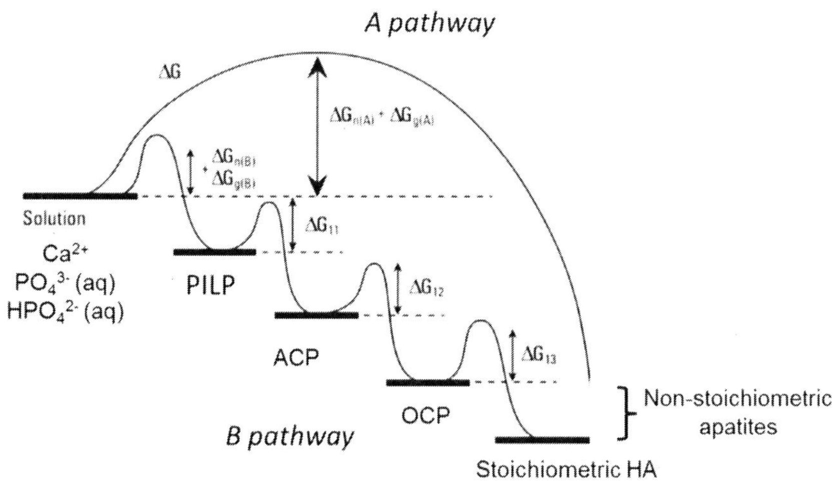

Figure 5. Gibbs free energy of activation associated with nucleation (n), growth (g) and phase transformation (T) through two main pathways: with (**B**) and without (**A**) intermediate phase(s) (adapted from Gower's review paper for the calcium phosphate system [107]).

Based on all of these considerations, several authors suggest that the amorphous precursor pathway could be widespread in the formation of biominerals [106,107,111,112]. The possible presence of transient amorphous mineral phases in mineralized vertebrate tissues has been a subject of debate over recent decades, and it still remains an open question. Furthermore, the introduction of an additional precursor step has been suggested by several authors, first in the case of the calcium carbonate system, which involves the formation of a polymer-induced liquid-precursor (PILP) phase before the amorphous phase formation step in the presence of anionic polymer (Figure 5) [104,107]. Such a phenomenon seems very similar to the adsorption effect schematized in Figure 3B. This PILP phase is considered to be a highly hydrated phase, which is more labile than the amorphous phase and could be formed of stabilized nuclei with subcritical radii. The mechanism proposed by Gower for collagen mineralization during bone formation involves a PILP phase [107], *i.e.*, liquid-like amorphous precursor droplets that are drawn into the gap zones of the collagen fibrils through capillary forces. Then, once these drops have infiltrated the gap zones and interfibrillar space of collagen, the PILP phase is transformed into the amorphous (solid) phase (ACP), which in turn crystallizes into apatite nanocrystals.

Veis and Dorvee [105] further examine the mechanism of bone and dentin mineralization through a two-step mechanism involving: (i) the formation of a dense liquid cluster (a dense liquid phase of first-layer water-bound hydrated calcium and phosphate ions); followed by (ii) the formation of a crystal nucleus inside this dense liquid cluster. They also highlight the role of polyelectrolytes/protein interfaces, which have similar dense, liquid, first-layer, water-bound surfaces that can interact with the dense, liquid calcium phosphate nucleation clusters and control crystallization within the bone and dentin collagen fibril matrix most probably in accordance with the mechanism proposed in Figure 3.

Enamel crystals form in direct contact with ameloblast cells in a row-type arrangement, forming aligned bundles of apatite crystals, which are often referred as prisms of apatite. A small deviation of crystal orientation across each prism can be observed in densely packed enamel apatite prisms. Small disorientation angles have also been identified between enamel and dentin crystals excluding epitactic relationships [113]. Several studies report that the ribbon-like shaped structures formed at the earliest stage of enamel formation include a non-crystalline mineral (ACP) and/or octacalcium phosphate (OCP) crystals, both potent precursors of enamel apatite [114,115]. Several authors have identified an amorphous calcium phosphate phase which transforms with time into the final apatitic crystalline mineral during enamel formation in murine incisor teeth [113,115,116]. Beniash *et al.* [116] hypothesize that this transformation can be triggered by proteolytic degradation of the enamel matrix proteins and suggest that the shape and organization of the mineral particles is determined very early, implying direct control of mineral shape and organization by the matrix proteins. Bodier-Houllé *et al.* [113] have also identified the presence of amorphous calcium phosphate at the interface of enamel and dentin (dentino-enamel junction).

7. MATURATION AND EVOLUTION

Mineral structures are often considered stable and unreactive entities. This is not the case for biomineral apatites, especially nanocrystalline ones, which with their hydrated surface layer, appear highly reactive. This reactivity is illustrated by the maturation and evolution of bone mineral after it has been formed, and that of enamel crystals during teeth formation.

7.1. Bone

The maturation and evolution of bone mineral is dependent on physico-chemical and biological events. The physico-chemical transformations are related to the structure of the bone mineral apatite and exhibits similar characteristics to that of synthetic biomimetic apatites as described inSection 3. On ageing, the following events have been reported [22,23,27,32,117,118]:

1. a decrease in the labile environments;
2. an increase in CO_3^{2-} and a decrease in HPO_4^{2-} with a constant Ca/(P + C) ratio; and
3. the development of the apatite domains and an increase in their dimensions and crystallinity.

The most important event to consider in regards to the biological alterations is the remodeling process that renews bone tissues at different rates, depending on the bone type, age and mechanical sollicitations. This biological process produces mature bone with its characteristic osteonal structure and high mechanical strength. This sustained bone formation involves the fresh precipitation of nanocrystalline apatite with a well-developed hydrated layer and a high ability to exchange with the bone extracellular fluid, and by extension with blood, allowing for the maintenance and control of homeostasis.

Several processes are involved in homeostasis: hormonal regulation implicating bone mineral dissolution and re-formation, and also fast equilibration with bone mineral crystals believed to involve calcium binding molecules [119]. However, physico-chemical data regarding nanocrystalline apatite suggest a possible control by direct equilibration with the hydrated layer of nanocrystalline apatites related to their specific solubilization mechanism [120] depending on the extend of dissolution and the age of mineral crystals. Considering the high levels of calcium and phosphate in body fluids only crystals with a well-developed hydrated layer can be involved in the homeostasis. Strongly packed old bone crystals would be excluded from this fast equilibration due to the reduced water content and packing limiting diffusion to and out of their crystals surface. It has been shown that when nanocrystals have come close to each other in synthetic biomimetic apatite preparations, surface ion exchanges are significantly limited [121]. Thus, remodeling, considered to occur mainly for preserving mechanical properties of the bone tissues, could also be a necessity for homeostasis function.

The mineralization of new osteons is believed to occur in two stages [122,123,124]: a fast primary mineralization leading to about 75% of the total mineral load achievable in bone tissue, followed by a slow secondary mineralization process lasting for several months and even years, leading to mineralization completion. A possible explanation for this mineralization process is that the crystals involved in the primary mineralization, with their hydrated surface layer and lower mineral density, occupy most of the space within the matrix. Several mineral ions stabilize the hydrated layer and/or slow down the development of the apatite domains (mostly Mg^{2+}, $CO_3^{2-}$and $P_2O_7^{4-}$). With ageing and the growth of the more stable apatite domains with higher density, the initial crystal volume decreases and the inter-crystalline space globally increases, leaving space for new mineral deposition. This could results in a partial reconstruction of the hydrated layer, offering new crystal growth possibilities. The secondary mineralization would then be determined by the growth of the apatite domains associated with the shrinkage of the mineral complex. Although this process leads to bone tissue with increased mineral content and mechanical strength [125], it progressively decreases the pool of exchangeable ions useful for homeostasis. Thus bone remodeling can be seen as a quest for equilibrium between the two main functions of the bone tissues, mechanical strength and homeostasis of essential elements.

An important aspect of remodeling is the partial re-use of mineral ions. Although it is linked with bone turn-over, each element has its own turn-over rate depending on its recycling ability in newly formed bone mineral. This recycling rate is often considered to be related to the effect of the element on apatite solubility. Thus, fluoride ions, which attach very easily to apatite and decrease its solubility, even at low concentrations [126], are very difficult to eliminate when they have entered bone mineral apatite and they have a very long half-life in bones evaluated to about 20 years [78]. As the effect of fluoride ions on apatite solubility has been shown to depend on its concentration [126], the clearance delay of fluoride should also depend on its concentration. As well as affecting apatite solubility, fluoride also has a biological effect on bone cells and remodeling [78]. Another aspect of fluoride incorporation is its effect on the hydrated layer. It has been shown that fluoride favors the development of apatite domains at the expense of the hydrated layer [127] and thus influences the probable biological function of the hydrated layer in calcium homeostasis [128]. This can lead, for moderate fluoride intoxications, to hypocalcaemia [129]. Other ions such as Pb^{2+} are also retained in bone for long periods, which is possibly related to their effect on apatite solubility [130]. On the contrary, ions like Sr^{2+}, which have been found to increase the solubility of Ca-Sr hydroxyapatites [131], can be easily removed from bone mineral within a few weeks after interruption of intake [132]. The possible location of Sr^{2+} within the hydrated layer has been hypothesized but has not been proven.

From a general point of view, mineral ions can be incorporated preferentially in the apatite domains or in the hydrated layer, although in most cases the partition coefficients are not known and seem to vary with the maturation stage [35]. For example, carbonate or magnesium ions are first incorporated in the hydrated layer during the precipitation of carbonate or Mg containing apatites [35,127]. However, during the maturation process two types of behavior can be distinguished depending on the nature of the mineral ion: carbonate ions are progressively incorporated into the growing apatite domains in type A and B positions whereas magnesium, although it could also belong to the apatite domains, remains preferentially in the hydrated layer and can be easily removed by ion exchange experiments. Strontium, like carbonate, penetrates progressively into the apatite domains from a surface location [35]. The behavior of most mineral ions during maturation is not known, although their preferential location could help to explain their biological effects. For example, infants are more sensitive to acute lead poisoning than adults and their forming mineralized tissues have been shown to easily incorporate this element; the consideration of the possible role of the non-apatitic hydrated layer in these events could be of interest for the improvement of models describing lead metabolism variations with age [133,134].

7.2. Enamel

During its formation enamel mineral is subject to several changes [18,135,136,137]. As already mentioned, little is known of the composition of the first ribbon-like, very elongated crystals. After this first secretory stage the crystals are analogous to those of embryonic bone or mineralizing cartilage, with a low carbonate and a high HPO_4^{2-} content and no discernible OH^- ions [30]. During its development however, a divergence with bone mineral evolution is clearly observed (Figure 2): although the amount of non-apatitic environments decreases, the Ca/(P + C) ratio begins to increase with increasing OH^- ion content. This process is similar to that observed for synthetic biomimetic apatites in carbonate poor solutions (Figure 2). In fact, the amount of bicarbonate ions in enamel fluid is very low (about 10 mM) [18] compared to that of bone fluid and blood (27 mM) [45]. It has been shown that the ribbon-like crystals first deposited grow in width and thickness by alternate accretion and maturation of mineral deposits modulated by ameloblast activity, alternating characteristic ruffle-ended and smooth-ended morphologies. These steps are accompanied by pH fluctuations between slightly acidic (6.1–6.8) and physiological (7.2–7.4) values [135]. The low pH values have been associated with mineral deposition and could be related to the development of a hydrated layer on the surface of existing crystals (appearing amorphous on de-hydration). The maturation period also seems associated to changes in the mineral, especially a loss of labile surface carbonate [18]. This evolution can be compared to that of biomimetic apatite precipitates in solutions with a physico-chemical maturation which is driven by the growth of apatite domains at the expense of the hydrated layer. This phenomenon leads to a loss of protons due to the compositional differences between the apatite domains (mainly PO_4^{3-} ions) and the hydrated layer (mainly HPO_4^{2-} ions) and can be related to the loss of labile carbonates. Thus, the rest period in enamel crystal growth corresponds to the development of apatite domains and a considerable decrease of the hydrated surface layer due to the removal of one of its stabilizing ions, carbonate. Other changes which may facilitate the apatite growth during the formation of enamel have also been postulated, such as the removal of adsorbed Mg^{2+} ions related to calcium transport during the maturation stage, and the effect of fluoride ions favoring the development of apatite domains [18].

A last developmental stage of enamel occurs after eruption. This post-eruptive maturation corresponds to the completion of mineral formation from saliva leading to a denser, harder enamel surface [138]. A change in the surface enamel composition is observed with decreasing Ca/P ratio, which can be attributed to a decrease in the surface carbonate content and an increase in the fluoride ion content, which in-turn improves the crystallinity of the apatite. These changes also improve the surface resistance of enamel to caries. The slightly acidic media of the mouth favors the release of carbonate ions, and the natural physico-chemical maturation in solutions poor in carbonate ions leads to an evolution of the apatite deposits towards stoichiometry. A physico-chemical factor rarely mentioned in the literature is the dominant orientation of the crystals in enamel, with the hexagonal unit-cell c axis perpendicular to the surface [15]. The c axis direction, in needle-like apatite crystals, corresponds to the fastest growing (and dissolution) direction. This orientation could favor a dissolution-remineralization processes and eventually, after successive cycles, the improvement of surface enamel crystal acid resistance. However, regarding their composition and possibly structure, the resulting crystals appear as heterogeneous entities, as evidenced by the well-established preferential dissolution of the central core of enamel crystals [139] with high HPO_4^{2-} and low OH^- content.

8. ADAPTABILITY OF APATITE BIOMINERALS TO BIOLOGICAL FUNCTION

Calcium phosphate biominerals are involved in different tissues with different biological functions. The example of bone and dental enamel illustrate the levels of adaptation of the mineral crystals to their biological function. Dental enamel is an acellular tissue that is subject to high compressive strains, abrasion and dissolution on contact with acidic foods or acids generated by oral bacteria. Bone is a cellular tissue (except in some fishes) with a self-repairing ability when damaged; it is involved in body architecture, protection of internal organs, attachment of muscle and mobility, and homeostasis. From a physico-chemical point of view, a first level of adaptation is the mineral-organic ratio, which is obviously different in bone and enamel (50%–85% mineral in bone and 97%–99% mineral in enamel), but also between other mineralized tissues, and even in bone of different species [125]. Mechanical properties and hardness are strongly related to this parameter but also to other characteristics of the tissues. The biological apatite crystals: composition, dimensions and surface properties can be distinguished as parameters of the biological adaptation to the function.

Apatite composition allowing the existence of different types of ionic lacunas in bone apatites can considerably alter the cohesion of the crystals [20] and their physico-chemical properties, especially their dissolution properties and solubility [120]. Apatites, which are closer to stoichiometry in enamel, are less soluble than bone apatites, which are more lacunar and far from stoichiometry. Less soluble enamel apatites are thus more resistant to acidic attacks from foods or oral bacteria, whereas more soluble bone apatites help facilitate the ion reservoir function of bone through remodeling. In addition to non-stoichiometry, ion substitutions can also specifically act on the solubility; the apatite stability record is thought to be held by shark teeth consisting of fluorapatite, which is among the less soluble apatites.

The crystal size is another variable of biological adaptability. Enamel is comprised of large apatite crystals with low specific surface area for which the dissolution rate is low. On the other hand, bone mineral is comprised of apatite nanometric platelet crystals with high specific surface areas, which favor homeostasis through dissolution phenomena but also directly through rapid surface equilibration reactions [119,128]. As mentioned earlier the orientation of crystals could also play a role in the biological adaptation. If the orientation of enamel crystals could enable improving partial remineralization from saliva, for dentine or bone platelet crystals, the orientation in parallel arrays along collagen fibers favors crystals packing and junction, and some mechanical properties but prevents surface exchanges and reactions and increases stiffness and fragility. An interesting aspect is that the total fusion of crystals is not observed. The limiting phenomenon could be the impossibility to grow an apatite lattice with the residual ions in the hydrated layer (Mg, carbonate and possibly pyrophosphate ions), the presence of adsorbed molecules like citrate ions, amino-acids (especially phosphorylated ones), polypeptides, and more generally any molecules able to bind on the surface of the platelet crystals.

Due to its ion exchange ability, the hydrated surface layer of bone apatite crystals participates in homeostasis. However with ageing, due to the decrease of the labile ion reservoir, which is related to the evolution of the hydrated surface layer and bone apatite crystal maturation [32], its ability to regulate "chemically" the concentration of mineral ions in blood decreases accordingly. This phenomenon parallels the increase of stiffness and fragility mentioned above and determines a cyclical renewal of bone mineral to preserve the ion reservoir and ion exchange properties of freshly formed apatite nanocrystals as well as the original mechanical properties of the bone tissue. This is especially true for well-vascularized epiphysis of long bone and physico-chemical studies in long, weight bearing bones have shown that bone mineral crystals appeared less mature in trabecular bone of the epiphysis than cortical bone of diaphysis, with larger amounts of non-apatitic environments especially in young animals [140].

9. DIAGENETIC EVOLUTION OF BIOLOGICAL APATITE FOSSILS

Fossils of hard tissues are used for dating purposes, and can also be used to identify life conditions, dietary habits, climate changes, and environmental pollution. It is quite difficult to evaluate the diagenetic alterations of these fossils and their corrections to garner desired information.

Several stages in the evolution of hard tissues after death can be distinguished (Figure 6). After cell death, the first alteration is related to the destruction of the organic matrix by microorganisms and hydrolysis [141,142,143,144,145]. This process is of course important for tissues with a high content of organic components, such as fish scales and bones, and seems rather limited in the case of mature enamel. This degradation can increase the porosity of the tissues, modify its hardness and favor its ulterior degradation. Several driving forces are involved in the evolution of the apatite mineral. Based on our experience on ageing of synthetic analogues in wet conditions, a fast alteration of the hydrated layer in a few months seems likely: the amount of labile ionic environments decrease [146] and can trap surface pollutants in contact with the tissue. However, this phenomenon does not seem to touch the internal apatite core. A second degradation process sees a change in the chemical composition, which is related to the non-stoichiometry of apatites and their relative solubility in media different from body fluids. Numerous elements can enter the apatite lattice at this stage such as fluoride and heavy elements. Considering that apatites offer large ranges of solid solutions, this re-equilibration process, provided it does not involve acidic media, only concerns the surface of apatite crystals in the first stage. Progressively, however, the more stable equilibrated apatite domains will grow through a generally slow dissolution re-precipitation processes driven by the relative thermodynamic stabilities of the apatite fractions, including foreign (non-biological) elements and Ostwald ripening. The crystals grow bigger and their composition can be strongly altered [147]. Although the ions released by dissolution can certainly be locally reused at the precipitation loci, similar to the *in vivo* remodeling of bone, the system around the fossil is largely open and several foreign elements stabilizing the apatite structure such as fluorine, silicon, heavy metals (Pb, Ba) and rare earth elements play a determining role in this recrystallization, modify the apatite composition and characteristics [113]. The kinetics of these processes depend on the presence of groundwater, dissolved mineral ions, pH and temperature and also on the fossil apatite characteristics (composition, crystal size) and fossil porosity [141,144,147]. Thus small bones and trabecular bone are more strongly altered than cortical, compact bone. Tooth enamel, due the large size of its apatite crystals and its relatively low porosity, is less affected by diagenetic evolution than dentine, although enrichment by foreign elements can be measured even in the case of enamel [142]. Very often isotopic ratios of different elements in fossils are used for paleodietary or paleoclimatology purposes [143,144]. Globally, the mineral ions in largest quantities in the fossil apatite structures (calcium and phosphate and to a lesser extent carbonate) are probably those which are the least affected by foreign contamination, simply because of their mass effect. Thus determinations made using these ions and their isotopes can probably be trusted whereas the use of isotope ratios of elements showing strong diagenetic variations appears uncertain. Other effects that may disturb the diagenetic evolution of fossils also have to be considered, most notably biological degradation involving microorganisms (yeast, fungi, and bacteria) in search of essential elements [141]. Unlike purely physico-chemical diagenetic evolution, driven by thermodynamics, the dissolution–reprecipitation phenomena driven by microorganisms can lead to unstable and reactive precipitates, which can experience a new cycle of diagenetic evolution. These biological alterations would only be limited by the presence of toxic ions for the living organisms involved. Degradation of apatite fossils by other living organisms such as insects or animals has also been observed [141].

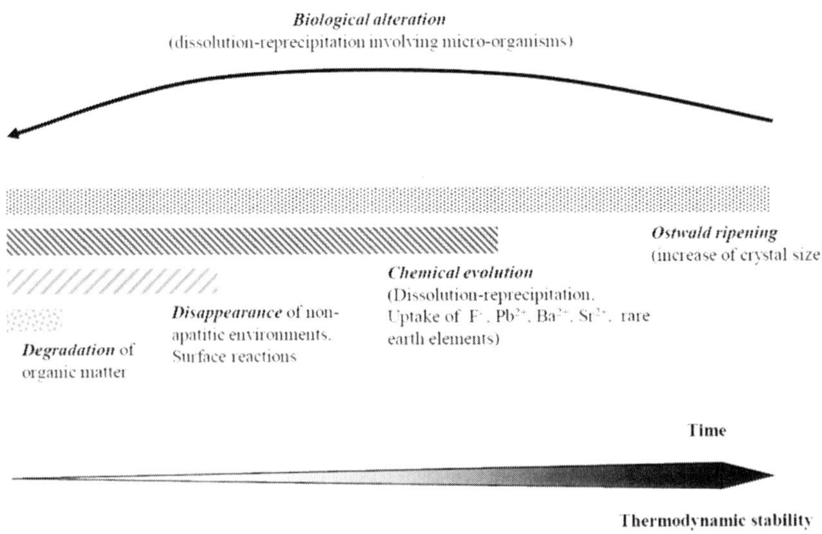

Figure 6. The various phenomena involved in the diagenetic evolution of biological apatites (adapted from [145]).

10. CONCLUSIONS AND PERSPECTIVES

Diversity and heterogeneity are the two words that may best qualify apatite biominerals. These might not show the original fine morphological details and lace-like patterns of silica and calcium carbonate biomineralizations, but their chemical richness, with large substitution facilities, acceptance of defects, and strong surface reactivity, have led to sophisticated materials with a large adaptability to different essential biological functions in many different animal species living in different environments. Although from an evolutionary perspective the storage of essential elements and detoxifications can be seen as the starting point of apatite biomineralizations, several other functions have emerged necessitating an improvement and a progressive control of all the possibilities given by the chemistry of apatites. The control of chemical composition and non-stoichiometry associated with a surface hydrated layer, allows fast mobilization of mineral ions and a fine tuning of homeostasis adapted to living environments. The control of crystals size, orientation and packing associated with an elaborated architecture participates to improving mechanical performance and lightweight. From the point of view of material science, such elaborated constructions are difficult to imitate although micro-fabrication processes seem to open perspectives. One of the most difficult problems is to manage the reactivity and ageing phenomena, which affect these structures and eventually change their properties. *In vivo*, the remedy is remodeling which, in some vertebrates, destroys and reconstructs new tissue with some energetic expenses, but allows a preservation of the bone-body fluids equilibria. In some animals, with no remodeling process, the deposition of new mineral is continuous affording at the very least a control of body fluids mineral content. Regarding biomaterials, the fabrication of bone substitutes analogous to the original has long be considered as an ideal. The way seems difficult and costly, in addition, the objective of a bone substitute is to favor bone tissue reconstruction and thus the conception has evolved towards the realization of scaffolds allowing cells recruitments, attachment, proliferation and differentiation and progressively disappearing as the reconstruction the tissue is advancing. The "chemical" imitation of bone tissue *in vitro* presents an interest as it allows the parameters involved in the mineralization process and mineral evolution and maturation to be assessed and the relations between mechanical properties and the tissue characteristics to be more precisely determined.

ACKNOWLEDGMENTS

The authors thank the french state funds managed by the ANR within the Investissements d'Avenir programme under reference ANR-11-IDEX-0004-02, and more specifically within the framework of the Cluster of Excellence MATISSE (MATériaux, Interfaces, Surfaces, Environment) led by Sorbonne Universités for covering the costs to publish this review paper in open access.

CONFLICTS OF INTEREST

The authors declare no conflict of interest.

REFERENCES

1. Lowenstam, H.A. Minerals formed by organisms. Science 1981, 211, 1126–1131.
2. Elliott, J.C. Calcium phosphate biominerals. Rev. Mineral. Geochem. 2002, 48, 427–453.
3. Bazin, D.; Daudon, M.; Combes, C.; Rey, C. Characterization and some physicochemical aspects of pathological microcalcifications. Chem. Rev. 2012, 112, 5092–5120.
4. Carlstrom, D. A crystallographic study of vertebrates otoliths. Biol. Bull. 1963, 125, 441–463.
5. Li, Z.; Pasteris, J.D. Tracing the pathway of compositional changes in bone mineral with age: Preliminary study of bioapatite aging in hypermineralized dolphin's bulla. Biochim. Biophys. Acta 2014, 1840, 2331–2339.
6. Sakae, T.; Nakada, H.; LeGeros, J.P. Historical review of biological apatite crystallography. J. Hard Tissue Biol. 2015, 24, 111–122.
7. Uskokovic, V. The role of hydroxyl channel in defining selected physicochemical peculiarities exhibited by hydroxyapatite. RSC Adv. 2015, 5, 36614–36633.
8. Xie, B.; Halter, T.J.; Borah, B.M.; Nancollas, G.H. Tracking amorphous precursor formation and transformation during

induction stages of nucleation. Cryst. Growth Des. 2014, 14, 1659–1665.

9. Omelon, S.; Ariganello, M.; Bonucci, E.; Grynpas, M.; Nanci, A. A review of phosphate mineral nucleation in biology and geobiology. Calcif. Tissue Int. 2013, 93, 382–396.

10. Boskey, A.L. Mineralization of bones and teeth. Elements 2007, 3, 385–391.

11. Gomez-Morales, J.; Iafisco, M.; Delgado-Lopez, J.M.; Sarda, S.; Drouet, C. Progress on the preparation of nanocrystalline apatites and surface characterization: Overview of fundamental and applied aspects. Prog. Cryst. Growth Charact. Mater. 2013, 59, 1–46.

12. Wopenka, B.; Pasteris, J.D. A mineralogical perspective on the apatite in bone. Mater. Sci. Eng. C 2005, 25, 131–143.

13. De Jong, W.F. La substance minérale dans les os. Recueil des Travaux Chimiques des Pays-Bas1926, 45, 445–448. (In French).

14. White, T.J.; Zhili, D. Structural derivation and crystal chemistry of apatites. Acta Crystallogr. Sect. B Struct. Sci. 2003, B59, 1–16.

15. Elliott, J.C. Structure and Chemistry of the Apatites and Other Calcium Orthophosphates; Elsevier: Amsterdam, The Netherlands, 1994.

16. Labarthe, J.C.; Bonel, G.; Montel, G. Structure and properties of B-type phosphocalcium carbonate apatites. Ann. Chim. 1973, 8, 289–301.

17. LeGeros, R.Z. Calcium phosphates in oral biology and medecine. Monogr. Oral Sci. 1991, 15, 1–201.

18. Aoba, T. Recent observations on enamel crystal formation during mammalian amelogenesis.Anat. Rec. 1996, 245, 208–218.

19. Fleet, M.E.; Liu, X. Coupled substitution of type A and B carbonate in sodium-bearing apatite. Biomaterials 2007, 28, 916–926.

20. Rey, C.; Combes, C.; Drouet, C.; Grossin, D. Bioactive ceramics: Physical chemistry. InComprehensive Biomaterials; Ducheyne, P., Healy, K.E., Hutmacher, D.W., Grainger, D.W., Eds.; Elsevier: Amsterdam, The Netherlands, 2011; Volume 1, pp. 187–221.

21. Legros, R.; Balmain, N.; Bonel, G. Age-related changes in mineral of rat and bovine cortical bone. Calcif. Tissue Int. 1987, 41, 137–144.

22. Wu, Y.; Glimcher, M.J.; Rey, C.; Ackerman, J.L. A unique protonated phosphate group in bone mineral not present in synthetic calcium phosphates. Identification by phosphorus-31 solid state NMR spectroscopy. J. Mol. Biol. 1994, 244, 423–435.

23. Rey, C.; Shimizu, M.; Collins, B.; Glimcher, M.J. Resolution-enhanced Fourier transform infrared spectroscopy study of the environment of phosphate ions in the early deposits of a solid phase of calcium phosphate in bone and enamel, and their evolution with age. I: Investigations in the ν4 PO4 domain. Calcif. Tissue Int. 1990, 46, 384–394.

24. Winand, L. Etude physico-chimique du phosphate tricalcique hydraté et de l'hydroxyapatite. Ann. Chim. 1961, 6, 951–967. (In French).

25. Montel, G.; Bonel, G.; Heughebaert, J.-C.; Trombe, J.-C.; Rey, C. New concepts in the composition, crystallization and growth of the mineral components of calcified tissues. J. Cryst. Growth 1981, 53, 74–79.

26. Rey, C.; Combes, C.; Drouet, C.; Sfihi, H.; Barroug, A. Physico-chemical properties of nanocrystalline apatites: Implications for biominerals and biomaterials. Mater. Sci. Eng. C2007, 27, 198–205.

27. Rey, C.; Renugopalakrishnan, V.; Shimizu, M.; Collins, B.; Glimcher, M.J. A resolution-enhanced Fourier Transform Infrared spectroscopic study of the environment of the CO32–ion in the mineral phase of enamel during its formation and maturation. Calcif. Tissue Int.1991, 49, 259–268.

28. Kourkoumelis, N.; Tzaphlidou, M. Spectroscopic assessment of normal cortical bone: Differences in relation to bone site and sex. Sci. World J. 2010, 10, 402–412.

29. Jäger, C.; Welzel, T.; Meyer-Zaika, W.; Epple, M. A solid-state NMR investigation of the structure of nanocrystalline hydroxyapatite. Magn. Reson. Chem. 2006, 44, 573–580.

30. Rey, C.; Hina, A.; Tofighi, A.; Glimcher, M.J. Maturation of poorly crystalline apatites: Chemical and structural aspects in vivo and in vitro. Cells Mater. 1995, 5, 345–356.

31. Nassif, N.; Martineau, F.; Syzgantseva, O.; Gobeaux, F.; Willinger, M.; Coradin, T.; Cassaignon, S.; Azaïs, T.; Giraud-Guille, M.-M. In vivo inspired conditions to synthesize biomimetic hydroxyapatite. Chem. Mater. 2010, 22, 3653–3663.

32. Cazalbou, S.; Combes, C.; Eichert, D.; Rey, C.; Glimcher, M.J. Poorly crystalline apatites: Evolution and maturation in vitro and in vivo. J. Bone Miner. Metab. 2004, 22, 310–317.

33. Eichert, D.; Sfihi, H.; Combes, C.; Rey, C. Specific characteristics of wet nanocrystalline apatites. consequences on biomaterials and bone tissue. Key Eng. Mater. 2004, 254–256, 927–930.

34. Eichert, D.; Combes, C.; Drouet, C.; Rey, C. Formation and evolution of hydrated surface layers of apatites. Key Eng. Mater. 2005, 284–286, 3–6.

35. Cazalbou, S. Echanges Cationiques Impliquant des Apatites Nanocristallines Analogues au Minéral Osseux. Ph.D. Thesis, Institut National Polytechnique de Toulouse, Toulouse, France, 2000. (In French).

36. Cazalbou, S.; Eichert, D.; Ranz, X.; Drouet, C.; Combes, C.; Harmand, M.F.; Rey, C. Ion exchanges in apatites for biomedical application. J. Mater. Sci. Mater. Med. 2005, 16, 405–409.

37. Rey, C.; Combes, C.; Drouet, C.; Lebugle, A.; Sfihi, H.; Barroug, A. Nanocrystalline apatites in biological systems: Characterisation, structure and properties. Materialwissenschaft und Werkstofftechnik 2007, 38, 996–1002.

38. Drouet, C.; Carayon, M.; Combes, C.; Rey, C. Surface enrichment of biomimetic apatites with biologically-active ions Mg2+ and Sr2+: A preamble to the activation of bone repair materials.Mater. Sci. Eng. C 2008, 28, 1544–1550.

39. Errassifi, F.; Menbaoui, A.; Autefage, H.; Benaziz, L.; Ouizat, S.; Santran, V.; Sarda, S.; Lebugle, A.; Combes, C.; Barroug, A.; et al. Adsorption onto nanocrystalline apatitic calcium phosphates. Applications to growth factors and drugs delivery. In Advances in Bioceramics and Biotechnologies; Narayan, R., McKittrick, J., Singh, M., Eds.; John Wiley & Sons: Hoboken, NJ, USA, 2010.

40. Ouizat, S.; Barroug, A.; Legrouri, A.; Rey, C. Adsorption of bovine serum albumin on poorly crystalline apatite: Influence of maturation. Mater. Res. Bull. 2000, 34, 2279–2289.

41. Rey, C.; Combes, C. Physical chemistry of biological apatites. In Biomineralization and Biomaterials, Fundamentals and Applications; Aparicio, C., Pau Ginebra, M., Eds.; Woodhead publishing: Swanston, UK, 2015; pp. 95–128.

42. Brown, W.E. Crystal structure of octacalcium phosphate. Nature 1962, 196, 1048–1050.

43. Eichert, D.; Drouet, C.; Sfihi, H.; Rey, C.; Combes, C. Nanocrystalline Apatite Based Biomaterials: Synthesis, Processing and Characterization; Eichert, D., Drouet, C., Sfihi, H., Rey, C., Combes, C., Eds.; Nova Science Publishers Inc.: New York, NY, USA, 2009.

44. Rey, C.; Combes, C.; Drouet, C.; Cazalbou, S.; Grossin, D.; Brouillet, F.; Sarda, S. Surface properties of biomimetic nanocrystalline apatites; applications in biomaterials. Prog. Cryst. Growth Charact. Mater. 2014, 60, 63–73.

45. Driessens, F.C.M.; Verbeeck, R.M.H. Biominerals; CRC Press: Boca Raton, FL, USA, 1990.

46. Iyengar, G.V.; Tandon, L. Minor and Trace Elements in Human Bones and Teeth; International Atomic Energy Agency: Vienna, Austria, 1999.

47. Teruel, J.D.D.; Alcolea, A.; Hernández, A.; Ruiz, A.J.O. Comparison of chemical composition of enamel and dentine in human, bovine, porcine and ovine teeth. Arch. Oral Biol. 2015, 60, 768–775.

48. Soares, M.A.B.; Adachi, E.M.; Saiki, M. INAA of enamel and dentine samples of a group of children and adults: A comparative study. J. Radioanal. Nucl. Chem. 2007, 276, 49–52.

49. Zenóbio, M.A.F.; Tavares, M.S.N.; Zenóbio, E.G.; Silva, T.A. Elemental composition of dental biologic tissues: Study by means of different analytical techniques. J. Radioanal. Nucl. Chem.2011, 289, 161–166.

50. Ishiguro, K.; Nakagaki, H.; Takeuchi, K.; Mukai, M.; Yoshika, I.; Miyauchi, K.; Robinson, C.; Weatherell, J.A. Distribution of fluoride in the dental tissues and their supporting mandibular bone from the same individual. Arch. Oral Biol. 1994, 39, 535–537.

51. Lakomaa, E.L.; Rytömaa, I. Mineral composition of enamel and dentin of primary and permanent teeth in Finland. Scand. J. Dent. Res. 1977, 85, 89–95.

52. Saiki, M.; Adachi, L.K.; Adachi, E.M. Elemental comparison in sound and carious human teeth by instrumental neutron activation analysis. Eur. Radiol. 2014, 24, 29–32.

53. Hendricks, S.B.; Hill, W.L. The nature of bone and phosphate rocks. Proc. Natl. Acad. Sci. USA 1950, 36, 731–737.

54. Lee, D.D.; Glimcher, M.J. Three-dimensional spatial relationship between the collagen fibrils and the inorganic calcium phosphate crystals of pickerel (Americanus americanus) and herring (Clupea harengus) bone. J. Mol. Biol. 1991, 217, 487–501.

55. Bala, Y.; Farlay, D.; Boivin, G. Bone mineralization: From tissue to crystal in normal and pathological contexts. Osteoporos. Int. 2013, 24, 2153–2166.

56. Smith, S.M.; Abrams, S.A.; David-Street, J.E.; Heer, M.; O'Brien, K.O.; Wastney, M.E.; Zwart, S.R. Fifty years of human space travel: Implications for bone and calcium research. Annu. Rev. Nutr. 2014, 34, 377–400.

57. Zylberberg, L.; Traub, W.; de Buffrenil, V.; Allizard, F.; Arad, T.; Weiner, S. Rostrum of a toothed whale: Ultrastructural study of a very dense bone. Bone 1998, 23, 241–247.

58. Rey, C.; Collins, B.; Goehl, T.; Dickson, I.R.; Glimcher, M.J. The carbonate environment in bone mineral: A resolution-enhanced fourier transform infrared spectroscopy study. Calcif. Tissue Int. 1989, 45, 157–164.

59. Elliott, J.C.; Holcom, D.W.; Young, R.A. Infrared determination of the degree of substitution of hydroxyl by carbonate ion in human enamel. Calcif. Tissue Int. 1985, 37, 372–375.

60. Penel, G.; Leroy, G.; Rey, C.; Bres, E. Micro-raman spectral study of the PO4 and CO3vibrational modes in synthetic and biological apatites. Calcif. Tissue Int. 1998, 63, 475–481.

61. Beshah, K.; Rey, C.; Glimcher, M.J.; Shimizu, M.; Griffin, R.G. Solid state carbon-13 and proton NMR studies of carbonate-containing calcium phosphates and enamel. J. Solid State Chem. 1990, 84, 71–81.

62. Combes, C.; Rey, C.; Mounic, S. Identification and evaluation of HPO4 ions in biomimetic poorly crystalline apatites and bone mineral. Key Eng. Mater. 2001, 192–195, 143–146.

63. Bohic, S.; Rey, C.; Legrand, A.; Sfihi, H.; Rohanizadeh, R.; Martel, C.; Barbier, A.; Daculsi, G. Characterization of the trabecular rat

bone mineral: Effect of ovariectomy and bisphosphonate treatment. Bone 2000, 26, 341–348.

64. Wang, Y.; Von Euw, S.; Fernandes, F.M.; Cassaignon, S.; Selmane, M.; Laurent, G.; Pehau-Arnaudet, G.; Coelho, C.; Bonhomme-Coury, L.; Giraud-Guille, M.-M.; et al. Water-mediated structuring of bone apatite. Nat. Mater. 2013, 12, 1144–1153.

65. Taylor, M.G.; Parker, S.F.; Mitchell, P.C.H. A study by high energy transfer inelastic neutron scattering spectroscopy of the mineral fraction of ox femur bone. J. Mol. Struct. 2003, 651–653, 123–126.

66. Cho, G.; Wu, Y.; Ackerman, J.L. Detection of hydroxyl ions in bone mineral by solid-state NMR spectroscopy. Science 2003, 300, 1123–1127.

67. Pasteris, J.D.; Wopenka, B.; Freeman, J.J.; Rogers, K.; Valsami-Jones, E.; van der Houwen, J.; Silva, M.J. Lack of OH in nanocrystalline apatite as a function of degree of atomic order: Implications for bone and biomaterials. Biomaterials 2004, 25, 229–238.

68. Rey, C.; Miquel, J.L.; Facchini, L.; Legrand, A.P.; Glimcher, M.J. Hydroxyl groups in bone mineral. Bone 1995, 16, 583–586.

69. Grossin, D.; Rollin-Martinet, S.; Estournès, C.; Rossignol, F.; Champion, E.; Combes, C.; Rey, C.; Geoffroy, C.; Drouet, C. Biomimetic apatite sintered at very low temperature by spark plasma sintering: Physico-chemistry and microstructure aspects. Acta Biomater. 2010, 6, 577–585.

70. Rey, C.; Renugopalakrishnan, V.; Collins, B.; Glimcher, M.J. Fourier transform infrared spectroscopic study of the carbonate ions in bone mineral during aging. Calcif. Tissue Int.1991, 49, 251–258.

71. LeGeros, R.; Trautz, O.R.; LeGeros, J.P.; Klein, E. Carbonate substitution in the apatite structure. Bull. Soc. Chim. Fr. 1968, 1712–1718.

72. Laurencin, D.; Wong, A.; Chrzanowski, W.; Knowles, J.C.; Qiu, D.; Pickup, D.M.; Newport, R.J.; Gan, Z.; Duer, M.J.; Smith, M.E. Probing the calcium and sodium local environment in bones and teeth using multinuclear solid state NMR and X-ray absorption spectroscopy.Phys. Chem. Chem. Phys. 2010, 12, 1081–1091.

73. Dykes, E.; Elliott, J.C. The occurrence of chloride ions in the apatite lattice of Holly Springs hydroxyapatite and dental enamel. Calcif. Tissue Res. 1971, 7, 241–248.

74. Burnell, J.M.; Teubner, E.J.; Miller, A.G. Normal maturational changes in bone matrix, mineral, and crystal size in the rat. Calcif. Tissue Int. 1980, 31, 13–19.

75. Landis, W.J.; Lee, D.D.; Brenna, J.T.; Chandra, S.; Morrison, G.H. Detection and localization of silicon and associated elements in vertebrate bone tissue by imaging ion microscopy.Calcif. Tissue Int. 1986, 38, 52–59.

76. Code, R.F.; Gelman, R.L.; Armstrong, R.S.; Hallsworth, C.; Lemaire, C.; Cheng, P.T. Field dependence of 19F NMR in rat bone powders. Phys. Med. Biol. 1990, 35, 1271–1286.

77. Grynpas, M.; Rey, C. The effect of fluoride treatment on bone mineral crystals in the rat. Bone1992, 13, 423–429.

78. Boivin, G.; Meunier, P.J. Fluoride and bone: Toxic effects and therapeutic role. In Therapeutic Uses of Trace Elements; Nève, J., Chappuis, P., Lamand, M., Eds.; Springer: Berlin, Germany; Heidelberg, Germany, 1996; pp. 283–295.

79. Nakagaki, H.; Koyama, Y.; Sakakibara, Y.; Weatherell, J.A.; Robinson, C. Distribution of fluoride across human dental enamel, dentine and cementum. Arch. Oral Biol. 1987, 32, 651–654.

80. Cazalbou, S.; Hina, A.; Rey, C. Interactions between trace elements and bone mineral matrix. In New Aspects of Trace Element Research; Abdulla, M., Bost, M., Gamon, S., Arnaud, P., Chazot, G., Eds.; Smith-Gordon: London, UK, 1999; pp. 58–62.

81. Glimcher, M.J. Bone: Nature of the calcium phosphate crystals and cellular, structural, and physical chemical mechanisms in their formation. Rev. Mineral. Geochem. 2006, 64, 223–282.

82. Landis, W.J.; Song, M.J.; Leith, A.; McEwen, L.; McEwen, B. Mineral and organic matrix interaction in normally calcifying tendon visualized in 3 dimensions by high-voltage electron-microscopic tomography and graphic image-reconstruction. J. Struct. Biol. 1993,110, 39–54.

83. Nikel, O.; Laurencin, D.; Bonhomme, C.; Sroga, G.E.; Besdo, S.; Lorenz, A.; Vashishth, D. Solid state NMR investigation of intact human bone quality: Balancing issues and insight into the

structure at the organic-mineral interface. J. Phys. Chem. C Nanomater. Interfaces 2012,116, 6320–6331.

84. Skwarek, E.; Janusz, W.; Sternik, D. Adsorption of citrate ions on hydroxyapatite synthetized by various methods. J. Radioanal. Nucl. Chem. 2013, 299, 2027–2036.

85. Davies, E.; Müller, K.H.; Wong, W.C.; Pickard, C.J.; Reid, D.G.; Skepper, J.N.; Duer, M.J. Citrate bridges between mineral platelets in bone. Proc. Natl. Acad. Sci. USA 2014, 111, E1354–E1363.

86. Rey, C.; Combes, C. What bridges mineral platelets of bone? BoneKEy Rep. 2014, 3.

87. Davis, E.C.; Jackson, D.C. Lactate uptake by skeletal bone in anoxic turtles, trachemys scripta. Comp. Biochem. Physiol. A Mol. Integr. Physiol. 2007, 146, 299–304.

88. Landis, W.J. The strength of a calcified tissue depends in part on the molecular structure and organization of its constituent mineral crystals in their organic matrix. Bone 1995, 16, 533–544.

89. Morris, D.C.; Vaananen, H.K.; Anderson, H.C. Matrix vesicle calcification in rat epiphyseal growth palte cartilage prepared anhydrously for electron microscopy. Metab. Bone Dis. Relat. Res. 1983, 5, 131–137.

90. Kim, H.-M.; Rey, C.; Glimcher, M.J. X-ray diffraction, electron microscopy, and fourier transform infrared spectroscopy of apatite crystals isolated from chicken and bovine cartilage. Calcif. Tissue Int. 1996, 59, 58–63.

91. McNally, E.A.; Schwarcz, H.P.; Botton, G.A.; Arsenault, L.A. A model for the ultrastructure of bone based on electron microscopy of ion-milled sections. PLoS ONE 2012, 7.

92. Nylen, M.U.; Eanes, E.D.; Omnell, K.A. Crystal growth in rat enamel. J. Cell Biol. 1963, 18, 109–123.

93. Frazier, P.D. Adult human enamel—An SEM study of crystallite size and morphology. J. Ultrastruct. Res. 1968, 22, 1–11.

94. Santo, A.R.E.; Line, S.R.P. The enamel organic matrix: Structure and function. Braz. J. Oral Sci. 2005, 4, 716–724.

95. King, J.C.; Shames, D.M.; Woodhouse, L.R. Zinc and health: Current status and future directions zinc homeostasis in humans 1. J. Nutr. 2000, 130, 1360S–1366S. [PubMed]

96. Burger, C.; Zhou, H.W.; Wang, H.; Sics, I.; Hsiao, B.S.; Chu, B.; Graham, L.; Glimcher, M.J. Lateral packing of mineral crystals in bone collagen fibrils. Biophys. J. 2008, 95, 1985–1992.

97. Rey, C.; Kim, H.M.; Gerstenfeld, L.; Glimcher, M.J. Characterization of the apatite crystals of bone and their maturation in osteoclast cell culture: Comparison with native bone crystals.Connect. Tissue Res. 1996, 35, 343–349.

98. Hodge, A.J. Molecular models illustrating the possible distribution of holes in simple systematically staggered arrays of type I collagen molecules in native-type fibrils. Connect. Tissue Res. 1989, 21, 137–147.

99. Christoffersen, J.; Landis, W.J. A contribution with review to the description of mineralization of bone and other calcified tissues in vivo. Anat. Rec. 1991, 230, 435–450.

100. Kokubo, T.; Takadama, H. How useful is SBF in predicting in vivo bone bioactivity?Biomaterials 2006, 27, 2907–2915.

101. Eidelman, N.; Chow, L.C.; Brown, W.E. Calcium phosphate saturation levels in ultrafiltered serum. Calcif. Tissue Int. 1987, 40, 71–78.

102. Combes, C.; Rey, C. Adsorption of proteins and calcium phosphate materials bioactivity.Biomaterials 2002, 23, 2817–2823.

103. Combes, C.; Rey, C.; Frèche, M. In vitro crystallization of octacalcium phosphate on type I collagen: Influence of serum albumin. J Mater. Sci. Mater. Med. 1999, 10, 153–160.

104. Olszta, M.J.; Cheng, X.; Jee, S.S.; Kumar, R.; Kim, Y.-Y.; Kaufman, M.J.; Douglas, E.P.; Gower, L.B. Bone structure and formation: A new perspective. Mater. Sci. Eng. R Rep. 2007, 58, 77–116.

105. Veis, A.; Dorvee, J.R. Biomineralization mechanisms: A new paradigm for crystal nucleation in organic matrices. Calcif. Tissue Int. 2013, 93, 307–315.

106. Meldrum, F.C.; Cölfen, H. Controlling mineral morphologies and structures in biological and synthetic systems. Chem. Rev. 2008, 108, 4332–4432.

107. Gower, L.B. Biomimetic model systems for investigating the amorphous precursor pathway and its role in biomineralization. Chem. Rev. 2008, 108, 4551–4627.

108. Colfen, H.; Antonietti, M. Mesocrystals and Non Classical Crystallization; John Wiley & Sons Ltd.: Chichester, UK, 2008.

109. Drouet, C. A comprehensive guide to experimental and predicted thermodynamic properties of phosphate apatite minerals in view of applicative purposes. J. Chem. Thermodyn. 2015, 81, 143–159.

110. Rollin-Martinet, S.; Navrotsky, A.; Champion, E.; Grossin, D.; Drouet, C. Thermodynamic basis for evolution of apatite in calcified tissues. Am. Mineral. 2013, 98, 2037–2045.

111. Weiner, S. Transient precursor strategy in mineral formation of bone. Bone 2006, 39, 431–433.

112. Nudelman, F.; Pieterse, K.; George, A.; Bomans, P.H.H.; Friedrich, H.; Brylka, L.J.; Hilbers, P.A.J.; de With, G.; Sommerdijk, N.A.J.M. The role of collagen in bone apatite formation in the presence of hydroxyapatite nucleation inhibitors. Nat. Mater. 2010, 9, 1004–1009.

113. Bodier-Houllé, P.; Steuer, P.; Meyer, J.M.; Bigeard, L.; Cuisinier, F.J. High-resolution electron-microscopic study of the relationship between human enamel and dentin crystals at the dentinoenamel junction. Cell Tissue Res. 2000, 301, 389–395.

114. Iijima, M.; Moradian-oldak, J. Control of apatite crystal growth in a fluoride containing amelogenin-rich matrix. Biomaterials 2005, 26, 1595–1603.

115. Takano, Y.; Hanaizumi, Y.; Ohshima, H. Occurrence of amorphous and crystalline mineral deposits at the epithelial-mesenchymal interface of incisors in the calcium-loaded rat: Implication of novel calcium binding domains. Anat. Rec. 1996, 245, 174–185.

116. Beniash, E.; Metzler, R.A.; Lam, R.S.K.; Gilbert, P.U.P.A. Transient amorphous calcium phosphate in forming enamel. J. Struct. Biol. 2009, 166, 133–143.

117. Spevak, L.; Flach, C.R.; Hunter, T.; Mendelsohn, R.; Boskey, A. Fourier transform infrared spectroscopic imaging parameters describing acid phosphate substitution in biologic hydroxyapatite. Calcif. Tissue Int. 2013, 92, 418–428.

118. Bonar, L.C.; Roufosse, A.H.; Sabine, W.K.; Grynpas, M.D.; Glimcher, M.J. X-ray diffraction studies of the crystallinity of bone mineral in newly synthesized and density fractionated bone. Calcif. Tissue Int. 1983, 35, 202–209.

119. Heaney, R.P. How does bone support calcium homeostasis? Bone 2003, 33, 264–268.

120. Baig, A.A.; Fox, J.L.; Wang, Z.; Higuchi, W.I.; Miller, S.C.; Barry, A.M.; Otsuka, M. Metastable equilibrium solubility behavior of bone mineral. Calcif. Tissue Int. 1999, 64, 329–339.

121. Eichert, D. Etude de la Réactivité de Surface D'apatites de Synthèse Nanocristallines. Ph.D's Thesis, Institut National Polytechnique de Toulouse, Toulouse, France, 2001. (In French).

122. Bala, Y.; Farlay, D.; Delmas, P.D.; Meunier, P.J.; Boivin, G. Time sequence of secondary mineralization and microhardness in cortical and cancellous bone from ewes. Bone 2010, 46, 1204–1212.

123. Busa, B.; Miller, L.M.; Rubin, C.T.; Qin, Y.-X.; Judex, S. Rapid establishment of chemical and mechanical properties during lamellar bone formation. Calcif. Tissue Int. 2005, 77, 386–394.

124. Boivin, G.; Meunier, P.J. The degree of mineralization of bone tissue measured by computerized quantitative contact microradiography. Calcif. Tissue Int. 2002, 70, 503–511.

125. Currey, J.D. The design of mineralised hard tissues for their mechanical functions. J. Exp. Biol. 1999, 202, 3285–3294.

126. Moreno, E.; Kresak, M.; Zahradnik, R.T. Fluoridated hydroxyapatite solubility and caries formation. Nature 1974, 247, 64–65.

127. Hina, A. Etude de la Réactivité en Milieu Aqueux, D'apatites Phosphocalciques D'intérêt Biologique. Ph.D. Thesis, Institut National Poytechnique de Toulouse, Toulouse, France, 1996. (In French).

128. Talmage, R.V.; Talmage, D.W. Calcium homeostasis: Solving the solubility problem. J. Musculoskelet. Neuronal Interact. 2006, 6, 402–407. [PubMed]

129. Shashi, A.; Singla, S. Parathyroid function in osteofluorosis. World J. Med. Sci. 2013, 8, 67–73.

130. Zhu, Y.; Zhu, Z.; Zhao, X.; Liang, Y.; Huang, Y. Characterization, dissolution and solubility of lead hydroxypyromorphite [Pb5(PO4)3OH] at 25–45 °C. J. Chem. 2015, 2015.

131. Christoffersen, J.; Christoffersen, M.R.; Kolthoff, N.; Bärenholdt, O. Effects of strontium ions on growth and dissolution of

hydroxyapatite and on bone mineral detection. Bone 1997, 20, 47–54.

132. Farlay, D.; Boivin, G.; Panczer, G.; Lalande, A.; Meunier, P.J. Long-term strontium ranelate administration in monkeys preserves characteristics of bone mineral crystals and degree of mineralization of bone. J. Bone Miner. Res. 2005, 20, 1569–1578.

133. Leggett, R.W. An age-specific kinetic model of lead metabolism in humans. Eviron. Heal. Perspect. 1993, 101, 598–616.

134. Flaherty, E.J.O. A physiologically based kinetic model for lead in children and adults.Environ. Health Perspect. 1998, 106, 1495–1503.

135. Simmer, J.P.; Hu, J.C.-C. Dental enamel formation and its impact on clinical dentistry. J. Dent. Educ. 2001, 65, 896–905. [PubMed]

136. Smith, C.E. Cellular and chemical events during enamel maturation. Crit. Rev. Oral Biol. Med.1998, 9, 128–161.

137. Robinson, C.; Kirkham, J.; Brookes, S.J.; Bonass, W.A.; Shore, R.C. The chemistry of enamel development. Int. J. Dev. Biol. 1995, 39, 145–152.

138. Lynch, R.J.M. The primary and mixed dentition, post-eruptive enamel maturation and dental caries: A review. Int. Dent. J. 2013, 63, 3–13.

139. Arends, J. Mechanism of enamel dissolution and its prevention. J. Biol. Buccale 1977, 5, 219–237.

140. Kuhn, L.T.; Grynpas, M.D.; Rey, C.C.; Wu, Y.; Ackerman, J.L.; Glimcher, M.J. A comparison of the physical and chemical differences between cancellous and cortical bovine bone mineral at two ages. Calcif. Tissue Int. 2008, 83, 146–154.

141. Fernandez-Jalvo, Y.; Sanchez-Chillon, B.; Andrews, P.; Fernandez-Lopez, S.; Alcala Martinez, I. Morphological taphonomic transformations of fossil bones in continental environments, and repercussions on their chemical composition. Archaeometry 2002, 44, 353–361.

142. Kohn, M.J.; Schoeninger, M.; Barker, W.W. Altered states: Effects of diagenesis on fossil tooth chemistry. Geochim. Cosmochim. Acta 1999, 63, 2737–2747.

143. Clementz, M.T. New insight from old bones: Stable isotope analysis of fossil mammals. J. Mammal. 2012, 93, 368–380.

144. Kohn, M.J.; Law, J.M. Stable isotope chemistry of fossil bone as a new paleoclimate indicator. Geochim. Cosmochim. Acta 2006, 70, 931–946.

145. Cazalbou, S.; Eichert, D.; Drouet, C.; Combes, C.; Rey, C. Minéralisations biologiques à base de phosphate de calcium. Comptes Rendus Palevol 2004, 3, 563–572. (In French).

146. Grunenwald, A.; Keyser, C.; Sautereau, A.M.; Crubézy, E.; Ludes, B.; Drouet, C. Novel contribution on the diagenetic physicochemical features of bone and teeth minerals, as substrates for ancient DNA typing. Anal. Bioanal. Chem. 2014, 406, 4691–4704.

147. Kocsis, L.; Trueman, C.N.; Palmer, M.R. Protracted diagenetic alteration of REE contents in fossil bioapatites: Direct evidence from Lu-Hf isotope systematics. Geochim. Cosmochim. Acta2010, 74, 6077–6092.

New Insights in the Ontogeny and Taphonomy of the Devonian Acanthodian Triazeugacanthus affinis From the MiguashaFossil-Lagerstätte, Eastern Canada

13

CHAPTER

Marion Chevrinais [1], Etienne Balan [2] and Richard Cloutier [1]

[1]Département de Biologie, Chimie et Géographie, Université du Québec à Rimouski, Rimouski, QC G5L 3A1, Canada

[2]Institut de Minéralogie, de Physique des Matériaux et de Cosmochimie, Sorbonne Universités, UPMC Univ Paris 06, CNRS UMR 7590, IRD UMR 206, MNHN, F-75005 Paris, France

ABSTRACT

Progressive biomineralization of a skeleton occurs during ontogeny in most animals. In fishes, larvae are poorly mineralized, whereas juveniles and adults display a progressively more biomineralized skeleton. Fossil remains primarily consist of adult specimens because the fossilization of poorly-mineralized larvae and juveniles necessitates exceptional conditions. The Miguasha *Fossil-Lagerstätte* is renowned for its Late Devonian vertebrate fauna, revealing the exceptional preservation of fossilized ontogenies for 14 of the 20 fish species from this locality. The mineralization of anatomical structures of the acanthodian *Triazeugacanthus affinis*from Miguasha are compared among larval, juvenile and adult specimens using Energy Dispersive X-ray Spectrometry. Chemical composition of anatomical structures of*Triazeugacanthus* reveals differences between cartilage and bone. Although the histology and anatomy is well-preserved, Fourier transform infrared spectrometry shows that the original chemical composition of bone is altered by diagenesis; the mineral phase of the bone (*i.e.*, hydroxyapatite) is modified chemically to form more stable carbonate-fluorapatite. Fluorination occurring in mineralized skeletal structures of adult *Triazeugacanthus* is indicative of exchanges between groundwater and skeleton at burial, whereas the preservation of larval soft tissues is likely owing to a rapid burial under anoxic conditions. The exceptional state of preservation of a fossilized ontogeny allowed us to characterize chemically the progressive mineralization of the skeleton in a Devonian early vertebrate.

Keywords: Acanthodii; biomineralization; fossilized ontogeny; paleontology; Devonian

1. INTRODUCTION

Fossilized fish ontogenies are rare, especially in the Paleozoic [1] because the mineralization of an animal's skeleton progresses during ontogeny and the preservation of poorly mineralized or ossified structures of immature (larval and juvenile) specimens requires exceptional fossilization conditions. Biomineralization of the skeleton during ontogeny is a condition shared by most vertebrates [2]. Extant jawless fish (hagfishes and lampreys) lack biomineralized tissues, which differs from their extinct relatives [1,2]. In extant cartilaginous fishes (Chondrichthyes; chimaera, sharks, skates and rays), the skeleton mineralizes as ontogeny progresses, at least superficially by means of tesserate calcified cartilage [3,4]. In extant bony fishes (piscine Osteichthyes; ray-finned fishes, coelacanths and dipnoans), young individuals show an unmineralized or poorly mineralized skeleton whereas older ones are highly mineralized [5,6,7]. In extinct jawed vertebrates, such as acanthodians and placoderms, this progressive biomineralization is assumed, rather than observed, because of the rarity of ontogenetic material but also because fossilization processes tend to favor the preservation of hard tissues, thus, frequently leaving an information gap during early ontogenetic stages. Since ontogenies have the potential to elucidate evolutionary developmental patterns and processes from the deep past [1,8], it is crucial to detect if fossilization biased our understanding of the early phase of biomineralization and how it could modify the preservation of the original tissues in fossils.

Vertebrate fossils predominantly consist of remains of mineralized skeletal structures [9], while soft tissues are only preserved exceptionally [10,11,12,13]. The quality and fidelity of tissue preservation depend, among other things, on the nature of the tissues in the organism, the amount of decay that precedes mineralization, and the availability of mineral-forming ions. It is however often difficult to determine whether the minerals observed in the fossils correspond to the original biomineralization of the tissues or result from diagenetic transformations of the fossils [14]. The comparison of chemical and mineralogical composition of anatomical structures during the ontogeny of an organism allows one to discriminate between these two mineralization processes. Indeed, as the biomineralization of skeletal structures progresses during ontogeny, or in relation to the size of an individual, the chemical and mineralogical signatures of skeletal structures change (e.g., carbon or calcium phosphate composition for cartilage or bone, respectively) [15,16]. On the other hand, an increase in exogenous ions (e.g., F ions) and modifications of the internal structure of skeletal tissues are expected with diagenetic mineralization, because it likely depends on local fossilization conditions.

The rapidity of consumption (*i.e.*, mechanical degradation by predators or scavengers) and decay (*i.e.*, degradation by bacteria) of an animal under oxic conditions is negatively correlated with the preservation of soft tissues in an aquatic environment [17]. The decomposition process of organic carbon by aerobic respiration roughly corresponds to the transformation of organic and dioxygen molecules into carbon dioxide and water molecules. Thus, anoxic conditions can be reached when the amount of organic carbon exceeds the oxygen supply. Specifically, the deposition of a large number of individuals (mass mortality) during a short period of time can locally reduce the level of dissolved oxygen which could create anoxic condition at the water/sediment boundary, decreasing scavenging and bioturbation [17]. These sporadic mass mortality events are characterized by a large number of fossil specimens in a given horizon. Fossiliferous sites and horizons yielding assemblages of mass mortalities are qualified as *Konzentrat-Lagerstätten*. Furthermore, when these fossiliferous environments present exceptionally well-preserved specimens, they are qualified of *Konservat-Lagerstätten*. It is likely that both, *Konzentrat-* and*Konservat-Lagerstätten*, favor soft tissues preservation.

Skeletal tissues showing early phases of biomineralization are prompt to be affected by two fossilization processes: the permineralization and the authigenic mineralization. Permineralization results from the early infiltration and permeation of tissues by mineralizing aqueous solutions [18]. In contrast, authigenic mineralization is often a product of biological tissues decay by bacterial activity; therefore, the types of minerals depend on the chemical composition of the sediment and the environmental conditions when bacterial decay occurred [19].

The Late Devonian Miguasha *Fossil-Lagerstätte* (Québec City, QC, Canada), a United Nations Educational, Scientific and Cultural Organization (UNESCO) World Heritage Site, yielded fossilized ontogenies for 14 out of the 20 fish species [20]. Among these 14 species, larval and juvenile specimens of the acanthodian *Triazeugacanthus affinis* have been reinterpreted recently [21]. The exceptional preservation of a large number of *Triazeugacanthus* within a large size range allowed the description of ontogenetic changes. This ontogenetic series is characterized by the simultaneous increase in the number of skeletal structures and the progressive extension of the squamation as a function of animal size [21]; thus, endoskeletal as well as exoskeletal mineralization proceed simultaneously.

In the present study, our main objectives are (1) to compare the chemical and mineralogical composition of larval, juvenile and adult specimens of *Triazeugacanthus*; and (2) to provide new information about the diagenesis of *Triazeugacanthus* from the Miguasha *Fossil-Lagerstätte*.

2. EXPERIMENTAL SECTION

2.1. Materials

The acanthodian *Triazeugacanthus affinis* Whiteaves, 1887 is common (more than 300 complete or almost complete specimens) in the middle Frasnian Escuminac Formation (Miguasha, QC, Canada) [22]. The distribution of *Triazeugacanthus* is uneven lithologically and stratigraphically.*Triazeugacanthus* is found predominantly in the laminites [22] and occurs primarily in two stratigraphic zones of the Escuminac Formation (*i.e.*, from 5 to 17 m and from 83 to 90 m from the base of the Escuminac Formation (Richard Cloutier, personal observation).

Three ontogenetic stages (larval, juvenile and adult) have been recognized in *Triazeugacanthus*based on the degree of squamation, the progression of the ossification and the size of the specimens [21]. Six complete or almost complete specimens, two larvae (Figure 1a–f), two juveniles (Figure 1j–o, and Figure 2a–f) and two adults (Figure 2h–s) and one thin section of the dermal skeleton of a juvenile (Figure 1g–

i) *Triazeugacanthus* preserved in laminites, have been used for the non-destructive scanning electron microscopy (SEM) coupled to the energy dispersive X-ray spectrometry (EDX) (Figure 1 and Figure 2). One partial adult specimen of *Triazeugacanthus*allowing destructive analysis has been used for Fourier transform infrared spectrometry (FTIR) and X-ray diffraction (XRD) (Electronic Supplementary Material Figure S1). Specimens are curated in the Musée d'Histoire naturelle de Miguasha (MHNM) (Nouvelle, QC, Canada).

For comparison we used juveniles of two extant species: the black dogfish, *Centroscyllium fabricii*(Chondrichthyes) and the Atlantic mackerel, *Scomber scombrus* (Osteichthyes). Specimens were frozen and stored at −20 °C. Specimens were thawed, dissected and dried in an oven at 35 °C for 72 h before EDX analyses. A series of anatomical structures were analyzed with EDX for both species: Meckel's cartilage, palatoquadrate, eye lenses, and vertebral centra of *C. fabricii* (ElectronicSupplementary Material Figure S2); exoccipital, dentary, basihyal and eye lenses of *S. scombrus*(Electronic Supplementary Material Figure S3).

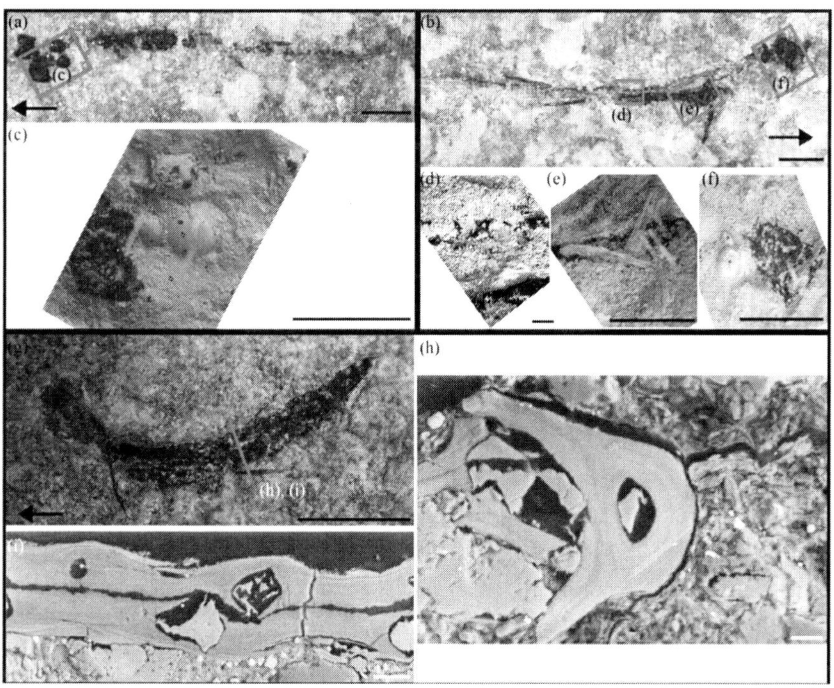

Figure 1. *Triazeugacanthus affinis* anatomical structures for energy dispersive X-ray spectrometry (EDX) analyses. (**a**) Larval specimen the Musée d'Histoire naturelle de Miguasha (MHNM) 03-440 #1 and scanning electron microscopy (SEM) close-ups of the (**c**) head region. (**b**) Larval specimen MHNM 01-440 #2 and SEM close-up of the (**d**) notochordal elements, (**e**) scapulocoracoid and pectoral fin spine, and (**f**) head region. (**g**) Juvenile specimen MHNM 03-398, red line shows the position of the histological section (**h**) of the anal spine and (**i**) the two lateral coverings of body scales in cross-sections. (**j**) Juvenile specimen MHNM 03-1252 and SEM close-ups of the (**k**) eye lenses, (**l**) otoliths, (**m**) notochordal elements (white arrows), (**n**) pectoral fin spine, (**o**) scales. Red squares show SEM close-up areas. Arrows point forward. Scale bars = 1 mm in (a) and (b); 600 μm in (c); 500 μm in (d); 800 μm in (e) and (f); 5 mm in (g) and (j); 10 μm in (h); 20 μm in (i); 200 μm in (k), (m), and (o); 100 μm in (l); and 500 μm in (n).

2.2. Scanning Electron Microscopy (SEM) Observation and Energy Dispersive X-Ray Spectrometry (EDX) Analysis

SEM microscopy has been performed using an environmental JEOL 6460LV SEM (JEOL USA, Peabody, MA, USA). The surface of samples was observed without conductive coating. Elemental composition analysis was performed using an INCA X-sight (Oxford Instruments, Austin, TX, USA) energy dispersive X-ray spectrometer; this analysis permits to identify the presence of chemical elements in the skeletal structures [23,24]. Each spectrum was acquired with a 10-μm spot size, for 100 s of lifetime (process time 5, spectrum range 0–20 keV, 2000 channels) at an accelerating voltage of 20 kV. The detection limits of chemical elements are about 1000 ppm or 0.1 wt %. System quantitative optimization was done using Cu as a standard.

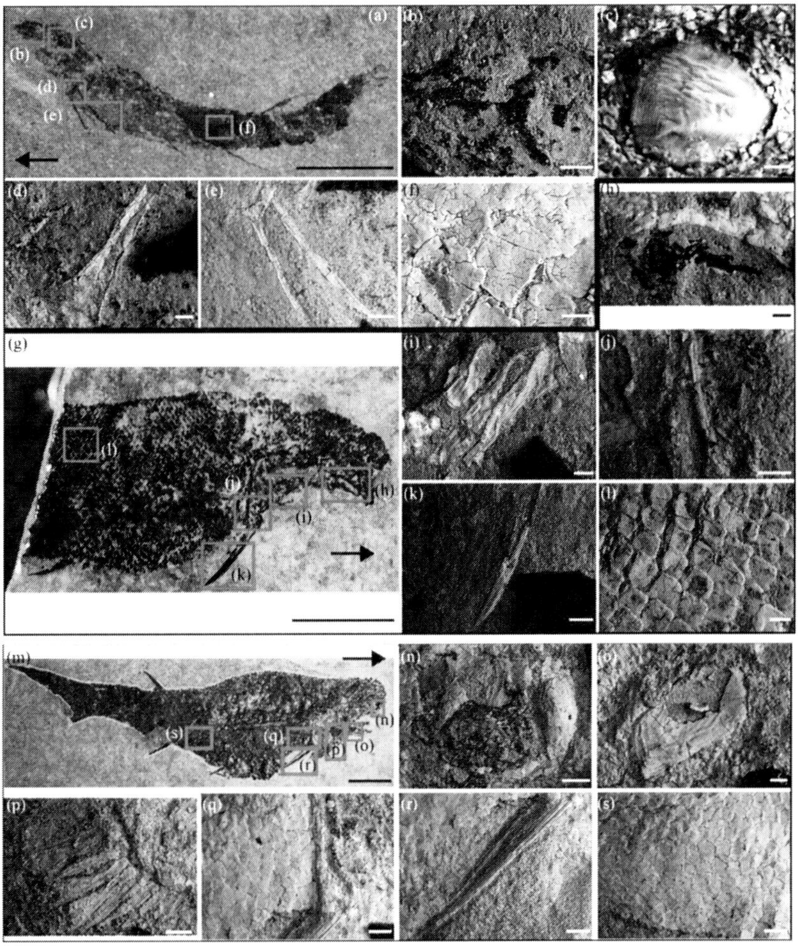

Figure 2. *Triazeugacanthus affinis* anatomical structures for EDX analyses. (**a**) Juvenile specimen MHNM 03-2684 and SEM close-ups of the (**b**) eye lenses, (**c**) otolith, (**d**) scapulocoracoid, (**e**) pectoral spines, and (**f**) scales. (**g**) Adult specimen MHNM 03-2669 and SEM close-ups of the (**h**) eye lenses and sclerotic plates, (**i**) branchiostegal rays, (**j**) scapulocoracoid, (**k**) pectoral fin spine and (**l**) scales. (**m**) Adult specimen MHNM 03-1497 and SEM close-ups of the (**n**) sclerotic plates, (**o**) palatoquadrate process, (**p**) branchiostegal rays, (**q**) scapulocoracoid, (**r**) pectoral fin spine and (**s**) scales. Red squares show SEM close-up areas. Arrows point forward. Scale bars = 5 mm in (a), (g) and (m); 200 μm in (b) and (h), 10 μm in (c), 100 μm in (d), (i), (o); 500 μm in (e), (j), (k), (n), (p), (q), (r), (s); 50 μm in (f).

2.3. Fourier Transform Infrared Spectrometry

The surface of one adult specimen of *Triazeugacanthus* (MHNM 03-1699; ElectronicSupplementary Material Figure S1) was scraped using a needle in order to obtain powder for Fourier transform infrared spectrometry (FTIR). Pellets for FTIR measurements were obtained by pressing a mixture of ~1.0 mg of gently ground sample diluted in 210 mg of dried KBr. Transmission IR spectra were recorded between 400 and 4000 cm^{-1}, by averaging 150 scans with a resolution of 1 cm^{-1}, using a Nicolet 6700 FTIR spectrometer (Thermo Fisher Scientific Inc., Waltham, MA, USA). The spectrometer was equipped with an Ever-Glo source, KBr beamsplitter and DTGS-KBr detector. The usual crystallinity index was calculated as the infrared splitting factor (IRSF): IRSF = (A605 + A565)/A590, where Ax is the measured absorbance at wavenumber *x*, assuming a straight baseline between 450 and 750 cm^{-1} [25,26].

2.4. X-Ray Diffraction

The mineralogical composition of the sedimentary rock was determined by powder X-ray diffraction (XRD) using a few mg of a sample obtained from the laminated matrix of*Triazeugacanthus* specimen MHNM 03-1699 (Electronic Supplementary Material Figure S1). The powder sample was deposited on a Si spinning sample holder. Analyses were performed using a PANALYTICAL® X'pert Pro apparatus (PANalytical, Almelo, The Netherlands) using the Co Kα radiation. Measurements were performed from 2° to 90° in 1000 s with a fixed slit size of 1°.

3. RESULTS

3.1. SEM Observations and EDX Analyses of Skeletal Tissues

3.1.1. Fossilized Tissues of *Triazeugacanthus affinis*

Chemical analyses of non-bony (eye lenses, otoliths), endochondral (palatoquadrate, scapulocoracoid, notochordal elements) and dermal elements (sclerotic plates, branchiostegal rays, pectoral spines, scales) have been performed for each ontogenetic stage (larval (Figure 3), juvenile (Figure 4) and adult (Figure 5)). Irrespective to the ontogenetic stages, (1) C and O are recorded in each skeletal element, (2) Ca and P are recorded in all dermal elements, and (3) the simultaneous presence of Ca and P is associated with F (with the exception of analyses performed on the internal structure of bones). Minor Al and Si contributions are detected and are likely related to the presence of aluminous clay minerals in the surrounding matrix (*i.e.*, illite and chlorite; Figure 3,Figure 4 and Figure 5).

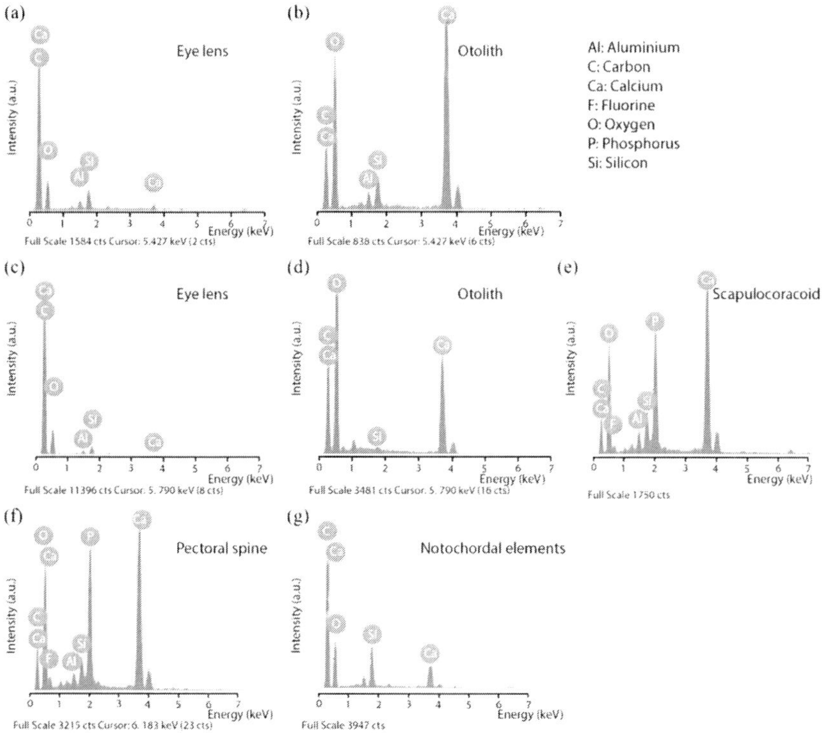

Figure 3. Representative spectra of larval *Triazeugacanthus*samples using EDX punctual microanalysis. (**a**) and (**b**) MHNM 03-440 #1. (**c**)–(**g**) MHNM 03-440 #2. Note that the intensity of the oxygen peak is non-significant and depends essentially on the vacuum level in the chamber of the environmental SEM.

In larval *Triazeugacanthus* (Figure 1a,b), notochordal elements and eye lenses are among the first elements to develop ([21]: Electronic Supplementary Material, Figure S3). Notochordal elements and eye lenses display C and O peaks with a minor contribution of Ca, Al and Si in the EDX spectrum (Figure 3a,c,g). The notochordal elements (visible only in rare specimens) show no sign of mineralization (Figure 3g). During the larval stage, otoliths form shortly after the previous elements ([21]: Electronic Supplementary Material, Figure S3). These non-bony elements display a strong Ca peak and lack P; these results are consistent with the expected calcium carbonate composition [27] (Figure 1c,f and Figure 3b,d). Among the first appendicular elements to form ([21]: Electronic Supplementary Material, Figure S3), the scapulocoracoid and the pectoral spine display clear Ca and P peaks associated to a F peak (Figures 1e and 3e,f). Even in the youngest occurrence of the scapulocoracoid (MHNM 03-440 #2; [21]: Electronic Supplementary Material,Figure S3), there is already a clear mineralized signature (Figure 3e); the pre-mineralized state of this endochondral element has not been sampled.

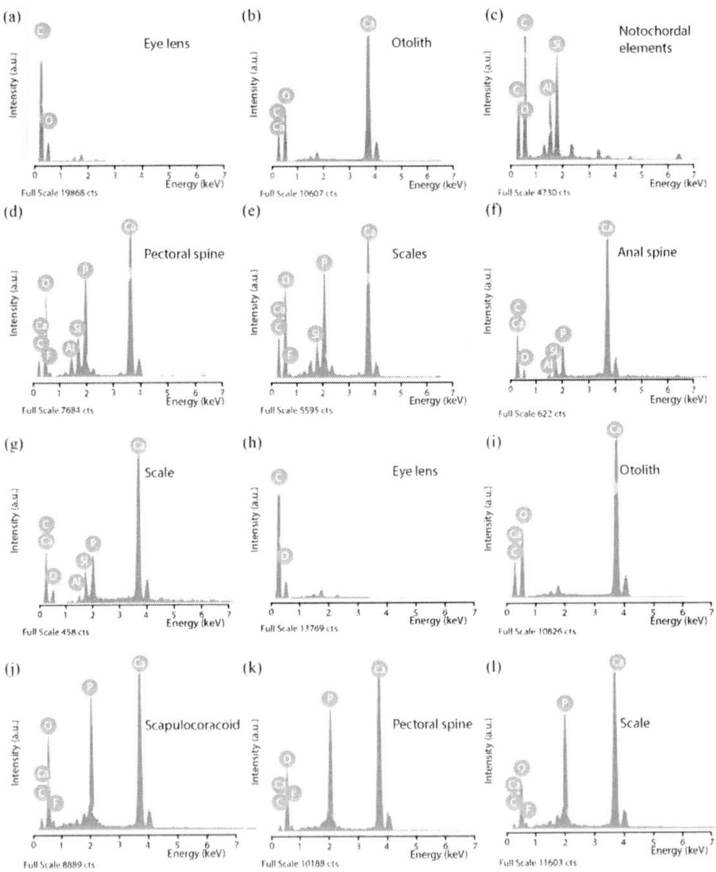

Figure 4. Representative spectra of juvenile *Triazeugacanthus*samples using EDX punctual microanalysis. (**a**)–(**e**) MHNM 03-1252. (**f**), (**g**) MHNM 03-398. (**h**)–(**l**) MHNM 03-2684. Note that the intensity of the oxygen peak is non-significant and depends essentially on the vacuum level in the chamber of the environmental SEM. See Figure 3 for chemical elements legend.

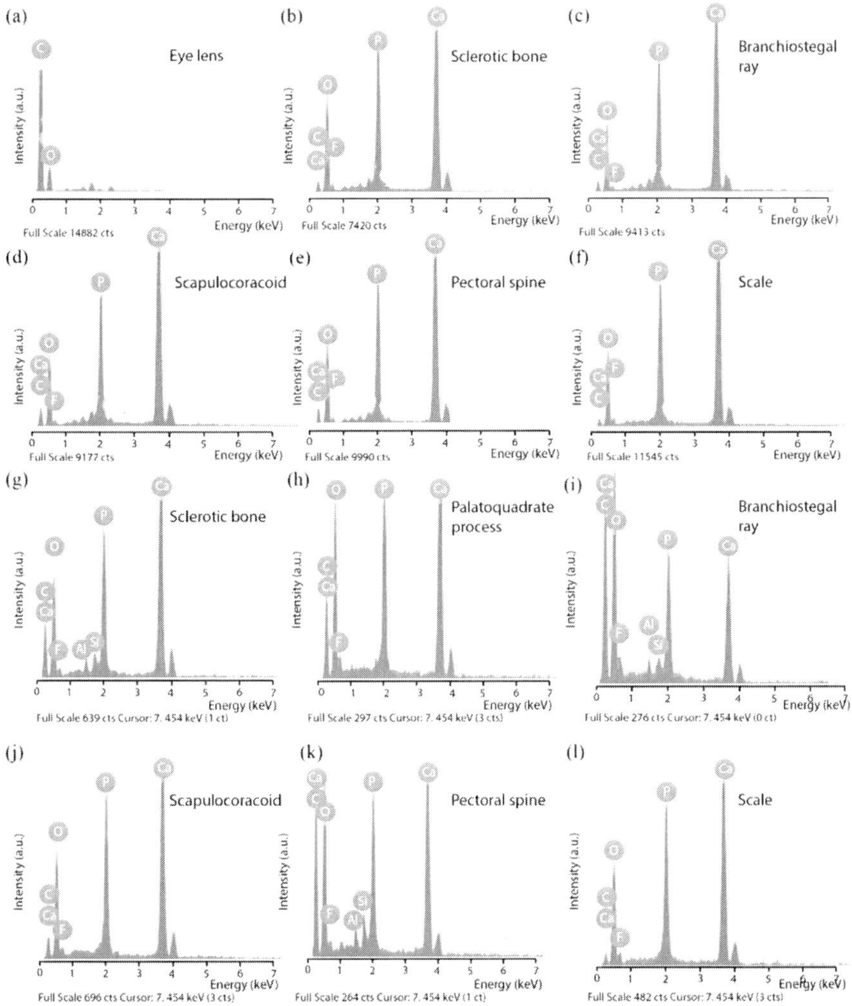

Figure 5. Representative spectra of adult *Triazeugacanthus*samples using EDX punctual microanalysis. (**a**)–(**f**) MHNM 03-2669. (**g**)–(**l**) MHNM 03-1497. Note that the intensity of the oxygen peak is non-significant and depends essentially on the vacuum level in the chamber of the environmental SEM. SeeFigure 3 for chemical elements legend.

Juvenile *Triazeugacanthus* specimens (Figure 1j and Figure 2a), characterized by the appearance and progression of the squamation [21], show dermal elements (scales: Figure 4e,g,l; spines: Figure 4d,f,k) with clear signatures of mineralization (Ca and P peaks). Elemental composition on the internal structure of dermal scales and spine (thin section; Figure 1h,i) also reveals the presence of Ca and minor P (Figure 4f,g), but the absence of F. Chemical composition of the eye lenses (Figure 4a,h), notochordal elements (Figure 4c), scapulocoracoid (Figure 4j) and otoliths (Figure 4b,i) remain similar to the larval condition.

In adult *Triazeugacanthus* specimens (Figure 2g,m), the chemical composition of previously formed skeletal structures remains similar to that of the immatures (Figure 5). The newly developed palatoquadrate (Figure 5h) is already mineralized. The dermal sclerotic plates (Figure 5b,g) and branchiostegal rays (Figure 5c,i) are also mineralized as suggested by the presence of strong Ca and P peaks (with similar relative intensities) associated with a F peak. Notochordal elements and otoliths are not observed in adults because they are overlaid by body and head scales, respectively.

3.1.2. Living Models: *Centroscyllium fabricii* and *Scomber scombrus*

The elemental composition analysis of the juvenile *C. fabricii* mostly consists in C, O and N, consistent with the absence of mineralization on the surface of the neurocranium, the eye lenses and the vertebral centra (Electronic Supplementary Material Figure S4). Tesserate mineralization indicated by the presence of P and Ca occurs in the mandibular and hyoid arches (ElectronicSupplementary Material Figure S4). The skeletal structures of the juvenile *S. scombrus* display mainly the presence of C, O and N in the eye lenses and Ca and P signatures in the exoccipital, the dentary and the basihyal (Electronic Supplementary Material Figure S5).

3.2. Mineralogical Analysis of Triazeugacanthus Skeleton

The FTIR spectrum of *Triazeugacanthus* skeletal structures displays the characteristic absorption bands of apatite (Figure 6a–c) [28]. These bands are related to the internal vibrational modes of PO_4 groups. The phosphate groups lead to intense bands at 565–576 and 605 cm^{-1} (PO_4 bending modes), and 1041 and 1096 cm^{-1} (PO_4 anti-symmetric stretching modes).

A weaker band related to the PO_4 symmetric stretching mode is observed at 965 cm^{-1}. The wavenumber of this band is correlated with F concentration and ranges between 960 in modern fish bone [29] to 965.6 cm^{-1} in a sedimentary carbonate-fluorapatite [28]. Its position in the present study thus corroborates the presence of a F concentration in apatite, consistent with the EDX measurements performed on specific skeletal structures of *Triazeugacanthus* (Figure 2i–n). The width of this band is dominated by the atomic-scale distortions of the apatite-structure [30] but overlapping with the strong PO_4 anti-symmetric stretching bands affects accurate analysis.

The splitting factor (IRSF) determined from the relative intensity of PO_4 bending bands is a practical way to assess the width of these bands, which is expected to reflect the apatite crystalline order but may also depend on the shape of apatite particles (Figure 6b). However, for samples with ordinary particle shapes, the full width at half maximum (FWMH) of the PO_4 symmetric stretching band and the infrared splitting factor (IRSF) parameter lead to similar information about apatite crystalline order [28]. In the present sample, the IRSF is 5.17, notably higher than that previously reported for bioapatite and slightly smaller than that reported for a sedimentary carbonate-fluorapatite sample [28]. For example, an IRSF of ~2.8 has been reported for vertebral centra of modern shark and ranging between 3.41 and 5.39 for Cenozoic fossil sharks [31,32]. Values of 2.6 and 3.2 have been reported for newly formed bones (fin rays) in the extant zebrafish*Danio rerio* [15]. Recent marine apatites also display low IRSF indices (3.0–3.6), whereas those of inland ancient apatites are higher (4.5–7.8) [26]. The IRSF of extant mammal bones exposed on tropical savannah grasslands varies from 2.7 to 3.5 and was found to be correlated to the mean length of apatite particles [33]. IRSF values increase to approximately 3.9 in Pleistocene bird and mammal bones [9]. The enamel of modern mammals displays a relatively high IRSF (e.g., 4.3 [24]). The comparatively high IRSF of *Triazeugacanthus* skeletal structures suggests that diagenetic transformations increased the apatite crystalline order with respect to that observed in extant and extinct biological materials.

Several bands related to vibrational modes of carbonate groups are also observed. Carbonate groups are responsible for the set of absorption bands at 1426 and 1453 cm^{-1} (anti-symmetric stretching) and the two overlapping signals at 865 and 874.4 cm^{-1} (out-of-plane bending modes). An additional band at 712 cm^{-1} indicates the presence of calcite.

Based on the signal related to out-of-plane bending modes, two main environments of carbonate groups occur in the sample. The band at 875 cm^{-1} can be mostly ascribed to calcite. The band at 865 cm^{-1} is related to carbonate groups incorporated in the apatite structure with a "francolite"-type environment [34]. This peculiar environment corresponds to an isovalent coupled substitution of the phosphate groups by clumped carbonate and F ions. Note that this type of environment does not imply any bonded interaction between the carbonate and the F ions but corresponds to a geometrical association, which ensures a local electrostatic charge neutrality.

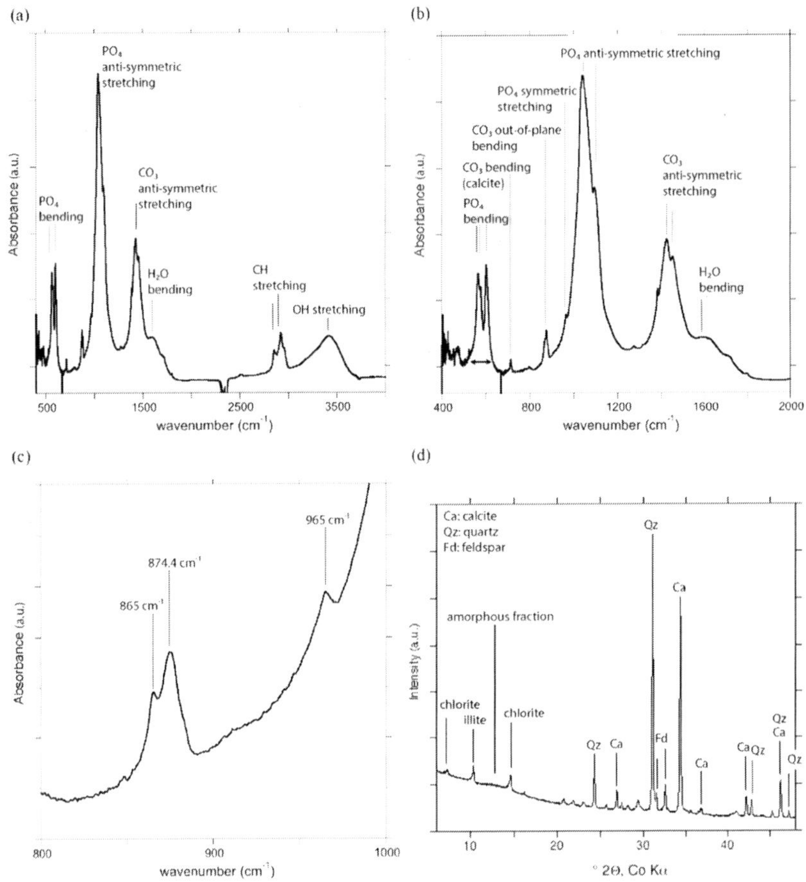

Figure 6. Powder transmission infrared spectrum of *Triazeugacanthus* and X-ray diffraction (XRD) pattern of surrounding sedimentary matrix. (**a**) Powder transmission infrared spectrum of *Triazeugacanthus* (MHNM 03-1699) showing PO_4, CO_3, OH, CH and H_2O bands and close-up (**b**) focusing on major CO_3 and PO_4 bands. Horizontal arrow points the line splitting of PO_4 bending modes. (**c**) Close-up of the Fourier transform infrared spectrometry (FTIR) spectrum focusing on the ν_2 CO_3 and ν_1 PO_4 signals. The band at 865 cm^{-1} is diagnostic of the presence of a francolite-type carbonate environment. (**d**) XRD pattern of sedimentary matrix surrounding the specimen of *Triazeugacanthus*.

The absorption bands related to the two anti-symmetric stretching modes of CO_3 groups, split by the site distortion, are observed at 1409–1440 cm^{-1} and 1450–1455 cm^{-1} (Figure 6b). These frequencies are consistent with the above interpretation. The absence of a marked absorption band at higher frequency (1550 cm^{-1}) rules out a major occurrence of carbonate groups in the structural channels of apatite (A-type carbonates). However, an uncertainty remains about the occurrence of carbonate groups in the usual B-type environment of hydroxyapatite because of its potential overlap with the calcite and "francolite"-type carbonate bands. The peculiar "francolite"-type substitution at the B-site of fluorapatite differs from ordinary B-type carbonate substitution. In biological apatite, the B-type carbonate substitution corresponds to the replacement of PO_4^{3-} by CO_3^{2-}, along with the substitution of Ca^{2+} by Na^+ or K^+ to preserve the electroneutrality.

Broad absorption features between 1560 and 1800 cm^{-1} are related to the presence of organic matter (carbonyl and amide functions) and water molecules (H_2O bending mode). Water molecules are also responsible for the broad OH-stretching band at 3420 cm^{-1}. In contrast, no stretching band related to OH groups in the structural channels of hydroxyapatite (at 3572 cm^{-1}) is observed.

Taken together, these results indicate that the mineral composition of the fossilized *Triazeugacanthus* skeleton mostly consists of carbonate-fluorapatite and calcite. As bioapatite usually consists in poorly ordered carbonate-bearing hydroxyapatite, the occurrence of carbonate-fluorapatite strongly suggests a post-depositional recrystallisation of the bioapatite [26].

Finally, the sedimentary matrix surrounding the fossil sample was analyzed by X-ray diffraction to better infer the conditions of preservation of *Triazeugacanthus* fossils (Electronic Supplementary Material Figure S1). The observed minerals are dominantly calcite and quartz with feldspar, chlorite and illite in lower abundance (Figure 6d). Note that the clay fraction could also contain smectite but its identification would require further physical separation and chemical treatments because of the overlap of the smectite with the chlorite (001) peak. The presence of an amorphous fraction is suggested by a weak broad band at 12° in the diffraction pattern (Figure 6d). Overall, these results are consistent with previously published data [35,36] showing the occurrence of quartz, calcite and smaller proportions of illite, muscovite, chlorite, plagioclase and amorphous organic matter. This mineralogical composition has been interpreted as resulting from shallow burial diagenesis and early formation of diagenetic calcite [35,36].

4. DISCUSSION

Here, we have shown that (1) both unmineralized (carbon signature) and mineralized anatomical elements (calcium and phosphorus signatures) are preserved depending on the ontogenetic stages of *Triazeugacanthus*; and (2) the carbonate-fluorapatite is the main mineralogical structure of *Triazeugacanthus* fossil skeleton.

4.1. Chemical Composition of Triazeugacanthus Biomineralized Tissues Depends on Ontogenetic Stages

Published spectrometry on *Triazeugacanthus* reported the dominant presence of Si with traces of Al, K, Mg, Ca and Fe for various skeletal elements (scales, "vertebral column" and "adjacent plates") [37]. Our observations do not confirm this previous report, which may correspond to an analytical artifact. It is unlikely that the specimen used for that study [37] would be indicative of silicification, while none of our specimens show such a chemical signature.

Our results indicate that both unmineralized (C signature) and mineralized anatomical remains (Ca and P signatures) are preserved differentially according to the ontogenetic stages of*Triazeugacanthus* and the histological origin of the elements. With the exception of the notochordal elements for which a non-mineralized signature has been found in the immatures, the pre-mineralized (cartilaginous) state of endochondral elements has not been observed in our coarse sampling of the ontogenetic series of *Triazeugacanthus*. The notochordal elements correspond either to the cartilaginous precursors of vertebral elements or to the notochord itself. The appearance and growth of dermal elements is always characterized by the presence of both Ca and P peaks. However, a difference has been noticed between the elemental composition at the surface of the bone and the internal structure of the bone; F has been recorded at the exposed surface (scales and spines), while it is not recorded internally.

While the larval individuals show the preservation of soft tissues (e.g., eye lenses, notochordal elements), juveniles and adults display notable concentrations of Ca and P, which are characteristic of bone chemical composition [38]. However, in the juveniles, the relative P/Ca content in the scales and fin spines is smaller than in the adults. Therefore, we interpret this progressive mineralization of skeletal elements as reflecting an ontogenetic transformation of the bony material.

EDX analyses and anatomical comparisons allowed us to identify properly the rounded anteriormost elements observed in the head of *Triazeugacanthus* (Figure 1a,b) [37,39,40]. These elements have been alternately interpreted as (1) optic plates [37]; (2) orbits [40]; and (3) eye lenses [21]. The use of the term "optic plates" was not justified in the original paper [37], as this term does not correspond to any anatomical element. The term "orbit" (="orbital cavity" [41]) corresponds either to the cavity that structurally supports the eyeball, or the bones (e.g., circumorbitals) forming the margin of the eye socket [42]. None of the known acanthodian neurocrania [43,44] display paired rounded elements as those seen in *Triazeugacanthus*. A series of four dermal sclerotic plates forms the margin of the eye socket of *Triazeugacanthus* (this study, [40,42]). Thus, the rounded elements seen in *Triazeugacanthus* do not correspond to the orbit. Anterior paired rounded elements have been identified as the eyes of the Carboniferous acanthodian *Acanthodes bridgei* [45]. Rods and pigments, as well as molecules common only to eumelanine, were found in the dark brown fragments from these fossil eyes. EDX analyses on the eye lenses of the extant *C. fabricii* and *S. scombrus* are similar to those of the *Triazeugacanthus* elements (this study). Characteristics of the crystalline lens of aquatic vertebrates (e.g., rounded-shape of the lens with low deformability [46]) and similarities shared with cartilage (abundance and chemical composition of crystalline proteins and collagen proteins [47] in the lens and the cartilage, respectively) might explain its potential for fossilization. We suggested that the rounded elements in *Triazeugacanthus* specimens could be interpreted as eye lenses. The anterior morphological position in the head region, their rounded shape, the absence of mineralized tissues (C and O peaks), and the similarities with acanthodian, chondrichthyan and osteichthyan anatomy (this study, [45]) corroborate this interpretation.

4.2. Preservation versus Recrystallization of Triazeugacanthus Tissues

In the stratigraphic sequence of the Miguasha *Fossil-Lagerstätte*, larval, juvenile and adult*Triazeugacanthus* are found in the same lithologies, primarily in laminites, as well as on the same bedding plane. The characteristic "laminites" facies is composed of an alternation of fine siltstones and shales [22] rich in organic matter interpreted as a tidal deposit [35]. This lithofacies is in agreement with an intertidal deposit likely in a wave-dominated estuarine environment, with periodic input of sediment by floods [22]. Miguasha *Konservat-* and *Konzentrat-Lagerstätte* horizons occur mostly in the transgressive phases and primarily towards the maximum paleodepth water [22].

Sedimentological and geochemical investigations have shown that the Escuminac Formation has been subjected only to a shallow burial diagenesis (depth < 1500 m, T < 430 °C) [35,36,48]. The clay mineral fraction is considered to be mostly detrital whereas calcite cement and fibrous calcite are attributed to early diagenesis of organic matter [35]. Thus, shallow burial allowing exchanges between groundwater and skeleton, and early diagenesis are required conditions for recrystallization of skeletal structures.

Despite the excellent state of preservation of bones at the microstructural [49,50,51], histological (Figure 1h,i) and anatomical levels, our results indicate a pervasive recrystallization of bioapatite in *Triazeugacanthus* skeletal structures. Modification of the mineral composition of fossil bones (e.g., presence of authigenic calcite) from the Escuminac Formation has been previously reported in the placoderm *Bothriolepis canadensis* and the osteolepiform *Eusthenopteron foordi* [36]. Bioapatites usually display small particle sizes and significant carbonate contents [15,52]. In addition, apatite particles in bones of living animals are coated by a poorly-ordered hydrous phosphatic phase [53] and associated with organic polymers (collagen), which transform [54] or decay after an animal's death [55]. In contrast, carbonate-fluorapatite is less soluble than bioapatite [26,54] and often represents an ultimate stage of recrystallization of biomineralized anatomical structures in geological environments [26,28,52,54,56]. Carbonate-fluorapatite also occurs in sedimentary environments as an authigenic mineral, likely produced from the bacterial decay of organic matter [57,58,59,60].

Transformation of bioapatite to carbonate-fluorapatite can occur in low-temperature environments and has been reported in marine or continental fossil deposits [14,28,52,61]. It however depends on the F supply. In the present case, the F source might be related to the early diagenetic transformation of carbonates. As a matter of fact, the early diagenesis of biogenic calcium carbonate minerals has been reported to be a major source of F in modern shallow sediments of the Florida Bay, USA [62]. The transformation of bioapatite in carbonate-bearing fluroapatite is, thus, still consistent with the rapid burial and early diagenetic transformations under anoxic conditions. Indeed, rapid burial and early diagenetic transformations limiting the rapidity of consumption and degradation, which are the most important bias in the preservation of soft tissues, allowed the preservation of these tissues such as those observed in *Triazeugacanthus*larvae [17,63].

Finally, bioapatite to carbonate-fluorapatite transformation of mammal enamel in late Tertiary continental environments has been shown to occur through a dissolution/precipitation mechanism [28]. However, such type of replacement modifies the structure of the biological material at a micrometric scale (e.g., presence of carbonate-fluorapatite infillings discordant with the biogenic Retzius striae in mammalian tooth enamel [28]). In the Miguasha *Fossil-Lagerstätte*, the preservation of the bone microstructures [49,50,51,64] could suggest a more subtle transformation mechanism, consistent with an early diagenetic stage, even though the formation of the "francolite"-type environment of substituted carbonates implies a transformation of apatite down to the atomic scale. Therefore, the respective contributions of early transformation and potential slow modifications [65] in the sedimentological environment of the Escuminac Formation have still to be determined.

5. CONCLUDING REMARKS

The preservation of fossil anatomical remains, especially the preservation of soft tissues, and pervasive chemical modification of hard tissues in the same sedimentary horizons, suggest specific conditions of burial and diagenesis. The investigation of ontogenetic changes of the Late Devonian acanthodian *Triazeugacanthus affinis*, using different methods for chemical identification, allowed us to better describe the growth in this fossil fish in documenting the progressive mineralization. The presence of soft tissues in larvae and the biomineralization of tissues in juveniles and adults throughout a growth series are described for the first time in a Paleozoic vertebrate.

ACKNOWLEDGMENTS

For the access to collections, we thank Johanne Kerr, Olivier Matton and France Charest (MHNM). We thank Claude Belzile (UQAR) for scanning electron microscopy (SEM) and EDX analyses. We thank members of the Maurice Lamontagne Institute (DFO-MPO) for providing us with juvenile specimens of living fishes. Cyrena Riley (UQAR) helped with the preparation of the black dogfish specimen. François Grégoire (UQAR) provided juvenile mackerels. We are grateful to Maxime Guillaumet (IMPMC), Benoit Baptiste (IMPMC) and Guillaume Morin (IMPMC) for their help in FTIR and XRD measurements. This project was supported financially by NSERC Discovery grant (RC), UQAR Research Chair in Paleontology and Evolutionary Biology (RC), EnviroNorth mobility grant (MC), and the team "Evolution et développement du squelette" (CNRS,

IPBS Paris). We also thank the Labex Matisse for organizing the biomineralization short course and proposing this special issue.

AUTHOR CONTRIBUTIONS

Marion Chevrinais, Etienne Balan and Richard Cloutier conceived and designed the experiments; Marion Chevrinais and Etienne Balan performed the analyses; Marion Chevrinais, Etienne Balan and Richard Cloutier analyzed the data; Marion Chevrinais and Etienne Balan conceived the figures; Marion Chevrinais wrote the first version of the paper; all authors contributed to the final version.

CONFLICTS OF INTEREST

The authors declare no conflict of interest.

REFERENCES

1. Cloutier, R. The fossil record of fish ontogenies: Insights into developmental patterns and processes. Semin. Cell Dev. Biol. 2010, 21, 400–413.
2. Donoghue, P.C.J.; Sansom, I.J. Origin and early evolution of vertebrate skeletonization.Microsc. Res. Tech. 2002, 59, 352–372.
3. Kemp, N.E.; Westrin, S.K. Ultrastructure of calcified cartilage in the endoskeletal tesserae of sharks. J. Morphol. 1979, 160, 75–101.
4. Dean, M.N.; Summers, A.P. Mineralized cartilage in the skeleton of chondrichthyan fishes.Zoology 2006, 109, 164–168.
5. Grünbaum, T.; Cloutier, R.; Vincent, B. Dynamic skeletogenesis in fishes: Insight of exercise training on developmental plasticity. Dev. Dyn. 2012, 241, 1507–1524.
6. Faustino, M.; Power, D.M. Osteologic development of the viscerocranial skeleton in sea bream: Alternative ossification strategies in teleost fish. J. Fish Biol. 2001, 58, 537–572.
7. Gavaia, P.J.; Sarasquete, C.; Cancela, M.L. Detection of mineralized structures in early stages of development of marine Teleostei using

a modified alcian blue-alizarin red double staining technique for bone and cartilage. Biotech. Histochem. 2000, 75, 79–84.

8. Johanson, Z.; Trinajstic, K. Fossilized ontogenies: The contribution of placoderm ontogeny to our understanding of the evolution of early gnathostomes. Palaeontology 2014, 57, 505–516.

9. Trueman, C.N.; Privat, K.; Field, J. Why do crystallinity values fail to predict the extent of diagenetic alteration of bone mineral? Palaeogeogr. Palaeoclimatol. Palaeoecol. 2008, 266, 160–167.

10. Morris, S.C.; Caron, J.-B. Pikaia gracilens Walcott, a stem-group chordate from the Middle Cambrian of British Columbia. Biol. Rev. 2012, 87, 480–512.

11. Arsenault, M.; Desbiens, S.; Janvier, P.; Kerr, J. New data on the soft tissues and external morphology of the antiarch Bothriolepis canadensis (Whiteaves, 1880), from the Upper Devonian of Miguasha, Quebec. In Recent Advances in the Origin and Early Radiation of Vertebrates; Arratia, G., Wilson, M.V.H., Cloutier, R., Eds.; Verlag Dr. Friedrich Pfeil: München, Germany, 2004; pp. 439–454.

12. Trinajstic, K.; Marshall, C.; Long, J.; Bifield, K. Exceptional preservation of nerve and muscle tissues in Late Devonian placoderm fish and their evolutionary implications. Biol. Lett. 2007,3, 197–200.

13. Cloutier, R. Great Canadian Lagerstätten 4. The Devonian Miguasha biota (Québec): UNESCO World Heritage Site and a time capsule in the early history of vertebrates. Geosci. Can. 2013, 40, 149–163.

14. Kolodny, Y.; Luz, B.; Sander, M.; Clemens, W.A. Dinosaur bones: Fossils or pseudomorphs? The pitfalls of physiology reconstruction from apatitic fossils. Palaeogeogr. Palaeoclimatol. Palaeoecol. 1996, 126, 161–171.

15. Mahamid, J.; Sharir, A.; Addadi, L.; Weiner, S. Amorphous calcium phosphate is a major component of the forming fin bones of zebrafish: Indications for an amorphous precursor phase. Proc. Natl. Acad. Sci. 2008, 105, 12748–12753.

16. Cambra-Moo, O.; Nacarino-Meneses, C.; Díaz-Güemes, I.; Enciso, S.; Gil, O.G.; Rodríguez, L.L.; Barbero, M.Á.R.; Antonio, H.; Martín, A.G. Multidisciplinary characterization of the long-bone cortex growth patterns through sheep's ontogeny. J. Struct. Biol. 2015, 191, 1–9.

17. Allison, P.A.; Briggs, D.E.G. Taphonomy of Nonmineralized Tissues. In Taphonomy: Releasing the Data Locked in the Fossil Record;

Allison, P.A., Briggs, D.E.G., Eds.; Plenum Press: New York, NY, USA, 1991; Volume 9.

18. Schopf, J.M. Modes of fossil preservation. Rev. Palaeobot. Palynol. 1975, 20, 27–53.

19. Briggs, D.E.G. The role of decay and mineralization in the preservation of soft-bodied fossils.Annu. Rev. Earth Planet. Sci. 2003, 31, 275–301.

20. Cloutier, R.; Béchard, I.; Charest, F.; Matton, O. La contribution des poissons fossiles du parc national de Miguasha à la biologsie évolutive du développement. Nat. Can. 2009, 133, 84–95. (In French).

21. Chevrinais, M.; Cloutier, R.; Sire, J.-Y. The revival of a so-called rotten fish: The ontogeny of the Devonian acanthodian Triazeugacanthus. Biol. Lett. 2015, 11.

22. Cloutier, R.; Proust, J.-N.; Tessier, B. The Miguasha Fossil-Fish-Lagerstätte: A consequence of the Devonian land-sea interactions. Palaeobiodivers. Palaeoenviron. 2011, 91, 293–323.

23. Gupta, N.S.; Cambra-Moo, O.; Briggs, D.E.G.; Love, G.D.; Fregenal-Martinez, M.A.; Summons, R.E. Molecular taphonomy of macrofossils from the Cretaceous Las Hoyas Formation, Spain. Cretac. Res. 2008, 29, 1–8.

24. Schweitzer, M.H.; Avci, R.; Collier, T.; Goodwin, M.B. Microscopic, chemical and molecular methods for examining fossil preservation. Comptes Rendus Palevol 2008, 7, 159–184.

25. Weiner, S.; Bar-Yosef, O. States of preservation of bones from prehistoric sites in the Near East: A survey. J. Archaeol. Sci. 1990, 17, 187–196.

26. Shemesh, A. Crystallinity and diagenesis of sedimentary apatites. Geochim. Cosmochim. Acta1990, 54, 2433–2438.

27. Schultze, H.-P. A new acanthodian from the Pennsylvanian of Utah, USA, and the distribution of otoliths in gnathostomes. J. Vertebr. Paleontol. 1990, 10, 49–58.

28. Yi, H.; Balan, E.; Gervais, C.; Ségalen, L.; Roche, D.; Person, A.; Fayon, F.; Morin, G.; Babonneau, F. Probing atomic scale transformation of fossil dental enamel using Fourier transform infrared and nuclear magnetic resonance spectroscopy: A case study from the Tugen Hills (Rift Gregory, Kenya). Acta Biomater. 2014, 10, 3952–3958.

29. Pucéat, E.; Reynard, B.; Lécuyer, C. Can crystallinity be used to determine the degree of chemical alteration of biogenic apatites? Chem. Geol. 2004, 205, 83–97.

30. Balan, E.; Delattre, S.; Roche, D.; Segalen, L.; Morin, G.; Guillaumet, M.; Blanchard, M.; Lazzeri, M.; Brouder, C.; Salje, E.K.H. Line-broadening effects in the powder infrared spectrum of apatite. Phys. Chem. Miner. 2011, 38, 111–122.

31. MacFadden, B.J.; Labs-Hochstein, J.; Quitmyer, I.; Jones, D.S. Incremental growth and diagenesis of skeletal parts of the lamnoid shark Otodus obliquus from the early Eocene (Ypresian) of Morocco. Palaeogeogr. Palaeoclimatol. Palaeoecol. 2004, 206, 179–192.

32. Labs-Hochstein, J.; MacFadden, B.J. Quantification of diagenesis in Cenozoic sharks: Elemental and mineralogical changes. Geochim. Cosmochim. Acta 2006, 70, 4921–4932.

33. Trueman, C.N.G.; Behrensmeyer, A.K.; Tuross, N.; Weiner, S. Mineralogical and compositional changes in bones exposed on soil surfaces in Amboseli National Park, Kenya: Diagenetic mechanisms and the role of sediment pore fluids. J. Archaeol. Sci. 2004, 31, 721–739.

34. Yi, H.; Balan, E.; Gervais, C.; Segalen, L.; Fayon, F.; Roche, D.; Person, A.; Morin, G.; Guillaumet, M.; Blanchard, M. A carbonate-fluoride defect model for carbonate-rich fluorapatite. Am. Mineral. 2013, 98, 1066–1069.

35. El Albani, A.; Cloutier, R.; Candilier, A.-M. Early diagenesis of the Upper Devonian Escuminac Formation in the Gaspé Peninsula, Québec: Sedimentological and geochemical evidence. Sediment. Geol. 2002, 146, 209–223.

36. Matton, O.; Cloutier, R.; Stevenson, R. Apatite for destruction: Isotopic and geochemical analyses of bioapatites and sediments from the Upper Devonian Escuminac Formation (Miguasha, Québec). Palaeogeogr. Palaeoclimatol. Palaeoecol. 2012, 361, 73–83.

37. Béland, P.; Arsenault, M. Scauménellisation de l'Acanthodii Triazeugacanthus affinis(Whiteaves) de la Formation d'Escuminac (Dévonien supérieur de Miguasha, Québec): Révision du Scaumenella mesacanthi Graham-Smith. Can. J. Earth Sci. 1985, 22, 514–524. (In French).

38. Young, R.A.; Brown, W.E. Structures of Biological Minerals. In Biological Mineralization and Demineralization; Springer: Berlin, Germany, 1982; pp. 101–141.

39. Graham-Smith, W. Scaumenella mesacanthi, gen. et sp. n., a peculiar organism from the Upper Devonian of Scaumenac Bay, Province Quebec, Canada. Ann. Mag. Nat. Hist. 1935, 16, 473–476.

40. Gagnier, P.-Y. Acanthodii. In Devonian Fishes and Plants of Miguasha, Quebec, Canada; Schultze, H.-P., Cloutier, R., Eds.; Verlag Dr. Friedrich Pfeil: München, Germany, 1996; pp. 149–164.

41. Franz-Odendaal, T.A.; Vickaryous, M.K. Skeletal elements in the vertebrate eye and adnexa: Morphological and developmental perspectives. Dev. Dyn. 2006, 235, 1244–1255.

42. Burrow, C.J.; Newman, M.J.; Davidson, R.G.; Den Blaauwen, J.L. Sclerotic plates or circumorbital bones in early jawed fishes? Palaeontology 2011, 54, 207–214.

43. Brazeau, M.D. The braincase and jaws of a Devonian "acanthodian" and modern gnathostome origins. Nature 2009, 457, 305–308.

44. Davis, S.P.; Finarelli, J.A.; Coates, M.I. Acanthodes and shark-like conditions in the last common ancestor of modern gnathostomes. Nature 2012, 486, 247–250.

45. Tanaka, G.; Parker, A.R.; Hasegawa, Y.; Siveter, D.J.; Yamamoto, R.; Miyashita, K.; Takahashi, Y.; Ito, S.; Wakamatsu, K.; Mukuda, T. Mineralized rods and cones suggest colour vision in a 300 Myr-old fossil fish. Nat. Commun. 2014, 5.

46. Levine, J.S. The vertebrate eye. In Functional Vertebrate Morphology; Hildebrand, M., Bramble, D.M., Liem, K.F., Wake, D.B., Eds.; Harvard University Press: Cambridge, MA, USA, 1985; pp. 317–337.

47. Goodenough, D.A. The crystalline lens. A system networked by gap junctional intercellular communication. Semin. Cell Biol. 1992, 3, 49–58.

48. Chidiac, Y. Paleoenvironmental interpretation of the Escuminac Formation based on geochemical evidence. In Devonian Fishes and Plants of Miguasha, Quebec, Canada; Schultze, H.P., Cloutier, R., Eds.; Verlag Dr Friedrich Pfeil: München, Germany, 1996; pp. 47–53.

49. Sanchez, S.; Dupret, V.; Tafforeau, P.; Trinajstic, K.M.; Ryll, B.; Gouttenoire, P.-J.; Wretman, L.; Zylberberg, L.; Peyrin, F.; Ahlberg, P.E. 3D microstructural architecture of muscle attachments in

extant and fossil vertebrates revealed by synchrotron microtomography.PLoS One 2013, 8.

50. Sanchez, S.; Tafforeau, P.; Ahlberg, P.E. The humerus of Eusthenopteron: A puzzling organization presaging the establishment of tetrapod limb bone marrow. Proc. R. Soc. B Biol. Sci. 2014, 281.

51. Downs, J.P.; Donoghue, P.C.J. Skeletal histology of Bothriolepis canadensis (Placodermi, Antiarchi) and evolution of the skeleton at the origin of jawed vertebrates. J. Morphol. 2009,270, 1364–1380.

52. Nemliher, J.G.; Baturin, G.N.; Kallaste, T.E.; Murdmaa, I.O. Transformation of hydroxyapatite of bone phosphate from the ocean bottom during fossilization. Lithol. Miner. Resour. 2004, 39, 468–479.

53. Wang, Y.; Von Euw, S.; Fernandes, F.M.; Cassaignon, S.; Selmane, M.; Laurent, G.; Pehau-Arnaudet, G.; Coelho, C.; Bonhomme-Coury, L.; Giraud-Guille, M.-M.; et al. Water-mediated structuring of bone apatite. Nat. Mater. 2013, 12, 1144–1153.

54. Kalvoda, J.; Novák, M.; Bábek, O.; Brzobohatý, R.; Holá, M.; Holoubek, I.; Kanický, V.; Škoda, R. Compositional changes in fish scale hydroxylapatite during early diagenesis; an example from an abandoned meander. Biogeochemistry 2009, 94, 197–215.

55. Sansom, R.S.; Gabbott, S.E.; Purnell, M.A. Atlas of vertebrate decay: A visual and taphonomic guide to fossil interpretation. Palaeontology 2013, 56, 457–474.

56. Pasteris, J.D.; Ding, D.Y. Experimental fluoridation of nanocrystalline apatite. Am. Mineral.2009, 94, 53–63.

57. Cosmidis, J.; Benzerara, K.; Menguy, N.; Arning, E. Microscopy evidence of bacterial microfossils in phosphorite crusts of the Peruvian shelf: Implications for phosphogenesis mechanisms. Chem. Geol. 2013, 359, 10–22.

58. Cosmidis, J.; Benzerara, K.; Gheerbrant, E.; Estève, I.; Bouya, B.; Amaghzaz, M. Nanometer-scale characterization of exceptionally preserved bacterial fossils in Paleocene phosphorites from Ouled Abdoun (Morocco). Geobiology 2013, 11, 139–153.

59. O'Hagan, D.; Harper, D.B. Fluorine-containing natural products. J. Fluor. Chem. 1999, 100, 127–133.

60. Murphy, C.D.; Schaffrath, C.; O'Hagan, D. Fluorinated natural products: The biosynthesis of fluoroacetate and 4-fluorothreonine in Streptomyces cattleya. Chemosphere 2003, 52, 455–461.

61. Ifrim, C.; Stinnesbeck, W.; Frey, E. Upper Cretaceous (Cenomanian-Turonian and Turonian-Coniacian) open marine plattenkalk deposits in NE Mexico. Neues Jahrb. Geol. Paläontol. Abh.2007, 245, 71–81.

62. Rude, P.D.; Aller, R.C. Fluorine mobility during early diagenesis of carbonate sediment: An indicator of mineral transformations. Geochim. Cosmochim. Acta 1991, 55, 2491–2509.

63. Allison, P.A. The role of anoxia in the decay and mineralization of proteinaceous macro-fossils. Paleobiology 1988, 14, 139–154.

64. Zylberberg, L.; Meunier, F.J.; Laurin, M. A microanatomical and histological study of the postcranial dermal skeleton in the Devonian sarcopterygian Eusthenopteron foordi. Acta Palaeontol. Pol. 2010, 55, 459–470.

65. Reynard, B.; Balter, V. Trace elements and their isotopes in bones and teeth: Diet, environments, diagenesis, and dating of archeological and paleontological samples.Palaeogeogr. Palaeoclimatol. Palaeoecol. 2014, 416, 4–16.

Index